Time's Urgency

The Study of Time

Founding Editor

Julius T. Fraser†

VOLUME 16

The titles published in this series are listed at *brill.com/stim*

Time's Urgency

Edited by

Carlos Montemayor
Robert Daniel

BRILL

LEIDEN | BOSTON

Cover illustration: BnF Manuscrit Français 28, Cité des Dieux Fol. 66v, Déluge.

The Library of Congress Cataloging-in-Publication Data is available online at http://catalog.loc.gov

Typeface for the Latin, Greek, and Cyrillic scripts: "Brill". See and download: brill.com/brill-typeface.

ISSN 0170-9704
ISBN 978-90-04-40823-4 (hardback)
ISBN 978-90-04-40824-1 (e-book)

Contents

PART 2
Urgency and Time Scales

Acknowledgments

Robert Daniel and Carlos Montemayor

The sixteenth conference celebrating the 50th anniversary of the founding of the International Society for the Study of Time took place at the University of Edinburgh, Scotland, UK, from June 26–July 2, 2016. We are very grateful to our hosts in Edinburgh for the memorable moments and for making our work easier with their generosity and hospitality.

During our conference, all the participants agreed that it was a very special event. It was, as is our tradition, in a beautiful location with a fascinating history, and the presentations spoke to the conference's theme *Time's Urgency* with unique depth. Thanks to Raji Steineck, Jo Alyson Parker, Thomas P. Weissert and Lanei Rodemeyer for being the driving force behind the organization of our 2016 conference. Special thanks to Raji Steineck for his invaluable help at the last stages of the editorial process for this volume.

Thanks are also due to Dennis Costa, Michael Crawford, Claudia Clausius, Paul Harris, Steven Ostovich, and Daniela Tan. One of the main reasons why we had a fantastic conference was because of their hard work in selecting the presentations for our 50th anniversary conference.

Finally, our gratitude to all the participants at the conference. The dialogue created at this event was interdisciplinary, original, and exciting. All the panels had a special energy, regardless of the disciplinary approach they had to the topic of time's urgency. The contributions to this volume certainly capture some of that energy. It is our hope that we maintain this enormously valuable way of studying time, with rigor, friendship and originality. This is exactly the way the founder of our Society, Julius Thomas Fraser, wanted us to engage with each other, with academia more generally, and with society at large.

Notes on Contributors

Julian Barbour

was born in 1937. He studied mathematics in Cambridge and physics in Munich, where he became very interested in basic questions relating to the nature of time and motion. Having completed a Ph.D. in theoretical physics in Cologne, he decided in 1968 to become an independent researcher. He has published about 50 scientific papers, two books, *The Discovery of Dynamics* (2001) and *The End of Time* (1999), and was the joint editor of the much cited conference proceedings *Mach's Principle: From Newton's Bucket to Quantum Gravity.*

Dennis Costa

teaches Comparative Literature at Boston University, specializing in medieval and Renaissance literature and intellectual history. He has published on Aeschylus, Virgil, Albertus Magnus, Bonaventure, Dante, Petrarch, Pico, Erasmus, Rabelais, Montaigne, Tommaso Campanella, Christopher Smart, and Federico Garcia Lorca. He serves as managing editor of ISST's scholarly journal, *KronoScope / Journal for the Study of Time.*

Kerstin Cuhls

has been working at the Fraunhofer Institute for Systems and Innovation Research ISI in Karlsruhe since 1992 as a scientific project manager. She took her degree in Japanese studies, Sinology and Business Administration at the University of Hamburg. In 1997, she was awarded a Ph.D. at the University of Hamburg (Japanology). Since 2010 she has been teaching Foresight methods and implementation at the FU Berlin. Cuhls was member of the European Forum for Forward-Looking Activities (EFFLA) of the European Commission, the expert group "Strategic Foresight," and the High Level Expert Group RISE.

Margaret K. Devinney

is Emerita Professor of German at Temple University. She is currently preparing the fifth edition of *Introduction to World Mythology: Contemporary Approaches to Classical and World Myth*, which reveals her long-time research interest in cultural anthropology, focusing on indigenous peoples. Her work on time expands her study of Mesoamerican, and specifically Mayan, culture and mythology by illustrating the continuing centrality of ancient beliefs in contemporary society.

Ileana da Silva

Her main passions in life involve travel, literature, and recreating Food Network's *Chopped* in her kitchen with any scraps she may find. Habitually late to everything and everywhere, except perhaps to class (which is her bread and butter – so as to not get *chopped* from her job) at San Francisco State University, where she mentors students in the art of self-expression and the process of identity formation through multimodal compositions. Assignments to be submitted on time, of course.

Sonia Front

is Assistant Professor at the Institute of English Cultures and Literatures, University of Silesia, Poland, where she teaches film and contemporary literatures in English. Her research interests include time and temporality as well as representations of consciousness in twenty-first-century literature and film. Her last book is a monograph *Shapes of Time in British Twenty-First Century Quantum Fiction* (2015).

Peter A. Hancock

D.Sc., Ph.D., is Provost Distinguished Research Professor in the Department of Psychology and the Institute for Simulation and Training, as well as at the Department of Civil and Environmental Engineering and the Department of Industrial Engineering and Management Systems at the University of Central Florida (UCF). At UCF in 2009 he became University Pegasus Professor and in 2012 was named 6th ever University Trustee Chair. He directs the MIT² Research Laboratories and is the Associate Director of the Center for Applied Human Factors in Aviation (CAHFA). Prior to his current position he founded and was the Director of the Human Factors Research Laboratory (HFRL) at the University of Minnesota where he held appointments as Professor in the Departments of Computer Science and Electrical Engineering, Mechanical Engineering, Psychology, and Kinesiology, as well as being a member of the Cognitive Science Center and the Center on Aging Research.

Paul A. Harris

Professor of English at Loyola Marymount University, is co-editor of the literary theory journal *SubStance*, and served as President of the International Society for the Study of Time 2004–2013. He maintains a rock garden and blog called "The Petriverse of Pierre Jardin," which informs his research, artwork and writings about stone. In 2016 he launched a SLOW LMU initiative, and designed "slow time zones" on campus (The Garden of Slow Time and The

Displacement Garden). He has exhibited stone installations at the National Gallery of Denmark, Chapman University, and the Arizona State Art Museum Project Space. He has created recent work in collaboration with jazz musician David Ornette Cherry, author David Mitchell, and artist Richard Turner, and is working with Turner and Thomas Elias on a book project, *Viewing Stones: Contemporary Approaches to Display.*

Rose Harris-Birtill

Ph.D., teaches at the University of St. Andrews, UK. Her academic monograph, *David Mitchell's Post-Secular World: Buddhism, Belief and the Urgency of Compassion*, was published in 2019 with Bloomsbury Academic. Rose holds the ISST's New Scholar Prize, the Frank Muir Prize for Writing, and a McCall MacBain Teaching Excellence Award. She serves as an editor for *KronoScope* and guest edited the David Mitchell special edition of *C21 Literature* (6.3, 2018), and is also Secretary of the British Association of Contemporary Literary Studies.

David Mitchell

is the award-winning and bestselling author of *The Thousand Autumns of Jacob de Zoet, Black Swan Green, Cloud Atlas, Number9Dream, Ghostwritten* and *The Bone Clocks*. Twice shortlisted for the Man Booker Prize, Mitchell was named one of the 100 most influential people in the world by *TIME* magazine in 2007. With K. A. Yoshida, Mitchell co-translated from the Japanese the international bestselling memoir, *The Reason I Jump.* He lives in Ireland with his wife and two children.

Carlos Montemayor

Ph.D. (2009) Rutgers University, is Associate Professor of Philosophy at San Francisco State University, USA. His research focuses on philosophy of mind and cognitive science. He is the author of *Minding Time: A Philosophical and Theoretical Approach to the Psychology of Time* (Brill, 2013), co-author (with Harry H. Haladjian) of *Consciousness, Attention, and Conscious Attention* (2015) and co-author (with Abrol Fairweather) of *Knowledge, Dexterity, and Attention: A Theory of Epistemic Agency* (2017).

Jo Alyson Parker

Ph.D., is Professor of English at Saint Joseph's University in Philadelphia. Her publications include *The Author's Inheritance: Henry Fielding, Jane Austen, and the Establishment of the Novel* (1998), *Narrative Form and Chaos Theory in*

Sterne, Proust, Woolf, and Faulkner (2007), and essays dealing with narrative and time. With Michael Crawford and Paul Harris, she co-edited *Time and Memory: The Study of Time XII* (2007), and, with Paul Harris and Christian Steineck, she co-edited *Time: Limits and Constraints: The Study of Time XIII* (2010). From 2014–18, she was Managing Editor of *KronoScope: Journal for the Study of Time.*

Katie Paterson

(born 1981, Scotland) is widely regarded as one of the leading artists of her generation. Collaborating with scientists and researchers across the world, Paterson's projects consider our place on Earth in the context of geological time and change. Her artworks make use of sophisticated technologies and specialist expertise to stage intimate, poetic and philosophical engagements between people and their natural environment. Combining a Romantic sensibility with a research-based approach, conceptual rigour and coolly minimalist presentation, her work collapses the distance between the viewer and the most distant edges of time and the cosmos. Katie Paterson has exhibited internationally, from London to New York, Berlin to Seoul, and her works have been included in major exhibitions including Hayward Gallery, Tate Britain, Kunsthalle Wien, MCA Sydney, Guggenheim Museum, New York, and The Scottish National Gallery of Modern Art, Edinburgh. She was winner of the Visual Arts category of the 2014 South Bank Awards, and is an Honorary Fellow of Edinburgh University.

Walter Schweidler

holds the chair of Philosophy at the Catholic University Eichstätt-Ingolstadt since 2009. From 2000 to 2009, he was Professor for Practical Philosophy at Ruhr-University Bochum. Among his research interests are contemporary and modern concepts of ethics and political philosophy, phenomenology, and bioethics. He is the author of *Die Überwindung der Metaphysik* (Stuttgart 1987); *Geistesmacht und Menschenrecht. Der Universalanspruch der Menschenrechte und das Problem der Ersten Philosophie* (Freiburg/München 1994); *Das Unantastbare. Beiträge zur Philosophie der Menschenrechte* (Münster 2001); *Der gute Staat. Politische Ethik von Platon bis zur Gegenwart* (Stuttgart 2004); *Das Uneinholbare. Beiträge zu einer indirekten Metaphysik* (Freiburg 2008); *Über Menschenwürde: Der Ursprung der Person und die Kultur des Lebens* (Wiesbaden 2012); *Transcending Boundaries*, ed. (Sankt Augustin 2015); *Wittgenstein, Philosopher of Cultures*, ed. w. Carl Humphries (Sankt Augustin, 2017); *Kleine Einführung in die Angewandte Ethik* (Wiesbaden 2017).

Raji C. Steineck

is Professor of Japanology at University of Zurich (UZH), president of the International Society for the Study of Time (ISST), and principal investigator of the ERC Advanced Grant project "Time in Medieval Japan" (TIMEJ). His research interests combine medieval and contemporary history of concepts in Japan and philosophy of culture. Recent publications include *Critique of Symbolic Forms II: Configurations of Mythology in Ancient Japan* (in German, 2017), *Concepts of Philosophy in Asia and the Islamic World* (co-edited with Ralph Weber, Elena Lange and Robert Gassmann; 2018), and "Chronography: an Essay in Methodology" (*KronoScope* 18/2).

Daniela Tan

is a Senior Lecturer at the Japanese department of the Institute of Asian and Oriental Studies of Zurich University. Her research focus is contemporary literature and literary criticism. She completed her studies in Kyōto, Ōsaka and Zurich and wrote her Ph.D. on Ōba Minako and female writers of the 20th century, and has been teaching classes on Japanese Literature since 2005. Currently she also serves as a research associate in the ERC project "Time in Medieval Japan," and works on body time – the conceptualization of the female body in medical, religious and literary texts.

Frederick Turner

his science fiction epic poems led to his being a consultant for NASA's long-range futures group, through which he met Carl Sagan and other space scientists. He received Hungary's highest literary honor for his translations of Hungarian poetry with the distinguished scholar and Holocaust survivor Zsuzsanna Ozsváth, won *Poetry*'s Levinson Prize, and has often been nominated for the Nobel Prize for literature. Born in England, raised in Africa by his anthropologist parents Victor and Edie Turner, and educated at Oxford University, he is also known as a Shakespearean scholar, a leading theorist of environmentalism, an authority on the philosophy of Time, and the poet laureate of traditional Karate. He is the author of about 40 books, ranging from literary monographs through cultural criticism and science commentary to poetry and translations. He has taught at UC Santa Barbara and Kenyon College, edited the *Kenyon Review*, and is presently Founders Professor of Arts and Humanities at the University of Texas at Dallas. Recent publications include *Light Within the Shade: 800 Years of Hungarian Poetry*, translated and edited by Frederick Turner and Zsuzsanna Ozsvath (2014); *Apocalypse: An Epic Poem*, e-book, hardback and paperback (2016); and *More Light: Selected Poems, 2004–2016* (2017).

Thomas Weissert

Ph.D., is Director of Technology at the Agnes Irwin School. His publications include *The Genesis of Simulation in Dynamics: Pursuing the Fermi-Pasta-Ulam Problem* (1997), and essays on tsunami travel-time charts, dynamical systems theory, dynamics and narrative, Jorge-Luis Borges, and Stanislaw Lem. He served as Executive Secretary of the International Society for the Study of Time from 1998–2016.

Marc Wolterbeek

is a Professor of English at Notre Dame de Namur University, where he teaches a variety of courses, including Epic, Romance, and Mythology; Graphic Novels and Manga; and Mythology of Heroes and Superheroes. He holds a Ph.D. in Comparative Literature and Medieval Studies, and he has published articles and delivered papers in sequential art as well as medieval literature.

Barry Wood

is Associate Professor of English at the University of Houston. Wood's publications began with a Canadian teaching edition of *Huckleberry Finn* (1968). He authored his first book during his M.A. program, his second during his doctoral program. He has edited *Malcolm Lowry: The Writer and His Critics* (1980) which earned an Ontario Arts Council Award; his manuscript *Fictional History, Fabricated Power*, is now under review for publication. Wood now has more than 60 books, articles, reprints, book chapters, and reviews, which address four areas: literature, the environment, big history, and education.

Introduction

Carlos Montemayor

The International Society for the Study of Time celebrates its 50th anniversary with this volume. Focused on the topic of *Time's Urgency*, it compiles papers and exchanges that emerged from our 16th triannual conference, in Edinburgh, Scotland, which generated many insightful and timely discussions. Although it is very difficult to capture such intellectual vitality on paper, this volume offers a glimpse of the contributions that made the conference so special.

Renowned physicist and theorist of time, Julian Barbour, opens the volume with his Founder's Lecture in honor of Julius Thomas Fraser. J. T. Fraser was the ultimate interdisciplinary timesmith, who received his early training in physics. Barbour's contribution presents an original approach to the pressing question of the arrow of time. Given that the laws of physics are time symmetric, he asks why we systematically experience an arrow of time that determines the decay, rather than the rejuvenation, of everything. Barbour's proposal is based on the statistical properties of non-confined systems, revising the traditional assumptions behind the second law of thermodynamics, frequently associated with the arrow of time. If one accepts Barbour's proposal, then traditional puzzles about the arrow of time are removed from physical theory. Thanks to Raji Steineck for organizing this insightful Founder's Lecture by Julian Barbour.

The second contribution, by Rose Harris-Birtill, is the winner of our 50th anniversary's conference New Scholar Prize. Harris-Birtill analyzes the literary work of British author David Mitchell, whom we were enormously fortunate to have as an active participant at our conference. Harris-Birtill's contribution draws on philosophy, spirituality and literary analysis, presenting a multifaceted examination and criticism of the dominant "end of history" narrative.

The papers have been divided into two sections: *Dialogues* and *Urgency and Time Scales*. "Dialogues" captures the vibrant discussion that unfolded during our conference. The first is a conversation between Paul Harris, former president of the ISST, with David Mitchell and Katie Paterson about the *Future Library* project. We are very grateful to the three of them for this wonderful contribution and for all their interventions at the conference. The second comprises the Presidential Address by Raji Steineck and a response by Carlos Montemayor. Their exchange concerns the centrality of the concept of time in framing large sections of our world without this entailing a single nature or essence of time. In a third, vivid exchange, Peter Hancock and Frederick

© KONINKLIJKE BRILL NV, LEIDEN, 2019 | DOI:10.1163/9789004408241_002

Turner engage the old debate about the existence of time. Hancock argues for an eliminativist view that Turner challenges, eliciting a response by Hancock. We are very grateful to longtime member and unmoved mover of our Society, Frederick Turner, for offering to write this response and to Peter Hancock for responding in detail to Turner.

Urgency and Time Scales contains contributions that address the issue of urgency and temporality across various perspectives and time scales. It opens with a paper by Jo Alyson Parker and Thomas Weissert, addressing the challenging problem of how to understand time loops and recursion within a narrative structure. The next contributions share the theme of narrative and urgency. Barry Wood explores this issue within the field of geology. Margaret Devinney explains the importance of narrative and cycles in Mayan cosmology. Daniela Tan analyzes the temporal and narrative significance of traumatic events, focusing on nuclear accidents in Japan. Marc Wolterbeek and Ileana da Silva investigate the notions of sequence and duration in visual narratives. Sonia Front focuses on narrative complexity and continuity in film. Kerstin Cuhls addresses the issue of foresight and urgency in short and long term decision making. Finally, two contributions offer an in depth analysis of the topic of urgency in literature: Dennis Costa's contribution is on urgency in the work of Dante Alighieri and Walter Schweidler's is on similar themes in the work of Marcel Proust.

Time will continue to puzzle humanity, even as humanity continues to influence the fabric of time – the "end of time" and the Anthropocene were a major topic of discussion at the conference. Showcasing the cross-disciplinary dialogue and engaged interdisciplinary research fostered by the International Society for the Study of Time, we renew our commitment to the belief that these constitute the best way to approach this complex topic. May the ISST live many more years. Its 50th anniversary gives us strong evidence to believe that this will be the case.

Introductory Essays

∴

CHAPTER 1

A New Theory of Time's Arrows

Julian Barbour

Abstract

For over 150 years theoretical physicists have been faced with a conundrum: with one exception that is far too small to affect the issue, the known laws of nature give no explanation of why all macroscopic phenomena in the universe unfold in a common temporal direction. Like the stars, we all get older together; we never meet anyone getting younger. This issue first came to prominence with the discovery of entropy in the 1850s and the demonstration that in a closed system it can never decrease and, in accordance with the second law of thermodynamics, has a general tendency to increase. It is widely claimed that, in agreement with the second law, the entropy of the whole universe is increasing and that the only explanation for this is some very special – and inexplicable – state in the past, probably at the big bang. I will argue that the failure to find a more satisfactory explanation arises from an inappropriate approach to the problem, which goes back to the discovery of thermodynamics through the study of steam engines. Critical for the working of those machines, which had such a great impact on the development of the industrial revolution, was confinement of the steam in a cylinder. This led to the development of the beautiful theory of the statistical mechanics of systems confined to a 'conceptual box.' But it seems hardly plausible that the universe is in a box. By a simple example I will show that the theoretical consideration of unconfined systems has the potential to remove all the puzzles surrounding the existence of time's arrows.

1   Introduction

Thermodynamics was discovered as a phenomenological discipline in the 1850s.[1] It was initially mainly concerned with the behaviour of gases and liquids

---

[1] Personal circumstances prevented me from giving the Fraser Lecture at the June 2016 meeting of the International Society for the Study of Time in Edinburgh; my collaborator David Sloan kindly deputized for me at short notice. This paper presents some of the key ideas of the talk. I am currently writing a book with the provisional title *The Janus Point. A new theory of time's arrows and the big bang*. It will go into much more detail than this paper, which can give only

in confined spaces and came into existence through the belated recognition of the brilliance of Sadi Carnot's 1824 booklet *Reflections on the Motive Power of Fire*. In it, in a key section of barely six pages, he laid virtually all the foundations of thermodynamics, the science of which Einstein said "It is the only physical theory of universal content which I am convinced that, within the framework of applicability of its basic concepts, will never be overthrown.'" Note the caveat about the framework of applicability. That is going to be critical.

Let me expand. Carnot's aim was to understand the extent to which the efficiency of steam engines could be maximized and to do this he created a conceptual model of one that eliminated everything except the essential features. One of these is that the working medium, consisting of water and steam, is always confined to a cylinder. In order to establish the maximal conceivable efficiency of any steam engine, Carnot made the theoretically possible but practically unrealistic assumption that the working medium is always 'nudged' reversibly between equilibrium states in which the temperature T and pressure V are, while gradually changing, uniform throughout the working medium. In his 1854 discovery of entropy, one of the greatest achievements in the whole of science, Rudolf Clausius added just one thing to Carnot's insights, namely that heat does not flow spontaneously from a colder to a hotter body.

Now Clausius was very adept at 'riding two horses at once.' On the one hand, he arrived at profound insights without ever making any hypotheses about the underlying nature of matter, whether in the form of gases, liquids or solids. Along with Carnot and William Thomson (later Lord Kelvin), he did most to lay the foundations of phenomenological thermodynamics, the science in which Einstein had such confidence. But he was also the first significant pioneer of statistical mechanics, whose practitioners sought to explain the experimentally observed thermodynamic phenomena by means of the atomic hypothesis – by postulating the existence of atoms and molecules in ceaseless more or less random motion. Clausius's work was soon taken further, above all by James Clark Maxwell and Ludwig Boltzmann. All three modelled the putative atoms and molecules in the simplest case by hard elastic spheres; more refined models allowed non-spherical shapes or used point particles that interacted through short-range forces. This work was crowned with considerable successes and was eventually critical in the first identification of quantum effects.

It also gave a first simple and intuitive explanation of the growth of entropy, a most important feature of which was retained from Carnot's conceptual

---

a gist of a whole complex of ideas developed by my collaborators and myself. The papers [1, 2, 3] and the book [4] cover some topics that will be included in the book.

model of a steam engine: a box. At this later stage of scientific theorizing, it no longer contained steam and water with properties determined by the putative heat substance caloric but instead atoms or molecules, which were invariably assumed to bounce elastically off the box walls. In this framework, there is a very simple and often given explanation of entropy growth, which is as follows.

One supposes an initial state in which all the particles are confined to and uniformly distributed within a small box in the corner of a significantly larger otherwise empty box. The particles are moving around with random motions with an average speed that determines their temperature. At some instant, the confining walls of the small box are abruptly removed and the particles are allowed to spread out over the whole of the much larger box. One readily believes, and experience confirms it, that the particles will soon be spread out uniformly throughout the complete box. In accordance with the definition of entropy given in statistical mechanics, this corresponds to a growth of entropy; one never observes occasions in which this does not happen. Moreover, if such experiments are made at many different, widely separated locations, they all unfold in the same way. This is one example of time's arrows.

As satisfactory as the above explanation of entropy growth may be, it suffers from two serious problems, both of which were recognized clearly near the end of the 19th century in a famous exchange between Boltzmann and the young mathematician Ernst Zermelo, who later went on to become a renowned set theoretician.

The first is related to the recurrence theorem proved by Henri Poincaré in 1890. This says that any system of particles that are subject to Newton's laws, are restricted to a finite region of space and whose energies cannot grow without limit is quasiperiodic, which means that such a system must return infinitely often arbitrarily close to any region it initially occupied, which may be much smaller than the complete region it can 'explore.' This means that the entropy of such a system will, if one waits long enough, start to decrease in conflict with the second law of thermodynamics. It must come back to near its original low value.

This is generally dismissed as an academic rather than a real problem because the time between recurrences is hugely long, far longer than the age of the universe for even a few hundred particles let alone the vast number in a cubic centimeter of water. However, this does not explain why, wherever we see localized systems anywhere in the universe, we see that their entropy is increasing in the same direction. That suggests there is some controlling force at work throughout the universe whose existence cannot be explained within the framework of statistical mechanics. Boltzmann was forced to grant that the only explanation for our existence as intelligent beings, which requires a low entropy, must be some special state in the past of the observable universe. Many theoreticians have found themselves forced, willy nilly, to the same conclusion.

The second problem is closely related to the first: how does it come about that so many parts of the universe around us are in states with entropy much lower than one should expect? In accordance with the expectations one obtains from the rules of statistical mechanics, their states are hugely improbable. Early in his exchange with Zermelo, Boltzmann was totally frank about the problem and said: "Naturally, we cannot expect from natural science an answer to the question – how does it happen that at present the bodies surrounding us are in a very improbable state – any more than we can expect from it an answer to the question why phenomena exist at all and unfold in accordance with certain given laws." Although like Boltzmann I suspect that any answer to his second question, if it exists at all, is far beyond our present ken, I do think there may well be a satisfactory answer to his first question.

To begin to provide it, I need to say something about equilibration and the conditions under which it can occur. The establishment of a uniform density of particles in space by dynamical evolution from a density concentrated in the small region of the box, as described above, is one example of the very common phenomenon of equilibration. Another common example is the diffusion of heat from an initial 'hot spot.' In this second example, the speeds of the particles, which are initially great in the hot spot, are equalized by collisions among themselves. Note the critical importance of the confining box in both these examples of equilibration. Without it, the particles would simply spread out through space having collided at most only a few times. They could never equilibrate and maximize their entropy; they would become essentially free particles moving inertially in accordance with Newton's first law of motion. The question then arises of whether we can, without a box, speak about entropy at all, let alone whether it increases.

Because entropy is essentially related to probability, I'll first say something about the conditions under which probabilities can be properly defined. Consider a fair die. It has six faces and the probability of it landing on one particular face is $1/6$. This is a meaningful statement because the die only has six faces. However, you would be mad to bet on the fall of a die with infinitely many faces. Any finite number of faces selected together would have zero chance that any one of them would come up. There are numerous difficulties associated in situations with actual infinities. They arise with particles in statistical mechanics as soon as the confining box is removed. The best you can do is pretend a box is there, but you must give it some size to be able to define probabilities and with it an entropy. However, any such size is entirely arbitrary. Proper science does not lie that way.

The literature on the problem of time's arrows and the true explanation of why "at present the bodies surrounding us are in a very improbable state" is huge. I have read quite a lot of it, much by eminent scientists. I have not seen a single suggestion that the way of thinking inherited from the great work of Carnot, Clausius, Maxwell and Boltzmann might need to be reconsidered. I am thinking here of the role of the confining 'box' in the ongoing attempts to resolve the problem. This is despite the fact that many authors, like Boltzmann already in the 1890s, conjecture that the universe as a whole must be taken into consideration. Moreover, since the early 1930s it has been known that the universe is expanding; current observations actually indicate that the rate of expansion is increasing. This poses an immediate question: is it right to assume that the universe is in a box? When Johannes Kepler realized that observations made by Tycho Brahe of a comet meant the planets could not be carried in their orbits by crystal spheres, he commented: "Henceforth the planets must find their way through the void like the birds through the air. We must philosophize about these things differently." The remainder of my paper is presented in this spirit. Like the crystal spheres, the box has played the role of a prop for too long.

In fact, an important result in dynamics obtained by the great mathematician Joseph-Louis Lagrange as long ago as 1772 gives a hint of how things can be changed in a system that is not confined by a box. Lagrange studied the problem of the motion of three bodies, idealized as point particles, that interact through Newton's law of universal gravitation. He was interested in the system formed by the earth, moon and sun; Newton had given up tackling the problem because he said it gave him headaches. To this day, remarkably little that is definitive and exhaustive is known about the three-body problem. The situation with more particles, the so-called N-body problem, is even less satisfactory.

However, Lagrange was able to obtain relatively easily what is called a qualitative result. This is a result that characterizes a property that can be found in a large class or perhaps all solutions of a system of dynamical equations. In fact, Lagrange found the very first such result in dynamics. Some decades later Carl Gustav Jacob Jacobi extended it to the N-body problem. The result concerns the behaviour of the *root-mean-square length* of the system. This is essentially a measure of the diameter of the system of a system of mass points that gives greater weight to the more massive points. To find it you calculate for each pair of particles the product of their masses and the square of the distance between

them, add up all the results, take the square root of the sum and finally divide
by the total mass. This means that the root-mean-square length depends only
on the ratios of the masses and the distances between the particles; its value is
largely determined by the greatest inter-particle distances.

The qualitative result that Lagrange found and Jacobi generalized was this: if
the total energy of the system is either zero or positive, the curve of this quan-
tity, plotted against the time, is concave upwards, as shown in the figure. What
is particularly interesting about the solutions is that at the unique minimal
value of root-mean-square length the configuration of the particles is typically
rather structureless and uniform but that with increasing distance in either
direction from the minimum the system tends to break up into clusters and
the configuration becomes structured. As I will shortly explain, this leads to an
interesting example, first noted in [1], of the emergence of an arrow of time.

It must first be pointed out that in the normal interpretation of solutions of
Newtonian dynamics the specified direction of time is purely nominal. There

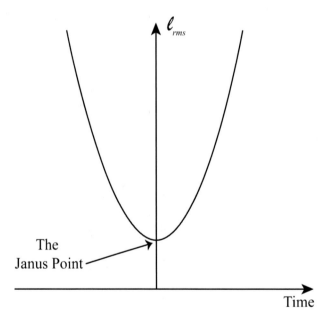

FIGURE 1.1   The root-mean-square length of the particles of the
N-body problem as a function of the time behaves
as shown in the diagram whenever the total energy
is either zero or positive. There is always a 'Janus
point' at which the size of the system, measured by
its root-mean-square length, has a unique minimum
value that divides the considered solution into two
qualitatively similar halves.

is nothing in the form of the equations that says the arrow in the figure must point to the right; it could just as well point to the left. The reason is that at any instant one could imagine all the velocities of the particles exactly reversed (the positions at that instant remaining unchanged). In accordance with the structure of Newton's equations, the particles of the system would then simply retrace their paths. This feature of Newton's laws is called time-reversal symmetry. If one thinks of complete solutions, it means that they come in pairs in which the paths of the particles through space are identical but traversed in opposite directions. If the system considered is just a small part of the universe, the two solutions of the pair are objectively distinct since there is a background direction of time defined by the universe as a whole by means of which they can be distinguished. But if the system of particles under consideration is taken to represent the entire universe, nothing intrinsic to any given solution would justify imposing an arrow such as the one shown in the figure. From where then does our extremely strong impression that there is a direction of time come? This has been a problem in physics since the discovery of thermodynamics in the 1850s.

The solution to the problem that my collaborators and I proposed in [1] and have since developed considerably is that the law which governs the universe is such that all of its solutions have a unique Janus point at which the size is minimal. I will explain later why this is not an unreasonable assumption; for the moment, suppose it is true. Then instead of trying to define an arrow of time that is the same on both sides of the distinguished point, the proposal is to assign *two*. They are to point in opposite directions away from the Janus point. This designation is, of course, suggested by the Roman god who looks in two opposite directions of time at once.

This proposal is justified by two things. First, our intuitive conviction that there is a definite direction of time arises from objective unidirectional change all around us. Everywhere we look, as far as we can see in both space and time, the processes we observe unfold in the same direction. As I said, we all get older together, just like the stars. Thus, to assign an arrow of time, we need universal unidirectional change in all that part of the universe which we can observe.

Now we come to the second justification for the proposal. Suppose that there are intelligent observers in a universe whose history is represented by a curve like the one in the figure. They must be on one or the other side of the Janus point. I mentioned that the configuration of the particles in its neighbourhood is typically rather structureless and uniform but that structured clusters tend to form with increasing distance either side of the special central point. Already at this level we can see that each half of the complete solution bears at least some resemblance to what we know about the history of the universe back to the big bang, where it was essentially uniform and hence structureless. From that time to our present epoch, the story of the universe is one of structure growth that, moreover, happens everywhere in more or less the same manner and direction.

The conclusion is this: if there are intelligent observers on both sides of the Janus point, the directions of time they identify from the objective change in the structure of the universe that they observe around them will, on either side, point away from the Janus point. For them, that special point will be effectively the beginning of time and a big bang. Although the observers will not realize it, such a scenario will resolve the mystery of how it can be that equations indifferent to the direction of time can nevertheless seem to break time-reversal symmetry by giving rise to arrows of time. The fact is that the solutions do all respect the time-reversal symmetry of the equations because the two halves of each and every one of them is symmetric about its Janus point. However, in each half there is a pervasive arrow of time away from a special state that looks like the big bang.

I think this is enough explanation of the key idea. However, it is bolstered by much more than the simple outline given here. Let me mention some of the most important things.

First, the assumption of a Janus point in the Newtonian N-body problem is not very restrictive. In fact, if one were to regard it as a model 'island universe' of particles in space, then as argued in [2], one would not only expect the energy to be exactly zero (which would ensure a Janus point) but also the angular momentum too. More important than such Newtonian 'toy' models of the universe is, of course, whether one can expect the existence of a Janus point in Einstein's general theory of relativity. Under not unreasonable assumptions, this is shown to be the case in [3]. Important for this is the transition to shape dynamics [4], about which I will say a few words.

It is a new way of describing the universe which grew out of the simple recognition that, by definition, the universe is everything; it is a contradiction in terms to suppose that there is anything outside it. It then follows that the description of the universe must be intrinsic, i.e., described in terms of purely relative quantities that do not rely on something like an external scale to define its size or Newton's absolute space to define orientations. Insights like these go back to Leibniz in his 1716 *Correspondence with Clarke* if not earlier and were strongly advocated by Ernst Mach in a famous book in 1883 that was a major stimulus for Einstein in his creation of general relativity. If these relational ideas are taken to their logical conclusion, they imply that observers within the universe can only determine its shape and the way it changes.

Shape dynamics bears on two of the issues mentioned above: the definition of probabilities in statistical dynamics and the existence of a Janus point in cosmological solutions of general relativity. Luckily the discussion of both can be illustrated by the difference between the shape and size of a triangle. The simplest non-trivial model of the universe that one can conceive in purely intrinsic terms is the three-body problem that Lagrange studied.

Let us first consider statistical considerations when we want to contemplate all possible sizes and shapes of triangles with the same mass points at their three vertices. The possible sizes, measured by the root-mean-square length, range from zero to infinity. Now conventional statistical mechanics only makes sense because the size of the box puts an upper bound on the value of the root-mean-square length. That is why one can meaningfully define probabilities. If one imagines that size of the complete universe is something physically real and the universe can expand without limit, probabilistic statements become impossible. It is like trying to bet on the fall of a die with infinitely many sides.

The situation is quite different if one considers only the shape, which is defined by two internal angles. They both have only bounded ranges analogous to those for latitude and longitude, which have ranges from -90 to +90 degrees and -180 to +180 degrees, respectively. Mathematically, the space of possible shapes is *compact* while the space of sizes is *non-compact*. This makes all the difference when it comes to defining probabilities. As explained in [2], one can develop a statistical mechanics of the shapes of possible universes, characterize the growth of structure purely in terms of shapes and show how the second law of thermodynamics arises away from the Janus point once subsystems that are effectively 'self-confined.' The statistics of shapes is a key part of the overall explanation of time's arrows that my collaborators and I are developing.

Our ideas can be applied in the framework of both Newtonian dynamics and general relativity, which, of course, provides a much more realistic picture of the universe. It is important that we do not question the basic correctness of general relativity. What we do is argue that key features of the theory have escaped notice because it has not been properly understood as a theory of the changing shape of the universe.

This brings me to the potential significance of the paper [3] and the way it questions the interpretation (but not the proofs or results) of the famous singularity theorems of Stephen Hawking and Roger Penrose first obtained in the 1960s. These showed that in black holes and at the big bang general relativity appears to 'predict its own demise' in that certain key quantities like the curvature of space and the density of matter become infinite. It is widely accepted by physicists that the appearance of such infinities is an indicator of the breakdown of a theory and the need for it to be replaced by a deeper theory. For half a century, it has been almost universally believed that classical general relativity will only be 'cured' of the singularity problems when it has been successfully unified with quantum theory.

It is, indeed, almost certain that general relativity and quantum theory, the two great achievements of the physics of the 20th century, must be unified. However, the main result of [3] suggests that classical general relativity, if properly interpreted, does not, at least for the case of the big bang, breakdown. The basis of the argument can be illustrated by the behaviour of certain solutions of the Newtonian three-body problem in which so-called total collisions occur. This is when all three particles collide at once in a violent unpredictable manner: the speeds become infinite as the size of the system, measured by the root-mean-square length, tends to zero. When described in Newtonian terms, it appears that the theory breaks down and it is not possible to say what will happen after the collision.

However, the time and size that appear in the Newtonian description are extrinsic in the sense that, to be determined, they presuppose clocks and rulers outside the system for their measurement. In a proper intrinsic description of the universe, one should only consider how its shape evolves. This changes things. The way the extrinsic time and size change then become irrelevant; only the shape of the system counts. It can be shown that it behaves in a perfectly predictable manner and 'passes through' what only extrinsically appears to be a singularity. In fact, far from a breakdown of the equations that determine how the shape changes, they actually become simpler. They are moreover autonomous. What this means is that one can first find how the shape of the system changes, obtaining a curve of successive shapes in the space of possible shapes of the system. Time and size play no role in this. However, once the curve in 'shape space' has been found, one can 'add' equations that determine change of size and time from the already found change of shape. This result is not restricted to the three-body problem; it holds for any (finite) number of particles greater than three.

It is shown in [3] that essentially the same thing happens in general relativity provided a certain form of matter is present. This is a not too restrictive condition: general relativity allows many forms of matter to exist and interact with the gravitational field. The one required in [3], a massless scalar field, may well be present.

Thus, the upshot of this 'shape-dynamic' approach to general relativity is that the big bang might well be a Janus point, dividing solutions of the theory into qualitatively similar halves in which arrows of time defined by the way in which the shape of the universe becomes ever more structured point in opposite directions. We, of course, exist on side of the big bang and, unable to observe its 'other side,' take it to be the birth of time. In a sense it is, but only for us.

References

Barbour, J., T. Koslowski and F. Mercati. "Identification of a gravitational arrow of time," *Physical Review Letters* 113, 181101 (2014). [1]

Barbour, J., T. Koslowski and F. Mercati. "Entropy and the typicality of universes" arXiv: 1507.06498v2 [gr-qc] 24 July (2015). [2]

Koslowski, T., F. Mercati and D. Sloan. "Through the big bang: Continuing Einstein's equations beyond a cosmological singularity," *Physics Letters B* 778, 339 (2018). [3]

Mercati, F., *Shape Dynamics. Relativity and relationism*, Oxford: Oxford University Press, (2018). [4]

CHAPTER 2

'Looking Down Time's Telescope at Myself': Reincarnation and Global Futures in David Mitchell's Fictional Worlds

(*Winner of the ISST New Scholar Prize*)

Rose Harris-Birtill

Abstract

This essay explores the trope of reincarnation across the works of British author David Mitchell (b. 1969) as an alternative approach to linear temporality, whose spiralling cyclicality warns of the dangers of seeing past actions as separate from future consequences and whose focus on human interconnection demonstrates the importance of collective, intergenerational action in the face of ecological crises. Drawing on the Buddhist philosophy of *samsara*, or the cycle of life, death, and rebirth, this paper identifies links between the author's interest in reincarnation and its secular manifestation in the treatment of time in his fictions. These works draw on reincarnation in their structures and characterization as part of an ethical approach to the Anthropocene, using the temporal model of "reincarnation time" as a narrative strategy to demonstrate that a greater understanding of generational interdependence is urgently needed in order to challenge the linear "end of history" narrative of global capitalism.

Keywords

David Mitchell – time – reincarnation – Buddhism – Anthropocene – samsara – capitalism – Nietzsche

In the second section of *The Bone Clocks* (2014a) – the sixth of seven novels to date by British author David Mitchell (b. 1969) – Cambridge student Hugo Lamb visits his headmaster's old friend, Brigadier Philby, in a nursing home.[1]

---

1 This essay also appears in *KronoScope: Journal for the Study of Time* 17.2 (2017): 163–81, and is reprinted courtesy of Brill. An extended version of this essay can be found in *David Mitchell's Post-Secular World: Buddhism, Belief and the Urgency of Compassion* (Rose Harris-Birtill, Bloomsbury Academic, 2019; 89–108).

The brigadier, formerly a "linguist and raconteur" who taught Hugo how to cheat at card games and get a fake passport, is now suffering from dementia, and is largely "non-verbal" (118). After he leaves, Hugo realises "[w]hen I look at Brigadier Reginald Philby, I'm looking down time's telescope at myself" (120). For Hugo, the metaphor of time's telescope reflects his anxiety about his own unavoidably finite lifespan (he notes, "the elderly *are* guilty: guilty of proving to us that our wilful myopia about death is exactly that"), a preoccupation that leads him to join the soul-stealing, immortality-seeking Anchorites later in the novel (116). However, as this essay will demonstrate, the notion of "looking down time's telescope at myself" as a form of future-facing temporality has far greater significance: both for Mitchell's literary approach to time across his oeuvre and also for the exploration of alternative cyclical temporalities that it prompts outside of his fictional world.

As an imaginary model, "time's telescope" makes visible the gaze of the present self at its own future, and by implication, the causal relationship between them. The telescope holder becomes both immediate viewer and future subject, a simultaneously cross-temporal spectator and actor. This is not just any telescope: it is "time's telescope," allowing the individual to see their future self as a separately observable entity. It is an image of the temporally-multiplied self that evokes the author's fascination with cyclical temporality, transmigration and reincarnation in his fictional world, "time's telescope" providing an imaginary model that makes visible the gaze of the present self at its own future, and by implication, the causal relationship between them. The sight line created is both linear and cyclical – a linear device that provides a means of cyclical self-observation. However, such self-observation does not form a perfect circle but a spiralling gaze which continues to move forward in time, creating a feedback loop: having seen their future self, the present self can choose to modify their actions in response, in a relationship that fosters a heightened awareness of personal responsibility and agency. After all, it is only by choosing to gaze down this metaphorical instrument that the actions of the future self become visible, magnifying a temporally distant self which is *other* to the self, but "othered" only by time, creating a relationship in which the actions of each self – present and future – can become mutually influential.

Building on this idea of "looking down time's telescope at myself," this essay explores the trope of reincarnation in David Mitchell's works as an alternative approach to linear temporality, whose spiralling cyclicality warns of the dangers of seeing past actions as separate from future consequences, and whose focus on cross-temporal interconnection demonstrates the importance of collective, intergenerational ethical action in the face of ecological and humanitarian crises.

1   "Time's Telescope" and Reincarnation

In a 2016 interview, David Mitchell notes:

> linguistically time is singular, but actually it's plural – there are so many
> different kinds of time. There's a lifespan [...] there's geological time [...]
> there's reincarnation time [...] these tiny, tiny moments where vast things
> can happen.
>
> "Interview by Rose Harris-Birtill," MITCHELL 2016b

Mitchell's works reflect this fascination with "many different kinds of time,"
often using the micro-temporal to depict the heightened significance of immediate present-time decisions as "these tiny, tiny moments where vast things
can happen." To give just one example, his short story "My Eye On You" (2016a)
depicts a break in conventional temporality to introduce a simultaneous dimension in which time is experienced at a twelfth of the speed of "normal"
time. Here, the use of a slowed timescale lengthens the lived present to introduce an alternative temporality embedded within our own, a form of reflective
time in which its narrator is able to make life-changing ethical interventions
to help others in need. As several critics have noted, this exploration of temporal plurality runs throughout Mitchell's works. To give just a few examples, in
"David Mitchell's Fractal Imagination: *The Bone Clocks*" (2015a), Paul A. Harris
rightly notes the "labyrinthine nature of time" portrayed with the novel's temporal shifts (149), while in "The Historical Novel Today, Or, Is It Still Possible?"
(2013), Fredric Jameson explores the layered temporalities of *Cloud Atlas*
(2004a) using the model of an elevator lurching through "disparate floors on its
way to the far future" (Jameson 2013, 303). Similarly, Patrick O'Donnell notes in
A Temporary Future: The Fiction of David Mitchell (2015) that Mitchell's depiction of temporality occupies "multiple domains" (16).

   Alongside this recognition of temporal plurality, several critics also note the
author's portrayal of cyclical timescales, often presented in conjunction with a
disrupted linear temporality. For example, in "Genome Time: Post-Darwinism
Then and Now" (2013), Jay Clayton suggests that *Cloud Atlas* displays a "para-
doxical combination of linear and cyclical perspectives on time" (58), while in
"'On the Fringe of Becoming' – David Mitchell's *Ghostwritten*" (2004), Philip
Griffiths discusses the novel's "disparate, discontinuous and cyclical view of
history" whose "linear chronology of events is broken down" (80–84). Similarly,
Marco de Waard, writing in "Dutch Decline Redux: Remembering New
Amsterdam in the Global and Cosmopolitan Novel" (2012), argues that *The
Thousand Autumns of Jacob de Zoet* (2010a) "articulates a cyclical conception

of history" (115).[2] In each case, a cyclical temporality is identified in Mitchell's works which disrupts readerly expectations of temporal linearity but does not necessarily overturn it altogether, suggesting that a model which combines the linear and the cyclical – the spiral – may be more appropriate.

In his 2016 interview, Mitchell also names "reincarnation time" as a specific form of temporality ("Interview by Rose Harris-Birtill," Mitchell 2016b). Read as a temporal model, "reincarnation time" can be seen to share the inherent spiralling cyclicality and future-facing self-awareness evoked by "looking down time's telescope at myself." Defined as "[r]enewed incarnation; the rebirth of a soul in a new body or form," reincarnation or rebirth forms part of many global belief systems, including Buddhist, Hindu, Native American, Inuit and Tibetan Bon traditions.[3] However, it is also defined as "revival, rebirth, or reinvention," providing the basis for a secular interpretation focused on causality, change, and rebirth. The concept of "reincarnation time" offers a flexible and specifically human temporal model, measured not by the mathematically-calculated and ultimately arbitrary second – a base unit calculated from a fixed rate of radioactive decay – but by the individual lifespan. The duration of a single human life provides a flexible temporal unit that is different for every individual, its finite length based not on assumed similarity, but on constant change. Read within a secular framework, "reincarnation time" provides a means of approaching the overlapping linearity of intergenerational interconnection, seeing each generation of species reproduction as not merely the beginning of another life, but as a form of rebirth in which past actions are unavoidably connected to – and visible in – the future lives of others.

The author has spoken on his interest in reincarnation in several interviews, as well as his interest in Buddhism, suggesting that the Buddhist model of reincarnation may be particularly relevant for a discussion of this trope in his works. For example, in an interview with Harriet Gilbert, Mitchell speaks of his "interest in Buddhism" (Mitchell 2010c), while in an interview for Shanghai TV he remarks, "of the great world religions, Buddhism [...] strikes the strongest

---

2 Several other critics also draw attention to the simultaneously linear and cyclical nature of Mitchell's works. In *The Cosmopolitan Novel* (2009), Berthold Schoene describes *Ghostwritten* (1999a) as an "oddly timeless and dislocated" work whose ending, linking back to its beginning, "dissolves the novel's linearity" altogether (Schoene 2009, 111). In "Cannibalism, Colonialism and Apocalypse in Mitchell's Global Future" (2015), Lynda Ng describes *Cloud Atlas* in terms of an ouroboric timescale which disrupts Western conventions of linear temporality (107, 118), while Peter Childs and James Green also note in "David Mitchell" (2013) that *Cloud Atlas'* structure is "at once linear and cyclical" (149).

3 "reincarnation, n." 2016. *OED Online*. Oxford University Press, June. http://www.oed.com/view/Entry/161570.

chord inside me, seems to suit me best" (Mitchell 2012). He later reveals in a 2013 interview with Andrei Muchnik, "I am a kind of secular Buddhist [...] Buddhism doesn't ask me to sacrifice my rationality or my common sense, it doesn't ask me to believe the impossible" (Mitchell 2013a). In an interview with Richard Beard he notes, "[t]he Buddhist model of reincarnation is particularly elegant" (Mitchell 2005b). In a 2004 interview by Eleanor Wachtel, although Mitchell states that he does not believe in literal reincarnation, he again notes that "it's an elegant, beautiful idea," adding that "there are more humans alive who believe there is such a thing as a soul [...] than humans alive who believe there isn't such a thing as a soul. So may I be wrong" [sic] (Mitchell 2004c). However, in a 2010 interview with Adam Begley he also emphasises the redemptive value of reincarnation in a secular, ecological sense: "[t]here is solace, however, in the carbon cycle, in the nitrogen cycle. Biochemically, at least, reincarnation is a fact" (Mitchell 2010b). Similarly, in a 2014 interview with Laurie Grassi he notes that there is "solace in the notion we can have another go and try and fix things [...] Reincarnation is a useful idea" (Mitchell 2014b). Throughout, the author discusses his interest in reincarnation in secular terms as a "useful" cross-transferrable model, a metaphysical concept which also has value for interpreting the physical world.

The concept of reincarnation resurfaces throughout Mitchell's body of fiction. As the author notes in a 2010 interview with Wyatt Mason, his works together form a growing "fictional universe with its own cast"; he asserts, "each of my books is one chapter in a sort of sprawling macronovel. That's my life's work" (Mitchell 2010d). Similarly referring to his continuous body of fiction as an "über-novel," Mitchell remarks in a 2013 interview with Jasper Rees that his libretti for the operas *Wake* (2010) and *Sunken Garden* (2013) are also part of this textual universe, noting "I like to think of everything I do as chapters in one bigger über-novel and the libretti are also chapters in the über-novel" (Mitchell 2013b). Many of Mitchell's short stories also continue the themes and characterization of his longer fictions: for example, characters that the thirteen-year-old protagonist of *Black Swan Green* (2006a) encounters in childhood resurface as adults in several short fictions, including "Acknowledgements" (2005a), "Preface" (2006b) and "The Massive Rat" (2009).

As opposed to a series of novels joined by a single setting, time, place, or protagonist, Mitchell's body of interconnected fictions cumulatively depicts not the world of a single individual or family, but that of a global community, in a macronovel that spans many continents, time periods, genres and art forms. For example, *Ghostwritten*'s (1999a) plot travels through Japan, Hong Kong, China, Mongolia, Russia, London, Ireland and New York; *Cloud Atlas* begins in the nineteenth century and moves forwards in time to a post-apocalyptic far future; *number9dream* (2001) is a coming-of-age novel; *The Thousand Autumns* is a historical novel; *Black Swan Green* a semi-autobiographical novel; *The Bone Clocks* is part-fantasy; and *Slade House* (2015b) is a supernatural horror story. Instead, these disparate works are drawn into an interconnected "macronovel" through their shared characters – such as Marinus, who reappears in several of Mitchell's novels and short stories, and both of his libretti – and their repeated cross-textual motifs and thematic preoccupations, from the mysterious "moon-grey cat" that appears in *Black Swan Green, The Thousand Autumns, The Bone Clocks* and *Slade House* (46; 228; 259; 9), to the recurring exploration of reincarnation.

The concept of reincarnation resurfaces throughout the macronovel in several different forms. Its shared characters are both metaphorically reborn, with characters reappearing from text to text, and literally reborn, with several characters depicted as being fully reincarnated, reborn into different bodies, as in *The Thousand Autumns, Cloud Atlas, The Bone Clocks* and *Slade House*. While the formal complexities of Mitchell's macronovel are beyond the scope of this essay – and further critical investigation is certainly needed into the author's entire corpus as an interconnected whole – the concept of reincarnation perhaps influences even the structure of these works. For example, along with *Cloud Atlas'* reincarnated characters,[4] the increasing effects of human predation are also "reborn" in different forms across different time periods and global geographies throughout the novel – as is the ongoing ethical problem of how to prevent such self-destruction. As will be discussed later in this essay, it is this ongoing exploration of how to collectively shape these shared global futures across many lifetimes, as explored through characters reborn across different times and texts, that lies at the heart of Mitchell's macronovel, drawing its disparate fictions into a continuous ethical worldview.

Reincarnated and transmigrated characters span the full length of Mitchell's writing, including: the disembodied spirit in his first novel *Ghostwritten* and short story "Mongolia" (1999b), both published in 1999; *Cloud Atlas'* reincarnated characters; the short story "Acknowledgements," in which its narrator discovers "psychomigration," or the ability of the mind to leave the body; and the reincarnated character of Marinus, as mentioned earlier, who appears in

---

4 In several interviews, Mitchell confirms the inclusion of the shared comet-shaped birthmark that links *Cloud Atlas'* narrators as evidence of their reincarnation. For example, in an interview with Harriet Gilbert, Mitchell notes, "they're the same person [...] it's the same soul, being reincarnated in different stages with just a dim, dream-level awareness that they've been here before and will be here again" (Mitchell 2010c).

three novels, two operas and two short stories to date.[5] However, as Caroline Edwards notes in "'Strange Transactions': Utopia, Transmigration and Time in *Ghostwritten* and *Cloud Atlas*" (2011), "transmigration" is "not a common trope in twentieth- and twenty-first-century literature" (191). While several critics have mentioned Mitchell's unusual use of reincarnation and transmigration, there is little sustained analysis of these across his macronovel. For example, in "'Gravid with the Ancient Future': *Cloud Atlas* and the Politics of Big History" (2015), Casey Shoop and Dermot Ryan briefly note that "the promise of reincarnation rises above the novel's bleak record of predacity" (105), while in "'This Time Round': David Mitchell's *Cloud Atlas* and the Apocalyptic Problem of Historicism" (2010), Heather J. Hicks goes further towards a discussion of reincarnation in the novel, suggesting that Sonmi's ascension is "the fabricant equivalent of the Buddhist state of Enlightenment," but ultimately concludes that "the complexities of Buddhist spirituality are beyond the scope of this essay." In "Food Chain: Predatory Links in the Novels of David Mitchell" (2015), Peter Childs also notes the "transcultural" influence of "Eastern philosophies" across Mitchell's works, observing "[t]he soul thus seems to represent the site and agency of reanimation for Mitchell," but again, the discussion is tantalizingly brief (190). Existing criticism tends to limit its mentions of reincarnation to *Cloud Atlas* and *Ghostwritten*, with no sustained discussion of this trope across his entire fictional universe to date.[6] Drawing on the Buddhist philosophy of samsara, or the cycle of life, death, and rebirth, this essay therefore investigates the macronovel's use of reincarnation as a transferrable temporal model, whose secular application provides a productive means of "looking down time's telescope at myself."

---

5   Read together, the novels *The Thousand Autumns of Jacob de Zoet* (2010a), *The Bone Clocks* (2014a) and *Slade House* (2015b), the operas *Wake* (2010) and *Sunken Garden* (2013), and the short stories "I_Bombadil" (2015a) and "All Souls Day" (2016c) form a seven-part subset of Mitchell's writing in which the reincarnated character of Marinus features more prominently, and is more fully realized, than any other character in the macronovel.

6   To give further examples, Caroline Edwards also comments on the "symbolic figure of transmigration" in *Ghostwritten* and *Cloud Atlas*, discussing the use of this trope as part of a construction of "a trans-historical community" that fosters "a forceful, oppositional agency" in the face of "colonial power" (Edwards 2011, 190). Berthold Schoene's *The Cosmopolitan Novel* also identifies the "eternal cycle of reincarnation" depicted in *Cloud Atlas* as a "general rather than specific symbol of humanity's potential for communal affiliation" (Schoene 2009, 115–16).

2   "Treading on Spirals": Reincarnation in the Macronovel

In Buddhist belief systems, reincarnation broadly refers to the process by which past actions influence future rebirths in an ethical model of cause and effect based on karma, defined in the Oxford English Dictionary as "[t]he sum of a person's actions [...] regarded as determining his fate in the next [life]; hence, necessary fate or destiny, following as effect from cause."[7] Compassionate actions are believed to result in being reborn as a "higher" lifeform, a process which repeats across many lifetimes until the individual breaks the seemingly endless cycle of life, death and rebirth, or samsara, to reach a state of Enlightenment, or nirvana.[8] As Richard Robinson and Willard Johnson note in *The Buddhist Religion* (1997), while reincarnation is not unique to Buddhism, its focus on karma offers a quintessentially Buddhist form of "moral causality" which is "intended to be experiential and concrete" (18–19).[9] The concept of "reincarnation time," in which this approach to reincarnation is considered as a secular temporal model, provides – as I have suggested – an alternative to linear temporalities. Its long timescale, focus on human causality, and potential for progression or regression all suggest an ethical means of approaching a new geological era defined by the shared consequences of compounded human actions over many lifetimes, while maintaining the possibility of positive change or dystopic decline. Its emphasis on causality also prompts a greater focus on individual ethical action and shared consequences. "Reincarnation time" shares important characteristics with the concept of "time's telescope"; applied to Mitchell's macronovel, it provides a model through which to approach its expansive temporal dimensions, its timescale "stretching back approximately seven millennia" from the reincarnated character Moombaki's

---

7   "karma, n." 2016. *OED Online*. Oxford University Press, June. http://www.oed.com/view/Entry/102561.

8   The Buddhist goal of Enlightenment or nirvana is variously described in Buddhist interpretations as a state of understanding, peace, and total release from the cycle of rebirth, rather than a fixed place (see "Nirvāṇa" in *The Buddhist Religion* (1997, 332) by Richard H. Robinson and Willard L. Johnson), described by John Snelling as "a Way Out of the system, one that did not lead to a temporary resting place" (see "The Buddhist World View" in *The Buddhist Handbook* (Snelling 1987, 48)).

9   As Marcus Boon notes in "To Live in a Glass House is a Revolutionary Virtue Par Excellence: Marxism, Buddhism, and the Politics of Nonalignment" (2015), "one might consider the doctrine of karma as an alternative form of general economy. Thanks to the 'kitschification' of this term by North American countercultures in the 1960s, the philosophical and religious meanings of the word have been obscured. [...] Karma is the universal law of cause and effect" (55).

distant past in *The Bone Clocks* (415), to several hundred years into the future in *Cloud Atlas*.

The concept of "reincarnation time" also addresses the author's fascination with temporal cyclicality. As Mitchell notes in a 2015 interview with Paul A. Harris, "metaphorically time can seem mighty circular or phase-like for something allegedly linear" (Mitchell 2015c, 9). David Mitchell further explores the concept of temporal cyclicality in *Cloud Atlas*, referring to the Nietzschean concept of the eternal return. The eternal return or eternal recurrence, described in Friedrich Nietzsche's 1882 text *The Gay Science* (2008) as the identical repetition of lived experience "innumerable times," as if the "eternal hourglass of existence [...] turned over again and again," suggests a potential alternative cyclical model for understanding Mitchell's treatment of temporality in the macronovel (194). The eternal return refers to a temporal cycle in which life is destined to endlessly repeat itself, as described in Robert Frobisher's suicide note in *Cloud Atlas*:

> Nietzsche's gramophone record. When it ends, the Old One plays it again, for an eternity of eternities. [...] my birth, next time around, will be upon me in a heartbeat. [...] Such elegant certainties comfort me.
>
> MITCHELL 2004a, 490

Frobisher's Nietzschean conception of time as an endlessly repeating set of events provides a particularly "elegant" possible temporal model within a book whose tribal warfare in its distant future disturbingly resembles that of its distant past. However, the eternal return is ultimately an ethical thought experiment which interrogates the possibility of acceptance of life as it is, and whose non-linear cyclicality refuses the possibility of meaningful change. As such, it suggests a paradoxically anti-temporal model of temporality, whose focus is not on the course of lived events as they unfold, but on the certainty of their endless recurrence. While Frobisher's imagining of an endlessly cyclical temporality may serve to "comfort" him at the moment of his death, this temporal model is not able to account for the vital moments of change that are at work across the macronovel. If such a future is unalterably destined to be repeated – as suggested with Frobisher's reference – then the long-term efforts of the reincarnated Horologists in *The Bone Clocks* are ultimately fruitless, while the narrator's urging of the reader to take up his compassionate lifelong course of helping others in "My Eye On You" is little more than an endlessly repeated plea. As *Cloud Atlas* demonstrates with Frobisher's suicide, the lived outcome of such temporal fatalism may be ideologically comforting, but practically futile. By contrast, the inclusion of reincarnation in the macronovel suggests an alternative model of temporal cyclicality that allows for ongoing change and progression, for better or worse, spiralling across unimaginably long time-scales and individual moments alike. Across this endlessly fluctuating fictional world, the depiction of "reincarnation time" in the macronovel reinforces that what is at stake is not whether change is possible, but whether such changes will result in ethical progression or regression.

Unlike the cyclical model of the eternal return, the spiralling model of "reincarnation time" is also visible across the macronovel's narrative structures, as well as in its content; a progressive narrative cyclicality is frequently embedded into Mitchell's works. For example, this is visible in *Ghostwritten's* interlinked beginning and ending (its narrative begins and ends with "[w]ho was blowing on the nape of my neck?" and "[w]ho is blowing on the nape of my neck?" respectively) (3, 436), in *Black Swan Green's* depiction of a single January to January year in its narrator's life, in *Cloud Atlas'* journey from the distant past into the distant future and back again, and in *Slade House's* repeated depictions of victimhood and soul-stealing, each successive generation of entrapment carried out in the same way, by the same people, in the same setting, for each of its characters. Yet even within these cyclical structures, a sense of ethical progression is maintained that prevents each narrative loop from being fully closed, allowing for change and movement within each text. For example, *Ghostwritten* returns the reader to the start of the novel, but it is with a new awareness of the fragility of shared human existence gained from witnessing its near-destruction by the sentient global security system in its final chapter. *Black Swan Green* returns to the beginning of another year for its narrator, but his life has been irrevocably altered with his parents' divorce as he leaves his childhood home at the end of the novel. The author's endings are often beginnings, rebirths in disguise: for example, the last words of *Black Swan Green* are Jason's sister Julia telling him "it's not the end" (371), while *The Bone Clocks'* final line is "[f]or one voyage to begin, another voyage must come to an end, sort of" (595). Similarly, the final words of *number9dream*, "I begin running," indicate a new beginning for its narrator (418), while *Slade House* ends with the rebirth of Norah Grayer as her soul leaves her dying body and transmigrates into "a foetal boy" (233). In each case, the narrative's structural cyclicality is accompanied by definitive narrative progression in its content, providing a spiralling temporal frame of reinvention and rebirth.

Nietzsche's *Thus Spoke Zarathustra*, first published in four parts between 1883 and 1885, also develops the theory of eternal recurrence. In it is the assertion that "[a]ll that is straight lies [...] All truth is crooked, time itself is a circle" (2006, 125). However, the macronovel's cyclical narrative structures refuse to provide the perfect temporal closure of the eternal return, maintaining the possibility of change and ethical progression; Frobisher's Nietzschean model of perfect temporal circularity is overturned in favour of a temporal model based on a simultaneously cyclical and linear "reincarnation time." In David Mitchell's essay "The View from Japan" (2007), he refers to the artistic process of "bending [...] time's false straight line into its truer shape, the spiral"; as opposed to a perfect circular temporality, "reincarnation time" maintains a spiralling possibility of forward and backward progression in a revaluing of the linear within the cyclical, even while conventional linearity itself is disrupted. In *The Bone Clocks*, we meet a benevolent race of compassionate reincarnated beings, called Horologists, who "live in this spiral of resurrections" (431). Significantly, their reincarnations are described as a "spiral," rather than the simpler circle that such rebirths would suggest. It would seem fitting, therefore, that when one of the novel's narrators, Crispin, hears a mysterious premonition that mentions spirals, it is only at the moment of his death that he finally realises the spirals were there all along, stitched into the carpet under his feet: "Not dots. *Spirals*. All these weeks. Treading on spirals. Look." (382).

The macronovel's inclusion of reincarnation ultimately presents the reader with a long view of the shared historical consequences of human action – or inaction. As Robinson and Johnson note, reincarnation is not unique to Buddhism, but is a belief "documented among archaic cultures throughout the world," identifying that "[t]he specifically Buddhist feature is the ethical correlation of good karma – intentional deeds – with happy births and bad karma with miserable ones" (1997, 18–19). Mitchell's reincarnated characters make visible these generations of human cause and effect, a narrative strategy that is particularly representative of Buddhist approaches to reincarnation, using a secular version of its experiential moral causality which depicts the far-reaching positive or negative consequences of individual actions across a vast scale, and portraying the macronovel's dystopic global future as caused by compounded human activity. As Holly realizes in *The Bone Clocks*, we are "leaving our grandchildren a tab that can never be paid," the ecological disaster in the novel directly caused by generations of damaging human behaviour in a secular depiction of the karmic model; she laments, "the regions we deadlanded, the ice caps we melted [...] the species we drove to extinction, the pollinators we wiped out, the oil we squandered [...] all so we didn't have to change" (533).

Cumulatively, Mitchell's secular depiction of "reincarnation time" suggests a form of collective experiential time with its own ethical foundations and consequences, based not on mathematical formulae or radioactive decay, but on the revisited consequences of human actions over many lifetimes, and the importance of individual action as the essential precursor to collective change. The measure of such a temporal system is what may be achieved – or destroyed

– for current and future generations at any stage in the cycle, each individual human lifespan providing a critical opportunity to make positive ethical decisions to help others. As Fredric Jameson notes in "The Historical Novel Today, or, Is It Still Possible?" (2013) "for better or for worse, our history, our historical past and our historical novels, must now also include our historical futures as well" (312). The macronovel's depiction of "reincarnation time" allows for such an incorporation of past and future temporalities. What is important in such a system is not a literal belief in reincarnation, but living *as if* the individual will be reborn to see their own behavioural consequences, and in doing so, "looking down time's telescope" not at an abstract future, but at future versions of "myself" – the lives of other finite yet interconnected beings who will inherit the future that we create, just as we inherit the consequences of others' past lives. Crucially, such a model uses the influences of Buddhist ethical philosophies to build a secular, humanist approach that is not dependent on the promise of a final redemptive heaven or nirvana, but also avoids the ethical stagnation of a self-centred postmodern nihilism. It suggests a temporal system rooted in humanist action, rather than the potentially damaging ideologies of religious doctrine or utilitarian sacrifice, by which the individual is marginalized for the greater good.

As a human-centred exploration of a spiralling cyclical temporality, "reincarnation time" suggests that a greater focus on generational interdependence and ethical causality are urgently needed in order to challenge the seemingly inescapable linearity of the "end of history" narrative of global capitalism. As Fredric Jameson observes in *The Cultural Turn* (1998), "[i]t seems easier for us today to imagine the thoroughgoing deterioration of the earth and of nature than the breakdown of late capitalism; and perhaps that is due to some weakness in our imagination"; such imagined "deterioration" seems all the more worrying in light of the proposed geological epoch of the Anthropocene, marking a new era defined by global human impact (50). Under the current geological system, the formal categorization of such an epoch will form a permanent addition within an irreversibly linear geological time scale. The challenge within such an era, then, will be to imagine other temporal counter-strategies which may help our species not merely to survive, but to regenerate; as the macronovel demonstrates, "reincarnation time" offers one such strategy.

3   "Reincarnation Time" and the Anthropocene

Mitchell's use of "reincarnation time" suggests the value of alternative temporalities in the face of the Anthropocene, a new geological epoch proposed to succeed the Holocene. As Simon L. Lewis and Mark A. Maslin note in "Defining the Anthropocene" (2015), "[r]ecent global environmental changes suggest that Earth may have entered a new human-dominated geological epoch" in which the "impacts of human activity will probably be observable [...] for millions of years into the future," and that consequently, "human actions may well constitute Earth's most important evolutionary pressure" (171). At the time of writing, no date for the beginning of such an epoch has yet been agreed on. As Lewis and Maslin recognise, "there is no formal agreement on when the Anthropocene began, with proposed dates ranging from before the end of the last glaciation to the 1960s" (171). However, these geological changes created by global human activity are already visible in the macronovel, a huge project that effectively traces the evolution of this human-centred epoch, from the early depiction of imperial conquest in *The Thousand Autumns*, to the near-future oil crises in *The Bone Clocks*, to the far-future civilisation breakdown in *Cloud Atlas*. For Mitchell's fictional world, the categorisation of this human-engineered global era is not under debate: it is already here.

The author's engagement with these concerns is by no means limited to the macronovel; Mitchell also notes the worrying ecological trajectory of globalisation in several interviews. To give just a few examples, in an interview with Wayne Burrows he notes that "[t]he logical extension of neo-capitalism is that it eats itself [...] corporate interests really will pollute the land that supports them, because they can make money by doing so" (Mitchell 2004b); in an interview with Adam Begley he observes that "[w]hat made us successful in Darwinian terms – our skill at manipulating the our environment – now threatens to wipe us out as a species" (Mitchell 2010b); and in an interview with Zahra Saeed he notes that "the increases in the standards of our living are being bankrolled by the standards of living of our children and our grandchildren" (Mitchell 2014c). In a 2015 essay, Paul A. Harris usefully recognizes that Mitchell is a "novelist of the Anthropocene" (Harris 2015b, 5); this is an author whose interconnected fictions, with their shared past and future, illustrate the shared causality – and consequences – of the Anthropocene in action. The macronovel forms a huge project that effectively depicts the evolution of this human-centered epoch, its earliest novel to date set in 1779 (*The Thousand Autumns*) during the age of imperial global navigation and trade. Read cumulatively, the macronovel shows the predatory human actions that contributed to the creation of such an epoch of Anthropocentric global impact, and warns of the potential futures created by such actions. However, while the categorization of a new epoch exists within an irreversibly linear geological temporality, the use of a spiralling "reincarnation time" in the macronovel complicates this with the possibility of a multiplicity of beginnings and endings at any stage, suggesting the value of the cyclical within the linear for envisaging productive strategies for change.

Mitchell notes in an interview with Richard Beard that "[t]he history of our species is made of endings and beginnings"; the macronovel's use of "reincarnation time" depicts this concept in action (Mitchell 2005b). For example, its oldest reincarnated character to date is Moombaki from *The Bone Clocks*, revealed not in terms of her age or birth date, but through the combined names of each of her previous selves which together form her "long, long, true name" (416). Listing rebirth after rebirth, Moombaki gradually reveals to Marinus that she has lived in 207 "previous hosts", and is "not thought of as a god [...] but as a guardian, a collective memory"; with no conventional age or date of birth given, it is Marinus who calculates that her lived history dates back "approximately seven millennia" (414–15). Moombaki's long name reimagines geological time's linearity in specifically human terms, each interconnected life becoming a unit of measurement, a literal depiction of Mitchell's envisaging of the "history of our species" through individual human "endings and beginnings." As Mitchell also notes in a 2016 interview, "[t]he atemporals are perhaps core samples [...] that record the human experience" ("Interview by Rose Harris-Birtill," Mitchell 2016b). Moombaki's long name offers a particularly human form of core sample, whose compacted layers do not comprise inert geological strata, but the grit of individual human lives.

Read through a purely linear sequence of events, the macronovel is heading toward periods of widespread dystopic decline. However, Mitchell's macronovel disrupts such linear temporalities with its shifts backwards and forwards in time, as in *Cloud Atlas*, but also in the publication of its narratives along a chronologically discontinuous timescale – for example, a 1980s period-piece (*Black Swan Green*) is followed by an eighteenth-century historical novel (*The Thousand Autumns*), which in turn is followed by a novel whose timescale envisages the near-future (*The Bone Clocks*). Such strategies deliberately complicate any attempt to read this body of works through a strictly linear chronology. Throughout, the possibility of meaningful change for each generation becomes a form of rebirth as another chance to alter the future, made possible through the ethical actions of individuals engaging anew with their responsibility to protect future generations within a global framework of impending ecological destruction. Such efforts can be in the form of big-scale interventions, such as Mo Muntevary's efforts to safeguard potentially devastating military technology in *Ghostwritten*, or in small-scale ethical decisions, such as Marinus' efforts to protect Holly's grandchildren in *The Bone Clocks* – regardless of their positions in the macronovel's chronological sequence of events. This is a narrative world infused with the possibility that "reincarnation time" brings, a human-centred model of a spiralling cyclical time which is not doomed to repeat the past, but finds the agency and determination to begin again in each successive generation, in what Heather J. Hicks (2010) describes as a "more hopeful deployment of a cyclical apocalyptic narrative."

If, as Berthold Schoene notes, "no one person or group of persons is ultimately in charge" of global events now unfolding, we must ask how humanity "can still hope to make a difference and shape such a world" in which "political agency is dilapidating into crisis management" (2009, 2). Mitchell's use of "reincarnation time" places intergenerational human action at the centre of this global "crisis." As he notes in a 2015 interview with Paul A. Harris, the reincarnated Horologists "are role models for me [...] I value the notion of reincarnation as a kind of metaphor for a single life" (Mitchell 2015c, 13). He adds:

> Horologists, then, are metaphors of mortals. They have repeated lives to slouch towards enlightenment, and we have just the one to scramble there as best we may, but the methods and the destination are the same.

The author's reference to supernatural beings that "slouch towards enlightenment" evokes William Butler Yeats' poem "The Second Coming," first published in 1920. Although in Yeats' poem it is a "rough beast" – and not a reincarnated Horologist – that "Slouches" towards "birth," the poem's opening line, "Turning and turning in the widening gyre," offers a structure for the immanent rebirth that it describes, again evoking the spiralling cyclicality of "reincarnation time" (Yeats 1997, 189–90). However, as "role models," the Horologists provide a collective example of the ethical actions that must be achieved in a "single life" if we are to avoid the same debilitating ecological and social meltdowns that they face, within a fictional world whose negative human impact remains unchecked. Although the Buddhist terminology of "enlightenment" is used, its dual meaning makes it particularly relevant to the macronovel, here referred to as the secular "action of bringing someone to a state of greater knowledge, understanding, or insight."[10] However, the difficult "scramble" towards such insight must be undertaken without the benefit of the Horologists' many lifetimes of lived experience; as such, the macronovel provides a shorthand depiction of the co-operative, compassionate actions that could be learnt through living in "reincarnation time." As in the macronovel, it is by no means certain whether such changes will be sufficient to avert global crises, but such collective action provides our only chance at redemption. The concept of reincarnation may

---

10   "enlightenment, n." 2016. *OED Online*. Oxford University Press, June. http://www.oed
      .com/view/Entry/62448.

be thousands of years old – but as Mitchell's macronovel demonstrates, it suggests a new temporal strategy whose interconnected past, present, and future reinforce the vital importance of intergenerational ethical action in imagining meaningful change.

Acknowledgements

My heartfelt thanks to David Mitchell for his time in interview, and to the members of the International Society for the Study of Time (ISST) for their feedback on an early version of this paper, presented at the ISST Sixteenth Triennial Conference at the University of Edinburgh, UK, in June 2016. Thank you also to the ISST for their award of the prestigious Founder's Prize for New Scholars 2016 for this research.

References

Boon, Marcus. 2015. "To Live in a Glass House is a Revolutionary Virtue Par Excellence: Marxism, Buddhism, and the Politics of Nonalignment." In *Nothing: Three Inquiries in Buddhism*, edited by Marcus Boon, Eric Cazdyn and Timothy Morton, 23–104. London: University of Chicago Press.

Childs, Peter. 2015. "Food Chain: Predatory Links in the Novels of David Mitchell." *Études anglaises* 68 (2): 183–95.

Childs, Peter, and James Green. 2013. "David Mitchell." In *Aesthetics and Ethics in Twenty-First Century British Novels: Zadie Smith, Nadeem Aslam, Hari Kunzru and David Mitchell*, 127–57. London: Bloomsbury.

Clayton, Jay. 2013. "Genome Time: Post-Darwinism Then and Now." *Critical Quarterly* 55 (1): 57–74.

de Waard, Marco. 2012. "Dutch Decline Redux: Remembering New Amsterdam in the Global and Cosmopolitan Novel." In *Imagining Global Amsterdam: History, Culture, and Geography in a World City*, edited by Marco de Waard, 101–22. Amsterdam: Amsterdam University Press.

Edwards, Caroline. 2011. "'Strange Transactions': Utopia, Transmigration and Time in *Ghostwritten* and *Cloud Atlas*." In *David Mitchell: Critical Essays*, edited by Sarah Dillon, 177–200. Canterbury: Gylphi.

Griffiths, Philip. 2004. "'On the Fringe of Becoming'–David Mitchell's *Ghostwritten*." *Beyond Extremes: Repräsentation und Reflexion von Modernisierungsprozessen im zeitgenösisschen britischen Roman*, edited by Stefan Glomb and Stefan Horlacher, 79–99. Tübingen: Gunter Nar Verlag Tübeingen, 2004.

Harris, Paul A. 2015a. "David Mitchell's Fractal Imagination: *The Bone Clocks.*" *SubStance: David Mitchell in the Labyrinth of Time* 44 (136): 148–53.

Harris, Paul A. 2015b. "David Mitchell in the Labyrinth of Time." *SubStance: David Mitchell in the Labyrinth of Time* 44 (136): 3–7.

Harris-Birtill, Rose. 2017. "'Looking down time's telescope at myself': reincarnation and global futures in David Mitchell's fictional worlds." *KronoScope: Journal for the Study of Time* 17.2: 163–81.

Harris-Birtill, Rose. 2019. *David Mitchell's Post-Secular World: Buddhism, Belief and the Urgency of Compassion.* London: Bloomsbury Academic.

Hicks, Heather J. 2010. "'This Time Round': David Mitchell's *Cloud Atlas* and the Apocalyptic Problem of Historicism." *Postmodern Culture* 20 (3) https://muse.jhu.edu/article/444704.

Jameson, Fredric. 1998. *The Cultural Turn.* London: Verso Press.

Jameson, Fredric. 2013. "The Historical Novel Today, or, Is It Still Possible?" In *The Antinomies of Realism*, 259–313. London: Verso.

Lewis, Simon L., and Mark A. Maslin. 2015. "Defining the Anthropocene." *Nature* 519: 171–80.

Mitchell, David. 1999a. *Ghostwritten.* London: Sceptre.

Mitchell, David. 1999b. "Mongolia." In *New Writing 8*, Vol. 8, edited by Tibor Fischer and Lawrence Norfolk, 514–61. London: Vintage.

Mitchell, David. 2001. *Number9dream.* London: Sceptre.

Mitchell, David. 2004a. *Cloud Atlas.* London: Sceptre.

Mitchell, David. 2004b. Interview by Wayne Burrows. "An Interview with David Mitchell on Cloud Atlas, Murakami, Money and the Sinclair Zx Spectrum." *The Big Issue in the North*, 2 March. https://serendipityproject.wordpress.com/2011/08/07/aug-7-2011-an-interview-with-david-mitchell-on-cloud-atlas-big-issue-in-the-north-2004/.

Mitchell, David. 2004c. Interview by Eleanor Wachtel. "'Cloud Atlas' with Author David Mitchell." *CBC Radio.* http://www.cbc.ca/player/play/2318715442.

Mitchell, David. 2005a. "Acknowledgements." *Prospect* 115, October.

Mitchell, David. 2005b. Interview by Richard Beard. "David Mitchell Interview." *RichardBeard.info.* Tokyo University, 12 January. http://www.richardbeard.info/2005/01/david-mitchell-interview/.

Mitchell, David. 2006a. *Black Swan Green.* London: Sceptre.

Mitchell, David. 2006b. "Preface." *The Daily Telegraph*, 29 April. *Nexis.* https://www.nexis.com/docview/getDocForCuiReq?lni=4JVD-GPV0-TX33-730B&csi=139186&oc=00240&perma=true.

Mitchell, David. 2007. "The View from Japan." *The Telegraph*, 14 October. http://www.telegraph.co.uk/culture/3668535/The-view-from-Japan.html.

Mitchell, David. 2009. "The Massive Rat." *The Guardian*, 1 August. https://www.the guardian.com/books/2009/aug/01/david-mitchell-short-story-rat.

Mitchell, David. 2010a. *The Thousand Autumns of Jacob De Zoet*. London: Sceptre.

Mitchell, David. 2010b. Interview by Adam Begley. "David Mitchell: The Art of Fiction No. 204." *The Paris Review*, Summer. http://www.theparisreview.org/interviews/6034/the-art-of-fiction-no-204-david-mitchell.

Mitchell, David. 2010c. Interview by Harriet Gilbert. "World Book Club: David Mitchell – Cloud Atlas." *BBC World Service*, April. http://www.bbc.co.uk/worldservice/arts/2010/06/100604_wbc_david_mitchell.shtml.

Mitchell, David. 2010d. Interview by Wyatt Mason. "David Mitchell, the Experimentalist." *The New York Times*, 25 June. http://www.nytimes.com/2010/06/27/magazine/27mitchell-t.html.

Mitchell, David. 2012. Interview by Shanghai TV. "2012 Shanghai Book Fair." *Reading Tonight*. Youtube, 21 August. http://www.youtube.com/watch?v=8Y1zPJ9xpFo&list=PLsMDsoSXlAFWQOlLiNdnDdbntJkMkyeRA&index=57.

Mitchell, David. 2013a. Interview by Andrei Muchnik. "David Mitchell Talks About Moscow, Literature and the Future." *Moscow Times*, 13 September. https://themoscowtimes.com/articles/david-mitchell-talks-about-moscow-literature-and-the-future-27620.

Mitchell, David. 2013b. Interview by Jasper Rees. "10 Questions for Writer David Mitchell." *Theartsdesk.com*, 10 April. http://www.theartsdesk.com/film/10-questions-writer-david-mitchell.

Mitchell, David. 2014a. *The Bone Clocks*. London: Sceptre.

Mitchell, David. 2014b. Interview by Laurie Grassi. "Best-Selling Author David Mitchell on His Biggest Wow Moment." *Chatelaine*. Chatelaine.com, 2 September. http://www.chatelaine.com/living/chatelaine-book-club/david-mitchell-interview-cloud-atlas-the-bone-clocks/.

Mitchell, David. 2014c. Interview by Zahra Saeed. "Cork-Based Cloud Atlas Author David Mitchell on Twitter, His Children, and Latest Novel." *Irish Examiner*. IrishExaminer.com, 4 September. http://www.irishexaminer.com/lifestyle/artsfilmtv/artsvibe/cork-based-cloud-atlas-author-david-mitchell-on-twitter-his-children-and-latest-novel-284747.html.

Mitchell, David. 2015a. "I_Bombadil." *Twitter*. https://twitter.com/i_bombadil.

Mitchell, David. 2015b. *Slade House*. London: Sceptre.

Mitchell, David. 2015c. Interview by Paul A. Harris. "David Mitchell in the Laboratory of Time: An Interview With the Author." *SubStance: David Mitchell in the Labyrinth of Time* 44 (136), 8–17.

Mitchell, David. 2016a. "My Eye On You." *Whirlwind of Time*. Kai Clements and Anthony Sunter. London: Kai & Sunny, 1–6.

Mitchell, David. 2016b. Interview by Rose Harris-Birtill. Personal Interview. 12 Feb.

Mitchell, David. 2016c. "All Souls Day." In *Jealous Saboteurs*, by Francis Upritchard, 78–86. Melbourne: Monash University Museum of Art. Originally published 2010.

Ng, Lynda. 2015. "Cannibalism, Colonialism and Apocalypse in Mitchell's Global Future." *SubStance: David Mitchell in the Labyrinth of Time* 44 (136): 107–22.

Nietzsche, Friedrich. 2001. *The Gay Science*. Translated by Josefine Nauckhoff. Edited by Bernard Williams. Cambridge: Cambridge University Press. Originally published 1882.

Nietzsche, Friedrich. 2006. *Thus Spoke Zarathustra*. Translated by Adrian Del Caro. Edited by Adrian Del Caro and Robert Pippin. Cambridge: Cambridge University Press. Originally published 1883.

O'Donnell, Patrick. 2015. *A Temporary Future: The Fiction of David Mitchell*. New York: Bloomsbury.

Robinson, Richard H., and Willard L. Johnson. 1997. *The Buddhist Religion*. 4th ed. California: Wadsworth Publishing Company.

Schoene, Berthold. 2009. *The Cosmopolitan Novel*. Edinburgh: Edinburgh University Press.

Shoop, Casey, and Dermot Ryan. 2015. "'Gravid with the Ancient Future': *Cloud Atlas* and the Politics of Big History." *SubStance: David Mitchell in the Labyrinth of Time* 44 (136): 92–106.

Snelling, John. 1987. "The Buddhist World View." In *The Buddhist Handbook: A Complete Guide to Buddhist Teaching and Practice*, 43–50. London: Rider.

Sunken Garden. 2013. Comp. Michel van der Aa. Lib. David Mitchell. English National Opera, London.

Wake. 2010. Comp. Klaas de Vries and René Uijlenhoet. Lib. David Mitchell. Nationale Reisopera, Enschede.

Yeats, William Butler. 1997. "The Second Coming." *The Collected Works of W. B. Yeats: The Poems*. Vol 1. 2nd ed. Edited by Richard J. Finneran. New York: Scribner, 189–90. Originally published in 1920.

PART 1

Dialogues

∵

CHAPTER 3

Archivists of the Future

Paul Harris, Katie Paterson and David Mitchell

David Mitchell and Katie Paterson presented work at the ISST's conference "Time's Urgency," at the University of Edinburgh in June 2016: Mitchell read three unpublished short stories, and Paterson talked about several of her artworks, and the two then engaged in a dialogue about Paterson's Future Library, in which a forest of trees will become an anthology of books to be printed in 100 years (2114), composed of works contributed by authors each year and held in trust. Mitchell was the second author invited to contribute a text; to date, Margaret Atwood and Sjón have also deposited works. "Archivists of the future," which evokes many aspects of Mitchell and Paterson's work, is a phrase from Mitchell's *The Bone Clocks* (2014) – Paul Harris.

Paul Harris: *Let's begin with Future Library as an archive of the future … Katie, what was the germination of the idea? Can you talk about it in relation to your sense of time?*

Katie Paterson: The idea to grow trees to print books arose for me through making a connection with tree rings to chapters, the material nature of paper, pulp and books, and imagining writer's thoughts infusing themselves, 'becoming' the trees, over an expansive period of time. As if the trees absorb the writer's words like air or water, and the tree rings become chapters, spaced out over the years to come.

The beginnings of *Future Library* could be anything from the small sketch of tree rings as chapters I made on a train several years ago, or the seedlings that have been planted, or the first words written. The idea for *Future Library* happened in a micro second where notions of time, growth, future, place, stories, pulp, matter, cells, smells, all collapsed into one.

Future Library has pillars, and an inherent ecology, parts interact to create the whole. The forest, the growth, the pulp, the soil, the light and all the natural ingredients that will bring the paper and books into being in the future. The texts, and the manuscripts written by 100 authors, over and through time. The ideas stored within the trees before harvest. The Silent Room, housed in an city landscape, that looks out to the forest and vice versa. *Future Library* exists in many spheres, many of them unseen (words written and read by unborn people, held in the current imagination). There are material and immaterial aspects.

Future Library is not a directly environmental statement, but involves ecology, the interconnectedness of things – those living now and still to come. It questions the present tendency to think in short bursts of time, making decisions only for us living now. It's an artwork made not only for us now, but for a future generation, in an unknown time and place.

Future Library connects with my wider art practice through its engagement with nature and time – long, slow time. Whilst previous works have dealt with time on geologic or cosmic scales (the billions of years light has travelled from an exploding star, the million year ages of ancient fossils) in many ways the human timescale of 100 years is more confronting. It is beyond many of our current lifespans, but close enough to come face to face with, to comprehend and relativize. Inside the forest time stands still. This place could have existed for one hundred, one thousand, one million, or even one hundred million years.

PH: *David, what was your response to the invitation and the project, and what was the challenge of it?*

David Mitchell: I wasn't quite sure how 'for real' Katie's proposal was when my agent first passed it on, even though came from Simon Prosser, a respected editor and trustworthy acquaintance. A hundred years? Nobody reads the manuscript? Really? I receive offers from half-baked to fully-baked to decidedly uncooked and I didn't know where to place this one. Then I learned that Margaret Atwood had taken the plunge the first year and, feeling a little foolish, I thought, *Fine, it's for real.*

I began to think about the beauty of launching a little bottle with my message in it onto the sea of time, and knowing that people not yet born will receive it, and read it, and think about the long-dead sender of the bottle. It's almost like a phone line to them: I can wave at them, and say 'Hi!' and see them wave back. It also feels like an affirmation that they, and books, and trees, and Norway, and civilisation, will still be there in a hundred years. After the last couple of years, I need that affirmation like Asterix needs his magic potion. In demonstrable everyday senses as well as some strands of Buddhist theology, the future is crafted by thought. The Future Library Project is a manifestation of this idea.

As you know, I can't tell you too much about my artistic response, i.e., what I submitted, because of the scroll Katie made me sign in blood during that weird ritual on a blasted heath with the witches. All I'll say is, I wrote something that engaged with the spirit of the project and played with its themes, particularly time and human memory and history. The title, which we are encouraged to make public, contains a few clues: *From Me Flows What You Call Time.* Which I nicked from Tōru Takemitsu, the Japanese composer.

The challenge ... was there a particular challenge? I'm not sure. When writers were writing fiction in 1917 they weren't thinking about what would and what wouldn't make sense to us, in 2017 – and good luck to any of them who may have tried, because their attempts would have been doomed. They were just trying to write as well as they could. I did the same. Nobody knows who the future writers are going to be – many of these, too, are not yet born – but it's a safe bet the standard of work is going to be very high, so when the whole library is published in 2114, I don't want mine to look too shoddy.

PH: *Each in your own way archivists of the future in your work. Can you talk about how that interest evolved? And, there is a specific temporality that is opened – the duration is different, its correspondence to reality is different....*

KP: *Future Library* could be seen as a future archive. Most of the project doesn't yet exist, it is archiving the unknown. It extends into the past and future, through layers of time and distance. It looks back and forward in time simultaneously: from the history of printed books on paper, to future authors who aren't yet born, trees that aren't grown, words yet unwritten and unimagined.

Future Library unfolds concurrently over long, slow time – a century – and the present moment – the diurnal, daily cycles of the trees, the seasons, and the author's yearly contributions and handover events. The momentary, imperceptible changes in the growth of the trees, the unread words in the author's manuscripts take shape year by year. The project is marked out by yearly demarcations like chapters in a book, which keep it fluid. It exists in the here and now, as well as for and into an unknown future whose temporality and correspondence to reality is distant and skewed. The duration of 100 years seems on the fringes of being graspable and knowable.

DM: I'm writing this three months after Katie wrote the above, so I get to read her answers just before I write mine. It's a little intimidating: God, she's so smart! Much more clod-hoppingly, then, I'm an archivist of the future insofar as I like to set stories there, and in order to do so I need to know what my characters know. Meaning, I have to know their pasts, and the recent history of their worlds – which lies, of course, in the future as viewed from the early twentieth century. Why do I like to set things in the future? Dunno ... Because I'm me? Because I read Isaac Asmiov and Ursula le Guin and William Gibson and watched *Star Trek* and *Doctor Who* when I was kid, and thought, *Wow, that works for me ... I'd like to do that too ...?* Could be.

"*Can you talk about how that interest evolved?*" Well, I hope I get better at doing it ... And my work needs to be different each time, otherwise I'm repeating myself, and that's rude to my readers and dull for me. My (nonverbal) son's autism makes me wonder about how he perceives time, and these wonderings are rearing their heads in a few recent pieces of fiction.

Similarly, my daughter has taught me a lot about DC and Marvel superheroes, some of whom are studies in atypical temporalities (normally I'd put that in quotes but in the present company I don't feel the need.) Quicksilver and The Flash, for example, whose superpower is moving about the place faster than a bullet: from their perspective, they're not fast – it's us, and everyone else, who is trapped in a much much slower stream of time. I wrote a story about this recently, called *My Eye on You.* ("The pram in the hallway is the enemy of art"? What a tawdry, opportunity-wasting way to think – my kids have enriched my creative life beyond all measure.) And, if this doesn't sound too sycophantic, my conversations with a certain Paul Harris and his neighbour Pierre Jardin have beefed up how I think about time, by adding 'time' to the list of singular nouns that ought really to be plurals. We live in a matrix of time*s*, with an 's.' Calendrical, life-span, diurnal, monthly, historic, geological, planetary, caffeine-hit-to-caffeine-hit, to name just those that spring to mind right now. The question is less, 'Why does temporality fascinate you?' and more 'How could temporality possibly not fascinate you?'

PH: *It seems no accident that you are both interested in the deep past in your work, as much as the future. What is the relation between stretching the boundaries of cultural and cosmic memory and looking to the future?*

KP: Astronomy is important to me and figures in so many of my artworks, mostly due to the revelations of the deep interconnections between humans and everything else. This can be evident through the geological connections in the strata on earth, to the remnants of supernovas flowing through our blood. I'm interested in how we attempt to contemplate ideas of being able to look back in time, to the beginnings of the universe. I was astonished to learn that all the matter that exists now in the universe is all that will ever exist, because everything – all the stars, galaxies and matter – is moving away from each other so quickly, it can no longer collide and produce new objects. For me, these are ideas that just can't be rationalized. How do we conceive of a time before the earth existed? Being able to look through powerful telescopes and witness eras in the universe so inexplicably distant can heighten awareness of deep time and stretch cosmic memory, from the past into the future.

You could say there is a cosmic memory engrained in the materials and phenomena I tend to work with: fractions of Saharan sand, the billion-year-old light from dying stars, ancient darkness from the edges of the early universe, to fossil beads from the first single-celled life on earth, and slides which chart the unknown evolving universe. These materials project the imagination forward and backwards through time and space. The beads could be micro future or past planets. Looking to the most distant past can feel like a journey into an infinite future. To experience time billions of years ago now, looking to a time proceeding the earth and ourselves, I find this almost as challenging as looking forward into an expansive future.

DM: Yeah, what she said.

I keep answering your questions with more questions, Paul – do the Germans have a word for that? – and here I go again: would it not be profoundly odd to be interested in the future but not the past, or vice versa? Time is all time, not just one stretch of it. Isn't every line infinite, and what we call a line not, in fact, a true line, but only a segment of a line?

PH: *You both stretch temporal boundaries in your work, and you both create work known for challenging artistic forms, genres, limits or conventions. Can you talk about the relation between these things?*

KP: A sense of limits, or impossibility, is a filter I don't think I was born with. Conventions rarely rule my creative life: curiosity, and making this kind of artwork, can be liberating. I am grateful that forms and genres are something I don't tend to dwell on – the idea at hand never does – whether that involves the colour of the end of time, working with nano-materials or carving out records made of glacial ice in a deep freezer. Stretching temporal boundaries may not be in my field everyday experience, but is wide open and malleable when it can take shape in the mind and imagination.

DM: Reality stretches temporal boundaries. Reality challenges artistic forms, genres, limit and conventions. I merely follow its lead.

Maybe what a temporal boundary or convention in my art – novels should be linear, for example, with no back-flashes or forward-flashes – has in common with genre boundary is its artifice, its arbitrariness, the fact that a human, or a critic (not that critics aren't human) decided the boundary is there. Thus these things are analogous to cartography. While cartography is sometimes connected to the geography it draws lines upon, geography for its part doesn't pay much attention to cartography. Cartography, to be sure, is an interesting discipline. Geography, however, is freakin' awesome.

PH: *Your work induces a kind of temporal vertigo, produces experiences of sudden unbounding of time, a lovely phrase for which is a "phoenix flash of reverie," from Gaston Bachelard. Have you had experiences of this kind? Or, do you seek to integrate such moments into your work or open that kind of experience to viewers/readers?*

KP: *A phoenix flash of reverie, catching hold of unbounding of time,* I would like to make the mandate of my creative practice. This is something I aspire to attaining with regularity, but time plays tricks. There might be a *flash,* and then three years of bringing an idea into being. I'm currently working on a book of *Ideas,* you could say a book of temporary flashes, a repository of reverie. It brings together short statements of intent, works that exist, that don't exist, that may exist, that are there simply to exist in the imagination. The *Ideas* all three-lined short, almost Haiku-like expressions, their subjects ranging from earth and the moon, to the very distant edges of light, unbounded time and space. The *Ideas* are mixed up with works that I have already brought into the world, blurring a fictional line. Some are way off scale, and are clearly things that you can't do in this world. I wanted to evoke and activate them in the reader's mind.

Working on lengthy research and busy production processes can contradict creativity so I found a way to work spontaneously, stripping years of work to only a few words, which can be communicated instantly with the reader. If flashes of reverie can be rich, does it matter if an experience exists for a nano-second or beyond?

DM: 'A phoenix flash of reverie' is indeed a lovely phrase: it is also highly interpretable. For me, it makes me think of the 'long moments' I may experience a handful of times a year when I feel like I'm removed from the ceaseless whitewater hurly burly of time, and I feel intensely aware of the present moment, of right now; and intensely conscious that this isn't how things usually are; and intensely calm, with the PAUSE button pressed on the monkey mind, mid-banana or mid-armpit scratch. I don't know what brings these moments on. I suspect they just find me, and I'm duly grateful, because they feel wonderful. The boundary between me and everything else feels membrane-light and transparent, at these times. The word 'vertiginous' applies: these moments make me catch my breath at the changed perspective. I think more clearly, or at least I think I do. It's like being in Rivendell. Then, suddenly, the reverie is gone and I'm out faffing with hobbits, the annoying ones, or getting impaled by orcs again, and I forget things were ever different until the next reverie comes along, days or weeks or months later – or until I think about them, which I'm doing now.

PH: *What associations do you have with our 2016 conference theme, "Time's Urgency"?*

KP: I associate this idea with our current sense of urgency in terms of acting and creating change, in relation to our lifetimes, the next generation's lifetimes, and the next. Beyond these coming centuries, time starts to stretch into the infinite and neverending, into geologic or cosmic stretches of time, where a sense of urgency falls away.

DM: I associate this idea with mortality. Time has all the time in the world, in the universe – I can't imagine Time ever really 'doing' urgency. This hairy skin bag of organs and tubes and inlets and orifices and nerves and secreting glands we fondly think of as 'Me,' however, most certainly does *not* have all the time in the world. Famously, and truly, our supply of time is spent at an accelerating rate as we age. Let's not waste the stuff on grudges, or angry arguments inside our heads with people who cannot hear us because they aren't there, or clickbait, or fretting over things we are wholly powerless to influence. The singer-songwriter David Crosby has a song called *Time is the Final Currency*. It's not a particularly great song, but it's a great title. It's true.

PH: On behalf of the ISST, I want to thank you both for enriching our conference with your presentations, and participating in this conversation. As a fan of both of your singular, inspiring bodies of work, I look forward to seeing how they develop, and hope you will attend a future ISST conference.

CHAPTER 4

Presidential Address: Should We Give Up "Time"?

Raji C. Steineck

1 After 50 Years: Time's Urgency, Again

Dear fellow time-smiths, on behalf of ISST's council I would like to express a warm welcome to you all to our 16th triennial conference, which also marks the 50th anniversary of our society.[1] Whether you are long-standing members or newcomers, or like myself, somewhere in between: I am happy that you have come to join, on this special occasion, to share in what is quite literally the core activity of our society: to exchange our knowledge, our questions, our doubts even, on matters of time, in order to invigorate our thinking on this seminal subject of human concern, and to bring new ideas and life to pertinent debates in our disciplines. "Time's urgency", the theme of this year's conference, is what drove the founding of our society 50 years ago. And while the times have kept changing, or precisely because of that, it is still, and yet again, what brings us together.

Let me quote J. T. Fraser on the initial motives of his quest for an integrated study of time, the quest that brought him to seek out the leading scientists and scholars on the subject and eventually led to the founding of ISST.

> The reasons that led to the founding of this Society had nothing to do with anyone's interest in the nature of time. They had to do with the puzzlement in the mind of a man of twenty-one who, in the autumn of 1944, found himself on a mountainside between two vast armadas. Behind him was the armed might of Nazi Germany, in front of him the immense masses of the Soviet Union. He knew that he was watching a struggle between two ideologies, each of which was convinced that it, and it alone, was destined to fight and win the final conflict of history. [...] Having been aware of both dogmas, I, [Fraser says,] came to wonder whether there does exist a final conflict in history.
>
> FRASER 2007, 383

---

1 This is a revised version of the speech given at the Edinburgh conference. I would like to thank Carlos Montemayor for his insightful comments on earlier versions of the draft.

Several pages later, speaking about the immediate antecedents of ISST's founding in 1966, he continues:

> I realized with a pleasant shock that the question I posed many years earlier, namely, whether there can exist a final conflict in history, was too crude to be fundamental. Namely, it is possible to imagine a world without mass murders but it is not possible to imagine humans who will not declare, in innumerably many ways, "Death, be not proud ..." because the conflict that gives rise to such a rhetorical command – the conflict between the knowledge of an end of the self and the desire to negate that knowledge – is at the very foundation of being human.
>
> FRASER 2007, 386

One of the lessons we were made to learn over the last 50 years is that the link between the constitutive conflict at the foundation of being human and the willingness of large parts of humanity to lose themselves to siren's songs of final solutions is stronger than we all would like to believe. Half-way through those 50 years, the fall of the iron curtain sparked hope in a world where global democracy and peaceful commerce would replace the clashing of vast armadas. The fact that this hope was heralded under the name of the "end of history" already foretold, to those who shared J. T. Fraser's insights, that it was ill-fated.

What probably only very few could foresee is how liberal democracies themselves would embrace more and more of the authoritarian repertoire once competition was removed with a system that had heralded the promise, however devious, to show a better way to human freedom; how they would shed restrictions on inequality and abandon large cohorts of their citizens and inhabitants to hopelessness and destitution; and how they would fuel the rise of figures fostering, and feeding on, resentment against the weak and vulnerable. These figures promise yet again a state of unimpeded glory once "we" – whoever "we" are – have finally got rid of "them" and installed a regime where the will of some higher authority reigns supreme. In the past years, we have had ample opportunity, in each of our own countries, to witness massive crowds abandon themselves to such sound and fury.

When assessing this development, it is helpful to keep in mind how J. T. Fraser's generally evaluated experiences of "timeless" extasy:

> What I suggest is that feelings said to be those of eternity, absolute rest, or infinite peace are descriptions of moods in which one or another lower temporality has become dominant. [...] they take us to archaic umwelts

of the brain, which are less evolved than the noetic. Such umwelts may be experienced as beautiful or terrifying, depending on whether they are perceived as aids or threats to the continuity of the self. The elation that may accompany the descent may be that of a person who has jettisoned the burden of individuation, including responsibilities for the future and regrets about the past; [...] But the experience may also be horrifying: the intense present-orientedness of love may be heavenly, but the intense present-orientedness of pain is hellish.

FRASER 1999, 130

On a social and political level, this Janus face of the ecstatic collapse into lower temporal umwelts is exposed in the correlated appearance of both its joyous and its painful side. One group's ascent into ecstatic community regularly leads to another group's descent into the hell of pain. This is even necessarily so when the elation of the first group is brought about by way of stepping on the back of the second.

ISST is not a Fraser society. But I maintain that it is our mission, in keeping with the intentions and the motives of our society's founder, to consciously embrace, and publicly stand up for, the complexities of what he called "the no-etic world" and its unlimited temporal horizons. Embracing the complexities means to champion the civility of reasoned dialogue, over and against inciting contempt, castigating sinners, and preaching to the faithful. It also means to empathize with the pain that fuels the drive for simple solutions. Such empathy needs to be sustained, however, by the unwavering awareness that human life is not a riddle to be solved once and for all. To maintain the noetic world and its complexities, the horizons of time have to be kept open. To think deeply and rigorously about time is, as J. T. Fraser rightly believed, a primary way to firmly establish that insight.

If such is, as I believe it is, the very heart of ISST's mission, it is one that has increased urgency in today's world. At the same time, and by the same token, we are facing more adversity than even a decade ago. And this is not simply a matter of intellectual environments. Programs that attempt to turn universities into power stations for the creation of economic value, and policies to foster public allegiance to some national or religious cause have diminished the room for endeavours such as ours. Some of us have experienced difficulties to obtain funding for travel to interdisciplinary conferences such as the one we are about to hold. Quite a number of papers accepted for this conference were in fact cancelled for such reasons. But this forms only a fraction of the pressures coming with the general restructuring of budgets in research and teaching. The open-ended exploration of the multifarious temporal dimensions, scales, and structures that make up our human world is viewed with less sympathy by many authorities in and out of university today. We will have to discuss what this means for the operations of ISST. I will call on your energy, intelligence, wit, and, let there be no doubt, on your time, to keep our society afloat. I will do so because ISST has an important contribution to make in the current world. It is our task to keep open, and even broaden, the exploration of time's horizons and dimensions in all directions, and to sustain the open-ended reflection on that mysterious thing called time.

2   Should We Give Up "Time"?

Indeed, the argument that I am going to propose in the remainder of this presentation is one in favour of radical openness. The question in the header to this paragraph may sound rhetorical in light of what I have been saying so far. But it is meant as a serious question. "Time" has long ceased to be accepted as referring to a given fundament of reality, as it may have appeared in the 19th century. The 20th century saw its universality contested, and its interpretation fragmented between disciplines and discursive traditions. Different modalities and shapes of time were explored in the various fields of the sciences and humanities. Several attempts were made to re-integrate the findings of such research, be it by reducing them to a logical or natural core, or by providing paradigms accounting for plural temporal modes and morphologies. J. T. Fraser's hierarchical theory of time is a formidable case of such over-arching synthesis. But there is continued and powerful scepticism whether what we call time may count as a fundamental feature of reality. The recent doubts put forth against time's character as a basic universal category call for close attention and rigorous inspection. It is for this reason that we invited Dr. Julian Barbour to present our first J. T. Fraser memorial lecture, in a presentation that was to explain his idea of "timeless physics" – an idea he succinctly formulated in a paper for the 2008 Foundational Questions Institute's essay contest on the "nature of time" (Barbour 2008; see also: Barbour 2000). Other contributions to said contest had a similar gist – one of them, by Carlo Rovelli, even bore the title "Forget time" (Rovelli 2008). Craig Callender has reviewed these discussions for *Scientific American*, under another heading that exposes the precarity of time: "Is Time an Illusion?" (Callender 2010).

But it is not just scientists – and among them typically physicists – who question the fundamental character of time. There are also prominent sociological analyses and postcolonial critiques of time as a social construct. In an essay for the journal *Public Culture*, historian Stefan Tanaka just recently proposed to consider "History without chronology" (Tanaka 2016). To be sure, that does not necessarily mean "history without time", but Tanaka does call into question the routine application of the current standard calendar, and more generallly the notion of time it is based on – the notion of a global system of coordination, marked by a continuous sequence of homogeneous quantitative units: the second, the minute, and so forth. More radically, Minoru Hokari in a book on Australian oral history has demanded that we seriously consider alternative ways to order and conceptualize historical events, ways that do not make use of time as a separate coordinate (Hokari 2004; Hokari 2011). As a Japanologist, I am reminded by Tanaka's and Hokari's calls of an ancient Japanese source called *Kojiki* ("Record of Ancient Matters"; Yamaguchi et. al. 1997; Antoni 2012; Heldt 2014) that supports their case. The *Kojiki*, an ultimately discarded draft for the first official imperial chronicle of Japan, makes use of time only as a secondary category. Its primary system of coordination is genealogy. This is a system of coordination that congeals, in its coordinates, what we would call temporal, spatial, and social aspects.[2] I believe *Kojiki* is a particularly pertinent case in this regard, because you cannot say, as classical modern anthropologists have said with regard to Australian aboriginees (cf. Dux 1992, 168–74 referring to Stanner, 1972), that it *lacks* a concept of abstract, continuous time. Instead, its authors made a conscious choice to treat time as a subsidiary of genealogy. It was a choice made in close connection to the social organization of the period, and to the negotiation of power (R. C. Steineck 2017, 89–92; Ooms 2008) – much as, or so Tanaka and others would argue, our preference for the concept of time as a homogeneous and metrical continuum is connected to the power relations and the mode of production of the modern, capitalist age (Tanaka 2016, see also Maki 2003; Postone 2006).

At this point you may feel prompted to object that the challenges I have mentioned so far are not directed at "time" as such, but at specific concepts of time. Indeed, I think this observation is both pertinent and correct. It holds not only for the historical and social criticisms mentioned above, but also for a large part of the scientific debate. Here, "time" regularly refers to the metrical, continuous, and independent variable "t" that was conceptualized by Newton. This variable "t", however, does not stand for time in its entirety, but for duration. Still, the question remains what is left of time once we take these challenges seriously and think rigorously through their consequences.

---

2   For a detailed discussion, see R. C. Steineck 2017, 220–25.

3   "Time": A Concept, Not an Illusion

Briefly said, I want to argue in favour of the following five propositions:
(1)   "Time" is best understood as *a concept*, built to make sense of realities we cannot ignore.
(2)   There is no concept of time that is both semantically rich and universally applicable.
(3)   That does not mean that time is an illusion, or otherwise obsolete.
(4)   The concept of time may not be meaningful for all segments of reality – some perhaps are "timeless".
(5)   It is necessary to explore the full relation of these "timeless" segments to the "time-rich" world we inhabit.

By and large, I take these propositions to be potential points of convergence in the recent discussion, or at least in the more sober parts of it. Both Rovelli and Barbour are not simply trying to explain time away. To the contrary, both expect that a good formulation of "timeless" physics (meaning, to wit, a physics in which the factor t = duration has no fundamental part) will also elucidate the emergence of time in a rich sense as a part of cosmic evolution (leading to propositions 2, 3, 4, 5). Similarly, neither Tanaka nor Hokari, even if they strongly emphasize the limited applicability of any rich concept of time (proposition 2), encourage us to completely discard the concept itself (propositions 1, 3, 5). They would not even do away with the standard modern notion of time as a universal, quantitative system of coordination. Last but not least, readers of J. T. Fraser will find that my propositions by and large are in concord with his work – with the notable exception of proposition 1, which conflicts with Fraser's naturalist realism.[3] Proposition 1 is also what I perceive to be a point of contention within pertinent recent discussions in the sciences and, to a lesser extent, the humanities. As a consequence, there is further disagreement on proposition 3 – the question whether time is an illusion. I will therefore focus on discussing these two propositions in the remainder of this presentation.

Let me first explain what I mean by saying "time is best understood as a concept" and how that relates to denying that it is an illusion. This will involve a somewhat technical discussion on the concept of concept.

When we speak of "concepts", many people will immediately make the association to "ideas", and consequently, something that exists, somehow, exclusively "in the mind". The danger of this association is that it easily leads to a

---

3   For an extensive discussion of Fraser's naturalism and its problems in terms of a critical theory of knowledge, see C. Steineck 2010.

dichotomy with "the material". In today's world, dominated by naturalism, that will connect to assumptions about the primacy of matter; the conclusion being that, if "time" is a concept, it is "merely an idea", and therefore, if not a straightforward illusion, then at least of secondary importance, because it is only the material that ultimately counts as real.

This chain of associations is, however, misguided. To counter it, it is useful to remind ourselves that concepts and ideas are not the natural endowment of an immaterial soul, as metaphysical thinkers of earlier ages believed. They are products of human activity, formed in the course of human history. In order to avoid misleading associations, it may therefore be best to understand concepts through a pragmatic definition of the term, such as the following:

> A *concept* is what we refer to when we relate to a common meaning operative in the use of either various instances of one symbol, or of various different symbols.

The term *symbol* here refers to a perceptible object we take to have been produced in order to indicate something else. The *meaning* of a symbol is its determined relation to that which it indicates.

Concepts may thus be understood as products of a specific class of human actions related to the use of symbols, much as, in the words of J. T. Fraser, the *mind* is the dynamic integral of the human activity of "minding" (Fraser 1999, 18). The point of using such pragmatic definitions of concept or mind is that they lead beyond the metaphysical dualism of mind and matter, without however discarding the distinction between the "ideal" and the "material".

Human activity obviously involves nerves, brains, muscles. But the activity in question here also entails the use of symbols. Symbols, again, come in many shapes – sound, printed ink on paper, stone tablets – all of which are undeniably material. The defining property of symbols is that they are produced to point *beyond* their material fabric and properties, and to some extent even beyond their perceptual qualities. This ingenious use of sensible matter as a means to transcend the boundaries of what may be immediately perceived enables human beings to re-organize increasingly large segments of reality, to shape them in accord with their needs and intentions.

Placing concepts firmly within the realm of human activity situates them in the context of a spectrum that ranges from the lower forms of conditioned or even instinctual behaviour via purposeful actions to cultural forms of agency such as technological invention, art or scholarship. Specific acts of forming and using concepts may build on each of these kinds of human activity, but, or so I argue, they will always be connected to the production of symbols, which is a necessary condition for the existence of concepts.[4] The formation and use of symbols, however, in turn is based on more pristine activities involving the perception and use of *signs*. (A sign is any sensible object that is taken to indicate something else.) It is easy to envision an evolutionary hierarchy of sign-related behaviour from the reaction to natural indicators (smoke for fire, traces for the previous presence of animals) to the active use of natural objects as communicative signs and finally, the purposeful creation and re-production of symbols.

In terms of time, this has the following, important implication: to say that "time is a concept" does not indicate that time is a mere mental figment, an illusionary construct disconnected from the material world. To the contrary, just as the formation and use of symbols builds on lower forms of sign-related behaviour (which may already be present in the non-human animate world), forming a concept of time has, as one of its preconditions, behaviour relating to, and actions accounting for, temporal features of the natural world – it has a tried and tested basis in 'material' actions and sensuous experience.

To elaborate, the concept of "time" may best be understood as a higher order concept that synthesizes results of various, more elementary functions of ordering and conceptualizing, functions that enable us to consciously experience, reference, and make sense of both stability and change, and to organize our behaviour accordingly.

In this respect, it is useful to distinguish between 1) operational temporal ordering (along with other "categorial functions"); 2) explicit (i.e., conceptual) reference to temporal order; 3) generalized concepts of time; and 4) the concept of "Time" – time with a capital T – the attempt to unite the manifold operational and conceptual objectifications of time.

1) By operational temporal ordering, we – along with other living organisms – structure what we perceive into different situations, events and processes, and we establish a specific type of relations *between* them. In fact, these two functions go hand in hand. What makes the specific way of identifying, localising and relating situations, events and processes *timelike* – to use an expression of Callender's (2010, § "Time the great storyteller") – is that they refer to the form of a fixed, causally related sequence.

2) A *temporal concept* (such as "at sundown", "new", "end") is formed when a set of such sequences crystallizes into a stable pattern that becomes an object of reference, indicated by a symbol or set of symbols. This may, on a basic level, be a concept of limited scope and applicability.

---

4   I have discussed this in some detail in R. C. Steineck 2014, 99–125.

3)   A *concept of time* arises once a plurality of temporal concepts have been
     formed, and reference is made to an integral that comprises them. It
     is important to note that, while such integration makes sense, it is not
     something that happens in and of itself. Time is not, as Immanuel Kant
     famously believed, a naturally given "pure form of sensitive intuition
     (reine Form der sinnlichen Anschauung)". (Kant, *KrV*, B 46–58, cf. Kant
     1998, 106–16) The formation of even a basic concept of time requires
     several layers of reflection and analysis, each involving the active search
     for commonalities between different types of events and sequential pat-
     terns, including, finally, the construction of a comprehensive order that
     unites them.

4)   Further reflection on different concepts of time then constitutes a search
     for "the concept of Time" (Time with a capital T). Such a search starts
     from a hypothesis: it presupposes a unified concept without necessarily
     being able to explain it in a well-defined manner. But even the search
     constitutes, or so I argue, a concept of Time.

When we speak of "Time" with a capital "T", when we try to understand how
"time in physics" relates to "time in Proust" or "legal time", we tentatively posit
such a concept – and we have good reason to do so. Its explication may turn
out, in the end, to be of book length, or even longer. I would suggest that a
full-fledged concept of time in fact requires a theory of culture – a theory of
how the sciences relate to the humanities, the arts, economy, law and all the
other symbolic forms we use to make sense of the world.[5] Such a theory would
allow for substantial differences between local concepts of time, but retain a
logical coherence between them. There are other options, such as a unified
naturalist theory of time as envisioned by J. T. Fraser. Even while we are un-
decided which course to take, it still makes sense to use the concept of Time,
since it enables us to meaningfully relate our temporal ideas, perceptions, or
regulations in one area of human knowledge, experience, or practice to those
in other fields.

Critical readers may feel that conceiving of time in this way brings it danger-
ously close to being an arbitrary construct, if not an outright illusion. Against
this objection, and in order to support my proposition 2, I would take recourse
to the time-honoured Kantian argument that if something is a logical condition of possibility of what we consider valid in our experience, it must, of necessity, be as valid as that experience itself (cf. Kant, *KrV*, A 96–97, cf. Kant
1998, 206–7). I would therefore say that time cannot be an illusion (proposition
3), because it has proven to be a category that is essential for the organisation of the largest parts of our experience. Without time, there is no law, no

---

5   On symbolic forms, see: Cassirer 1953; Bayer and Cassirer 2001; R. C. Steineck 2014.

technology, no art, no music, no literature, no religion, no politics, but also no chemistry, no biology, no psychology – and even no physics, for reasons to be explained below. There is furthermore no relating these various fields of experience to each other. Time is one of those categories that enable us to unite widely different realms of reality, and hugely diverging perspectives. It allows us to move, in a critical appropriation of Fraser's parlance, beyond our *umwelt* and to have a world, that is, an open-ended horizon of experience (C. Steineck 2010). We can ask questions like: "Is time an illusion?", precisely because we find ourselves as both curious and fundamentally insecure "strange walkers" (Fraser 1999, 5–20) in a world that is never entirely ours, and not as cognitively self-contented inhabitants of a closed *umwelt*. The symbolic functions of representation and mutual reference enable us to move beyond our perceptive *umwelt* by forming, in Ernst Cassirer's terms, substance concepts and function concepts.[6] But they turn out, at the same time, to be the tools of fate that place us in the inextricable situation of "excentric positionality", to borrow a term from Helmuth Plessner's classic in philosophical anthropology (Plessner 1975 [1928]): once we know that the world may be different, and larger, than what we actually perceive, and even than everything we have ever and will ever perceive, we can never go back to embeddedness in a closed loop of action and perception.

Now, time is one of the constituents of this symbolically mediated world that both unites and transcends manifold *umwelts*. Try to imagine a physics laboratory without shared protocols involving what to do first and what to do next, and without participants in research who agree that *now* they start this part of an experiment and that *after* a defined period of time, once it is finished, they will analyse the results – in other words, a physics laboratory that does not possess, and employ, the concept of Time. The frenzy caused by such a lack might be a compelling sight to a participating observer, but it would not produce anything that we call physics. As Ernst Cassirer has famously argued with reference to Bohr and Heisenberg, even 'timeless' quantum mechanics requires 'time' and 'causality' as a condition of possibility of its measurements (Cassirer 1994 [1936], 265).

4   Time as Mission

That said, "time" may not be a universally applicable way of identifying and ordering elements of experience. There may be atemporal segments of reality, as J. T. Fraser was quick to admit (cf. e.g., Fraser 1999, 60–61). In other segments,

---

6   For further elucidation, please see R. C. Steineck 2014, 50–59.

temporal ordering may be applicable, but not without alternatives. There is no problem in conceding such possibilities. Whether the alternatives to time can be used to construct a world as large and rich as that built on time, remains to be seen – the above quoted *Kojiki*, for example, systematically precludes any politically neutral view on questions of cosmology, geography, history, or ethics and thus shapes a world with very much restricted possibilities. In any case, the challenge is to find out how such segments and forms of order relate to our time-imbued world (cf. proposition 4). It is conceivable that the most convincing description of that relationship requires a re-evaluation of the concept of time. That is what Barbour and Rovelli seem to believe. They assume that timeless physics will eventually explain the phenomenon of time, and thereby, reduce it to secondary rank. In contrast, J. T. Fraser, facing discussions in physics on the so-called "theory of everything" in the late 1990s, remained sceptical about the potential of this mode of bottom-up inferential reasoning (Fraser 1999, 35, 132, 258 n. 32). His hierarchical theory of time demonstrates how one may accept that time does not denote a universal feature of reality while insisting that it is fundamental to those segments of the world that we inhabit and experience directly. Whether we choose to follow Fraser or to develop a theory of culture to integrate the atemporal segments of reality with the temporal ones, or whether we seek yet another way: if faced with challenges to the applicability of the concept of time, we need to rigorously enquire what precisely is called into question, on what grounds, and how this impinges on our understanding of reality as a whole.

As an academic society, ISST will continue to actively promote open discussion of such questions and the challenges posed by and to the concept of time. We can do so with some confidence. We can call on 50 years of robust experience, a series of 16 triennial conferences, documented in so many volumes of the *Study of Time* series, and on 17 volumes of our journal *KronoScope*, all of which bring divergent views and theories in dialogue with each other. Let's continue that journey together, into the second half of ISST's first century.

References

Antoni, Klaus. 2012. *Kojiki: Aufzeichnung alter Begebenheiten*. Frankfurt: Verlag der Weltreligionen im Insel Verlag.

Barbour, Julian. 2000. *The End of Time: The Next Revolution in Physics*. Oxford University Press.

Barbour, Julian. 2008. "The Nature of Time." Essay. Foundational Questions Institute. http://fqxi.org/data/essay-contest-files/Barbour_The_Nature_of_Time.pdf?phpMyAdmin=0c371ccdae9b5ff3071bae814fb4f9e9.

Bayer, Thora Ilin, and Ernst Cassirer. 2001. *Cassirer's Metaphysics of Symbolic Forms: A Philosophical Commentary*. New Haven: Yale University Press.

Callender, Craig. 2010. "Is Time an Illusion?" *Scientific American* 302 (6): 58–65. https://doi.org/10.1038/scientificamerican0610-58.

Cassirer, Ernst. 1953. *The Philosophy of Symbolic Forms*. Translation by Ralf Manheim. New Haven: Yale University Press.

Cassirer, Ernst. 1994 [1936]. "Determinismus und Indeterminismus in der modernen Physik." In: idem. *Zur modernen Physik.* 7th ed. Darmstadt: Wissenschaftliche Buchgesellschaft.

Dux, Günter, ed. 1992. *Die Zeit in der Geschichte : Ihre Entwicklungslogik vom Mythos zur Weltzeit*. Frankfurt am Main: Suhrkamp.

Fraser, J. T. 1999. *Time, Conflict, and Human Values*. Urbana: University of Illinois Press.

Fraser, J. T. 2007. *Time and Time Again: Reports From a Boundary of the Universe*. Brill.

Heldt, Gustav. 2014. *The Kojiki: An Account of Ancient Matters*. New York: Columbia University Press.

Hokari, Minoru. 2011. *Gurindji Journey: a Japanese Historian in the Outback*. Sydney: University of New South Wales Press.

Hokari, Minoru 保苅実. 2004. *Radikaru ōraru hisutorī: Ōsutoraria senjūmin Aborijini no rekishi jissen* ラディカル・オーラル・ヒストリー：オーストラリア先住民アボリジニの歴史実践. Tōkyō: Ochanomizu Shobō.

Kant, Immanuel. 1998. *Kritik der reinen Vernunft*. Edited by Jens Timmermann. Philosophische Bibliothek 505. Hamburg: Meiner.

Maki, Yūsuke 真木悠介. 2003. *Jikan no hikaku shakaigaku* 時間の比較社会学. Iwanami gendai bunko. Tokyo: Iwanami Shoten.

Ooms, Herman. 2008. *Imperial Politics and Symbolics in Ancient Japan: The Tenmu Dynasty, 650–800*. Honolulu: University of Hawaii Press.

Plessner, Helmuth. 1975 [1928]. *Die Stufen des Organischen und der Mensch : Einleitung in die philosophische Anthropologie*. 3rd edition. Berlin: De Gruyter.

Postone, Moishe. 2006. *Time, Labor, and Social Domination: A Reinterpretation of Marx's Critical Theory*. Cambridge: Cambridge University Press.

Rovelli, Carlo. 2008. "Forget time": Essay written for the FQXi contest on the Nature of Time. 24 August 2008. http://fqxi.org/data/essay-contest-files/Rovelli_Time.pdf?phpMyAdmin=0c371ccdae9b5ff3071bae814fb4f9e9.

Stanner, W. E. H. 1970. "The Dreaming." In *Cultures of the Pacific*, edited by Thomas G. Harding and Ben J. Wallace, 304–14. Simon and Schuster.

Steineck, Christian. 2010. "Truth, Time, and the Extended Umwelt Principle: Conceptual Limits and Methodological Constraints." In *Time: Limits and Constraints*, edited by Jo Alyson Parker, Paul A. Harris, and Christian Steineck, 350–65. Study of Time 13. Leiden; Boston: Brill.

Steineck, Raji C. 2014. *Kritik der symbolischen Formen I: Symbolische Form und Funktion.* Stuttgart: frommann-holzboog.

Steineck, Raji C. 2017. *Kritik der symbolischen Formen II: zur Konfiguration altjapanischer Mythologien.* Stuttgart: frommann-holzboog.

Tanaka, Stefan. 2016. "History without Chronology." *Public Culture* 28 (178): 161–86.

Yamaguchi, Yoshinori 山口佳紀, and Kōnoshi Takamitsu 神野志隆光. 1997. *Kojiki* 古事記. Tōkyō: Shōgakkan.

CHAPTER 5

Time as an Open Concept: A Response to Raji Steineck

Carlos Montemayor

This paper elaborates on Raji Steineck's thesis that time can be understood as a concept with unique characteristics – a crucial and productive symbolic entry point for interdisciplinary and historical dialogue. I agree with much of his approach, especially with the claim that time, as a concept, does not entail the view that time is an illusion, and shall respond to his proposals on the basis of my own research. Time as such, as a symbolic structure, is not reducible to the typical dichotomies of subjective/objective or apparent/real. It seeps through them, or transcends them. But how exactly? I expand here on key ideas presented by Raji Steineck in his Presidential Address, which might help finding an answer to this question.

These key ideas are that time, as a concept, cannot be a uniquely symbolic structure applicable universally to all disciplines in the same way. Time, as a concept, is compatible with time being real at different levels of analysis, without the need of having a dominant conception of time. A central theme I shall explore is that by giving up the specialized and reductive notions of time that are used to allegedly explain the essential nature of *time as such*, one opens the possibility of a conception of time that balances the realms of the eternal and the ephemeral, the permanent and the organic, the epistemic and the moral. I shall focus mostly on this normative task: the "should" of the search for a balanced conception of time.

Time may serve as a symbolic horizon for historical encounters, scientific breakthroughs, social transformation and political reform. However, unlike the metric or scientific conception of time, this conception of time does not necessitate any of these social and political changes. Therefore, one must ask the question: which conception of times should we adopt to make sense of such diversity without reducing one concept of time to the other? Why should we adopt a rich conception of time rather than simply get rid of the notion of time? What justifies this "should" that is very much at the heart of the International Society for the Study of Time, namely, why must we study such a rich conception instead of either giving it up or letting it be, as a scattered notion that must be defined by the scientific or humanistic specialties? We have

© KONINKLIJKE BRILL NV, LEIDEN, 2019 | DOI:10.1163/9789004408241_007

the conviction that based on all the ways of representing time, we *should* emphasize how the very concept of time allows for interrelations that would not be possible without it. But why? Why not think, instead, that such a broad conception of time is truly illusory, or even a bad idea?

To a first approximation, there is a strictly epistemic and instrumental justification for this "should." We should preserve the disciplinary definitions of time conceptions with a regulative ideal: we should favor the interconnections between those conceptions without trying to reduce one conception to the other and leave what is useful about them to help experts in a discipline develop their field. The concept of time, therefore, cannot be like any other concept even from this perspective, including scientific concepts, such as "energy" or "acceleration" or non-scientific ones, such as "sun" and "justice" (both folk and technical concepts). It is in the very nature of the concept "time" that one can navigate from discipline to discipline without necessarily loosing rigor. Thus, time is a basic concept in logic, semantics, epistemology, perception, sociology, history, physics and biology and, at the same time, the symbolic common ground among disciplines. Some conceptions may depend, or even "emerge" from others (Montemayor, 2010); others may be more fundamental to understanding the metric features of time, and other conceptual aspects of time (Montemayor, 2013).

These are all plausible ways of satisfying the "should" behind the regulative task of identifying, as carefully as possible, the main commonalities among different conceptions of time. However, these are all *epistemic* solutions and strategies, in the sense that they all aim at arriving at certain types of knowledge about time. Knowledge from one discipline may shed light on knowledge from another discipline, and the different commitments and representations of time may shed light on larger debates concerning time (e.g., its linearity, continuity or tensed structure). This is the typical way we proceed when we engage in interdisciplinary or cross-disciplinary studies of time, and much has been learned from these studies. The difficulty here is how to think of the dependencies among notions of time. Suppose the most fundamental understanding of time comes from physics (as many think). Then if, as Raji says, physicists come to the conclusion that time is unnecessary to describe the universe, then one must conclude that the very notion of time must be abandoned. That is a serious problem indeed.

But there is a different kind of "should" behind some of the remarks endorsed by Raji, particularly in reference to the work of Julius Thomas Fraser (1999, 2007). This "should" is most directly related to the normative force of *morality*, and even *aesthetics*. One should seek the good and the beautiful in our search for the conception of time. In investigations of time, this "should" is much more difficult to grasp than the epistemic notion. Although one can distinguish epistemic from moral agency in philosophy (see for instance Montemayor and Haladjian, 2015; Fairweather and Montemayor, 2017), how to exactly articulate this distinction in studies of time remains an elusive task. Here I shall sketch some possible approaches to these two types of "should" that might be at work behind Raji's insightful remarks concerning the nature of time as a general concept, focusing on the moral dimensions of this concept.

The epistemic and moral "should" cannot be explained by merely descriptive statements, such as "time is this in physics, and that in literature; or this according to this author, that according to this theory." Such a list, no matter how accurate and comprehensive, could never justify the way in which one should conceptualize time, either in epistemology or morality. How one should conceptualize time across the disciplines and how should one be guided, morally or aesthetically, by such conceptualization, are questions that demand the kind of general time-concept envisioned by Raji. But precisely because of its normative guidance, this cannot be merely a concept. Rather, it resembles what Kantians call a *regulative idea* – its main purpose is to provide guidance, even if it is ultimately false. However, as Raji emphasizes, the ideal conception of time should avoid the Kantian implication that it is somehow an ineffable, unknowable, or somehow illusory concept. All this means is that the ideal conception of time cannot be known simply by contrasting the disciplinary notions of time.

Thus, the regulative conception of time transcends the disciplinary notions of time and the descriptions of how they contrast with each other. Time is clearly a critical concept in many disciplines, but the regulative conceptions of time transcend any specific conception in at least two ways: as a measure and as a hierarchical structure. As a measure, what makes possible the use of time as a symbolic common ground among the sciences is its scale, from the very long to the very brief, but this feature of time requires a degree of realism about time, namely, that its existence must transcend the specific conceptions and scales required within a specific scientific discipline (again, a very Kantian notion of transcendence is at work here, as condition for the possibility of all measurement). Time measured is not merely time conceived (Montemayor and Balci, 2007) but clearly, time measured does not exhaust, or even begin to capture, all the normative dimensions of time. So this kind of transcendence, even if it fully resolves the normative imperatives of epistemology, as in Kant's transcendental deduction, cannot account for the moral and aesthetic dimensions of time.

The concept of time can serve as a symbolic common ground among the humanities as well, especially in relation to narrative. As a hierarchical structure that has genetic, organic, productive, and integrative features, the concept of time opens possibilities that guide human action towards the good and beautiful. Conflict here means something more than merely a "measure" of any kind. This conception of time has a powerful grip on our lives; it defines the human condition, stretching it and shrinking it, making it a month or a millennium of social explorations and historical encounters that cannot be defined in purely metric terms. But here too, time experienced and acted upon is not merely time conceived. The encounters defined by narrative must have real human depth, which is critical for Raji's proposal of "radical openness," namely, the suggestion that one must give up time as a given or datum and think of it instead as a concept, of a very peculiar kind, as I shall explain in what follows.

Suppose that time should be understood as a concept, not a multiplicity of concepts but a single ideal conception. This concept should be, in order to qualify as the ideal conception, a very unique type of concept: one that works as a regulative idea, guiding us towards the permanent values of the truth, the good and the beautiful, as J. T. Fraser emphasized. It is, therefore, not merely a concept or a conception, but a horizon of possibilities; a "radically open" regulative idea. There is the "should" of epistemology: we should measure time in this or that way; we know that time has this or that characteristic because we have measured and explained it; we all should accept these explanations. But there is also the "should" of morality and aesthetics: this way of conceiving the time of this narrative (mine, or my society) serves us better as guidance than other alternatives; it includes more forms of conflict and helps solve them better; it is more comprehensive than other alternatives.

Where does this narrative structure come from, if it is not explicable in terms of mere conceptions and measurements? I submit that its origin is the structure of memory itself (see Montemayor, 2016, 2018). Narrative depth and the nesting of narratives provide the structure for the kind of radical openness needed here. The possible and the necessary, determined by the logical and metaphysical structure of the universe, becomes flexible and meaningful at the individual and social scales. There is no micro or macro to be explained or reduced here at all: it all "fits in" and it *should* fit in; the historical and the personal, the individual and the collective. This is the possible and the "necessary" that bends and morphs to the complexities of memory, which is teeming with significance. This conception of radical openness does not entail that everything is possible. Rather, the concept of time as a regulative idea, neither as destiny nor as chaos, is what makes all personal meaning possible. It constrains but it also brings in freedom to the "should" of morality and aesthetics, and it also determines the way in which we encounter the world.

This conception of time, as the open thread of narrative, provides guidance and satisfies moral and aesthetic needs. It satisfies conditions that must be in place for the proper functioning of our species, based on the value we attribute to narratives and memories. The conceptions of time as measure and as open narrative have so far being discussed in rather abstract terms. My interest, for the remainder of this response to Raji's Presidential Address, lies in the complex functions and perplexities of the narrative conception of time.

1   Time, Memory, and History

Human memory is a complex and widely encompassing capacity, and its relation to history is an inexhaustible topic. Providing a detailed account of the relation between memory and history is beyond the scope of even entire volumes of work. My sole purpose here is to illustrate an aspect of time, as a concept that guides and regulates our behavior, which articulates the collective with the individual.

We inherited the basic capacity for recollection, like other species, from thousands of years of evolution. Animals must, to different degrees, rely on memories in order to guide themselves and make decisions. Certainly, we share central aspects of our memory capacity with other species. We also inherited the capacity to empathize with others, to jointly attend to objects and features for collective purposes, and to build social structures for mutual benefit. But human memory is unique because of the interaction between different forms of external and internal monitoring functions, that greatly expand the context for social interactions with the past and the future.

In particular, collective human memory, as Halbwachs (1992) explained, is layered in ways that integrate an individual into groups, as crucial parts of her identity. Memories of our family are constantly shared, monitored, revisited, based on constant exchanges, rituals and reunions. Family rituals that consolidate and solidify collective memories include religious gatherings, cultural activities, and celebrations. Social class affects how these memories are integrated, differing from social status to ethnic affiliation, all through careful negotiation. The rituals involved and the social sequences that become collective expectations, articulate a mesh of temporal structures, some of which are recent and secular, while others are ancient and deeply spiritual – all interlaced into the social fabric. These social and cultural symbolizations of time even force themselves upon the more ancient biological rhythms, generating an articulation of temporalities with unprecedented complexity.

The narrative richness of this symbolic and hierarchical conceptualization of time generates significant stress in our lives, but it also opens endless possibilities. Collective memory helps monitoring events that are important because of the groups we belong to, which transcend our existence. Humans give enormous value to collective memories partly because these are externally shared with groups that shape who we are. But we also cherish memories we find personally valuable, because of their autobiographical individuality – they define who we are at a personal and deeply private level. These two uniquely human aspects of memory (i.e., the historical and collective and the autobiographical) pull in opposite directions and they justify an analysis of human memory in terms of tradeoffs concerning accuracy and narrative value (Montemayor, 2016; Montemayor and Haladjian, 2015).

The tradeoff between accuracy and narrative value has a normative dimension because we have the possibility of relating deeply personal memories with widely shared collective structures through group membership and association, perhaps at the cost of forming more "objective" memories. By sharing memories collectively, we can reinforce mechanisms for the reliability of information, following the norms of good epistemic practices. But by being part of groups that inform our values, we also create collective memories that guide our actions and help us navigate the moral and aesthetic landscapes of society, independently of considerations of accuracy alone.

A contrast related to this tradeoff, but in the context of collective memory, is the role of official institutions in promoting a certain version of the past, many times in opposition to the activities of civil society groups.[1] In a democracy, collective memories should not be imposed by official institutions; on the contrary, they should be informed by the collective acknowledgment of the past that only groups in civil society can engender. This is a normative notion; an imperative that belongs to the moral domain: the civil society is *justified* in resisting a version of the past that does not match their experience of the past, especially the past that is relevant to a group and which the official institutions deny to that group. This is clearly not only an issue of reliability or accuracy – although clearly accuracy does play a role. It is essentially an issue of *recognition* and moral status. It is also an issue that touches on justified forms of forbidden knowledge. Should we know everything about the past? How aggressively? To what extent?

Our position in time is socio-politically mediated (and according to some, even determined) and yet, the struggle for one's life, as we experience it

---

1 For the importance of memory activism, and of how it puts into question this simple dichotomy, see Wüstenberg (2017).

through specific socio-political encounters, is deeply personal. Our position in time stretches not only into the infinite past and future, it also sublimates itself into eternity, through the "now" of the thinking self, exactly the way Hannah Arendt (1978) describes it in her book *The Life of the Mind.*[2] According to her, the ideal graphical representation of time should not be simply linear or cyclical, it should be a "parallelogram of forces" in which it is the human thinker that is *deflecting* either the simple chain of changing events or the ultimate mutual annihilation of the past and future (1978, 208–9). This is a "should" of the ultimate importance. The human mind should deflect time for it to be free and also for it to think in an "eternal now" in order to think thoughts that are ultimately eternal: a violent deflection that repudiates the passage of time.

Based on a parable by Franz Kafka, Arendt explains why this complex, almost paradoxical conception of time, is the only characterization that truly captures the "now" of the mind. I quote Kafka (as cited by Arendt) and Arendt at length in what follows, because of the clarity with which the moral and aesthetic dimensions of time are portrayed. Here is Kafka's parable, part of a collection of aphorisms under the title "HE":

> He has two antagonists; the first presses him from behind, from his origin. The second blocks the road in front of him. He gives battle to both. Actually, the first supports him in his fight with the second, for he wants to push him forward, and in the same way the second supports him in his fight with the first, since he drives him back. But it is only theoretically so. For it is not only the two antagonists who are there, but he himself as well, and who really knows his intentions? His dream, though, is that some time in an unguarded moment – and this, it must be admitted, would require a night darker than any night has ever been yet – he will jump out of the fighting line and be promoted, on account of his experience in fighting, to the position of umpire over his antagonists in their fight with each other.

As Arendt proceeds to explain, this parable captures the time of freedom, and also of despair: the "now" of a thinker, trapped between two infinities. She writes: "Man lives in this in-between, and what he calls the present is a life-long fight against the dead weight of the past, driving him forward with hope, and the fear of a future (whose only certainty is death), driving him backward

---

2 Although many aspects of this brilliant book by Arendt are relevant for the present discussion, I focus only on her interpretation of Kafka's parable, for the sake of conciseness and because it suffices to illustrate my main point.

toward 'the quiet of the past' with nostalgia for and remembrance of the only reality he can be sure of" (1978, 205). This remarkable passage – a profound and insightful comment on Kafka's parable – is followed by a lucid discussion of why the continuity of time depends on the spatiality of our bodily and daily activities, and why the proper time of the mind is the "now," which is, when it comes to thought and even to the activity of thinking, outside of time. Arendt says the following about the dream of the man trapped between the opponents:

> What are this dream and this region but the old dream Western metaphysics has dreamt from Parmenides to Hegel, of a timeless region, an eternal presence in complete quiet, lying beyond human clocks and calendars altogether, the region, precisely, of thought? And what is the "position of the umpire," the desire for which prompts the dream, but the seat of Pythagoras' spectators, who are "the best" because they do not participate in the struggle for fame and gain, are disinterested, uncommitted, undisturbed, intent only on the spectacle itself? It is they who can find out its meaning and judge the performance. (1978, 207)

We want this dream to become true, but we cannot step outside the present. The forces of the past propel us collectively, through memories, and through the contingency of our very own individuality, by the experienced impulse of an ephemeral origin pointing forcefully towards destiny. This is not, as Arendt emphasizes, a mere "chain of events". It is essentially a constant struggle, literally a fight, and although we machinate and idealize our exit through the trick of thinking a realm of eternal peace, in reality we cannot escape the combat of the present: our immediate socio-political, biological, bodily, spatially concrete and viscerally relevant present. The unsoiled indifference of the Pythagorean spectators is infinite in its capacity for timeless reflection, but it ultimately leads to delusion, and even tragedy. We all live in the present, and the present is a battle. The harmony of time as measurement, stretching mathematically into infinities, is nowhere to be found here. On the contrary, what one finds is oblique and conflictive deflection:

> Without doing too much violence to Kafka's magnificent story, one may perhaps go a step further. The trouble with Kafka's metaphor is that by jumping out of the fighting line "he" jumps out of this world altogether and judges from outside though not necessarily from above. Moreover, if it is the insertion of man that breaks up the indifferent flow of everlasting

change by giving it an aim, namely, himself, the being who fights it, and if through that insertion the indifferent time stream is articulated into what is behind him, the past, what is ahead of him, the future, and himself, the fighting present, then it follows that man's presence causes the stream of time to deflect from whatever its original direction or (assuming a cyclical movement) ultimate non-direction may have been. (1978, 207)

The insertion of humanity in time provides an aim, a purpose, both social and individual, non-measurable, and in constant conflict. One may call this insertion, without exaggeration, *the turmoil of humanity*. Social and historical struggles have, therefore, a similar structure to the one described by Arendt. This *aim*, this "now" of the thinker, is also the "now" of human freedom and precisely because of this, it is the source of moral and aesthetic value. One never escapes the moral demands of the present by dreaming a Pythagorean spectacle.

This is why collective memory is so important in a thorough conceptualization of time. Documenting any crime in the life of a democracy corresponds to the right of its citizens not only to know the truth, but also to evaluate what kind of moral standards their nation has; their collective right to gauge, as it were, the moral worth of their community at the present moment. This is a real struggle, repeated constantly. Narrating our group's identity allows us to hold on to the value of our past, in order for us not to be denied of what we find valuable in it; our way of *identifying* with it. The turmoil of humanity is a balance between freedom and order, truth and moral value, knowledge and ignorance. Should the moral guidance a collective gives us prevent us from learning truths about our past, and vice versa? Should we have specific conceptions of how the past should influence the future? We are pushed and pulled, free and determined, flowing in time and thinking outside of time – against time. A conception of time as a *parallelogram of forces* obliges us to conceive of time as conflict, but not just any conflict; time under this conception is a conflict with unique normative demands: the truth, the good, and the beautiful – the result of the constant struggle between the fixed past and the undetermined future.

A full version of the parallelogram of forces shows that collective memory, history and a collective understanding of the past and potential future are crucial components of the energy distribution that creates the turmoil of humanity. "HE" could easily be replaced by "WE" in Kafka's parable – the parallelogram of forces becomes a complex fractal representation of several billion struggles in the present, each living person at the present moment is her own turmoil: a constant clash of forces. It suggests, moreover, that there might be a *normative dissociation* in Arendt's interpretation of Kafka because the more we focus on concepts of time that abstract from the present, the less inclined we are to respond to immediate socio-political pressures, including the moral needs of others. This correlates, in many interesting ways, with the dissociation between consciousness and attention (Montemayor and Haladjian, 2015).

The correlation is as follows. On the one hand, the present moment is defined by our conscious awareness – the vital, visceral presence of our emotions, feelings, and thoughts. Moral and aesthetic values are associated with this vivid, visceral reactions, which can be dissociated form attention. The cognitive functions of attention, on the other hand, while not necessarily conscious, play an important epistemic role (Fairweather and Montemayor, 2017): they provide knowledge and justified beliefs that are salient for solving problems and reasoning. This is, in Arendt's words, the realm of thought, and it naturally expands into past and future, resisting the visceral demands of the "now". Moral values need to enter more into the open conception of the possibilities of time, and one way of doing this, is by providing Arendt's conception with this understanding of the cognitive foundations of the dissociation between consciousness and attention.

There is not enough space here to give a fully detailed account of this idea. All I wish to have established is that there is a vast possibility for research that needs to be explored if we decide to understand time as a concept or set of conceptions that are necessary to expand the dimensions of human dignity. This goal, justified by Arendt's parallelogram of forces is, I believe, one way in which we shall be, in Raji's words, radically open to new formulations of the concept of time. A collective form of moral agency, involved in the advancement of history towards the eternal values of beauty, truth and moral goodness seems paradoxical. But we may be guided by this ideal, and even succeed one day in arriving at a powerful new conception of time.

Perhaps, as Raji indicates, we must be radically open to give up time as a feature of the universe. To take it as concept, or a conception, requires that it itself is radically open, or torn, the way Arendt describes. We should be open to the idea of not talking about time *as such*, but as a construct that allows for conflicts, without which we could not be free, have moral standards or lead a meaningful life. Giving up time, in this sense, means turning our attention to the present needs determined by our social realities and collective conceptions of time, instead of exploring the mostly abstract formulations of time that Arendt associates with the Pythagoreans. It is a complex and difficult task, but it is one that is at the core of the mission of the International Society for the Study of Time from its inception; a conception that was, as Raji points out, very much in the mind of its founder.

References

Arendt, Hannah. 1978. *The Life of the Mind*. New York: Harcourt Brace & Company.

Fairweather, Abrol and Montemayor, Carlos. 2017. *Knowledge, Dexterity, and Attention*. Cambridge University Press.

Fraser, J. T. 2007. *Time and Time Again: Reports From a Boundary of the Universe*. Leiden, The Netherlands: Brill.

Fraser, J. T. 1999. *Time, conflict, and human values*. Urbana: University of Illinois Press.

Halbwachs, Maurice. 1992. *On Collective Memory*. Translated by Lewis A. Coser. Chicago: University of Chicago Press.

Montemayor, Carlos and Haladjian, H. H. 2015. *Consciousness, Attention, and Conscious Attention*. Cambridge, MA: MIT Press.

Montemayor, Carlos. 2018. "Consciousness and Memory: A Transactional Approach." *Essays in Philosophy* vol. 19, iss. 2: article 5. https://doi.org/10.7710/1526-0569.1612.

Montemayor, Carlos. 2016. "Memory: Epistemic and Phenomenal Traces." In *Time and Trace: Multidisciplinary Investigations of Temporality*, edited by Sabine Gross and Steve Ostovich, 215–31. Leiden, The Netherlands: Brill.

Montemayor, Carlos. 2013. *Minding Time: A Philosophical and Theoretical Approach to the Psychology of Time*. Leiden, The Netherlands: Brill.

Montemayor, Carlos. 2010. "Time: Biological, Intentional and Cultural." In *Time: Limits and Constraints*, edited by J. A. Parker, Paul Harris, and Christian Steineck, 37–64. Leiden, The Netherlands: Brill.

Montemayor, Carlos and Balci, Fuat. 2007. "Compositionality in language and arithmetic." *Journal of Theoretical and Philosophical Psychology* vol. 27, no. 1: 53–72.

Wüstenberg, Jenny. 2017. *Civil Society and Memory in Postwar Germany*. Cambridge: Cambridge University Press.

CHAPTER 6

Zero-Time Theory

Peter A. Hancock

Abstract

We must each write our own truth. The validity of that truth is only ostensibly tested by time. However, here, I seek to write the truth of time itself. The eventual conclusion of such an inquiry is that time is a necessary but delusional narrative for all living systems. Tragically then, the apparent necessity for time in life proves eventually to be, in the end, vacuous. But if we are willing to sacrifice our iatrogenic reification of living systems and our own hubristic pre-occupation with ourselves, albeit only for a brief span, we can eventually titrate and distill the truth of this proposition of delusional time. However, the degree of humility required of such a step is virtually anathema to the very nature of the human condition. Thus, I conclude that time's urgent delusion is unlikely to be recognized or dispelled, at least within any foreseeable 'future.' Further, there is of course, a moral dilemma in seeking to reveal this excoriating truth of time. For, our social civilization is fundamentally founded upon time's persistent mirage. Removing this foundation would not simply threaten society but perhaps actually assure anarchy. Nevertheless, I here present some keys to unlock the bounden hands of time in the hope of its final redemption.

1 Introduction

It is not simply hard, but for most humans it is virtually impossible, to comprehend a world without time (McTaggart 1908).[1,2] Time is woven so deeply, even into the very fabric of the languages through which we communicate, that to question time appears to be utterly absurd. Indeed, to negate time seems to be to question the very nature of reality itself. Most often such inquiries or possibilities are dismissed out of hand as futile. The question is not merely

---

1 This work is founded upon ideas presented in the paper Hancock, P. A. and Hancock, G. M. Time's urgent delusion: Implications of the phenomenological super-symmetricality of sentience in space-time. Presented at the Triennial Conference of the International Society for the Study of Time, Edinburgh, Scotland, June, 2016.

2 I think it worth quoting McTaggart's (1908) first line here when he observed: "It doubtless seems highly paradoxical to assert that Time is unreal, and that all statements which involve its reality are erroneous."

© KONINKLIJKE BRILL NV, LEIDEN, 2019 | DOI:10.1163/9789004408241_008

ruled inadmissible but simply preposterous on its face. While arcane philosophers are permitted to speculate upon the nature of time and associated conscious reality, few outside such a select cadre ever cogitate extensively upon it.
Perhaps even fewer ever consider the ramifications of such a proposition (but
see Barbour 1999; Russell 1915). For the 'man in the street,' (the American "Joe
Sixpack," or the English, "Man on the Clapham Omnibus"), time is simply a fact
of life.[3] Time is a given, an immutable element of our environment. In consequence, most individuals pragmatically conclude that rather than to keep
considering time, it is best simply to get on with life. When someone asked
St. Augustine what God was doing before he invented time, his answer was that
God was creating a special hell for people who asked such questions like that.
His comedic assertion serves actually to deflect the question, but crucially,
it does not answer it (and also see comments by Fraser 1987). St. Augustine's
other response about knowing what time is until asked still resonates well with
the vast majority of individuals today. For such diversion and avoidance prove
effective strategies if one wishes to defer engaging in apparently intractable
problems. Such responses prove to be misdirection, not solution, nor even a
step towards solution.

Indeed, so befuddling are the issues of time, which seem immediately and
spontaneously to emerge, that even "timesmiths" are often forced to retrench
to St. Augustine's second observation, noted above, which articulates the issue
but also does not move us toward any resolution at all.[4] It is quite feasible to
continue this whole essay in the same vein commenting on the behemoth
of time, its titanic origins and its evident cognitive intractability. Such works
are legion.[5] But what I look to provide here are a number of keys to time; to
emphasize the criticality of each, and to sequentially unlock the shackles
that bind our conceptual hands to time. I hope that, at the end of this exercise, we shall better "know" both time and ourselves. However, not even the
Delphic Oracle promised that such knowledge would prove to be of either a
salutary or heroic nature when finally secured. It turns out such knowledge is

---

3 By this I do not wish to infer that many, if not most human beings do not give some passing
   consideration to the nuances and subtleties of time; assuredly they do. However, it is just that
   it is most frequently a 'passing' consideration and not one upon which they act in any meaningful manner. Shortly after considering the bizarre nature of time and our relative ignorance
   of it, they rapidly return to 'real' life where the ramifications of what they have considered are
   overwhelmed by the force of the accepted narrative.

4 If you are not aware of Augustine's most famous pronouncement on time, it may be advisable
   to read about it at this juncture. As you will witness, his aphorism serves to deflect the argument about time without providing any resolution whatsoever. As Thomas More might have
   noted, a clever man indeed!

5 Indeed, we might reference a whole panoply of leading philosophers who have participated
   in this escapade, including but not limited to Spinoza, Kant, Hegel, and Schopenhauer.

not simply nihilistic but challenges our very conception of all reality. To begin such a daunting odyssey, let us proceed from a starting point of what is, apparently, a minor and rather esoteric observation. From this first initial crack in the temporal edifice, I shall elaborate upon other inconsistencies, aiming toward apparently evermore profound issues. However, each discrete issue, in actuality, represents differing faces of the same fundamental concern; the self-delusional nature of time.

2   The Psychophysics of Time

At first blush, psychophysics is a rather obscure point of departure for any such sequence of primarily philosophical argument. Psychophysics is a sub-discipline of psychology that has an extended, but rather specialized, history (Fechner 1860). In brief, psychophysics seeks to plot the subjective (or affective) response of an individual to a stimulus, against the physical magnitude of that stimulus. An example will help here. Suppose I ask you to estimate a distance of one metre. I could do this by presenting some sticks of differing lengths and asking you to choose the stick that you think is closest to one metre in length. This is often referred to as a "forced choice" situation, since you have to pick from the sticks that I, the experimenter, present or force on you. Alternatively, I could ask you to make two separate chalk marks on a table that you thought were one metre apart. This would not be a forced choice but rather a "magnitude estimation" approach in which you now generate your own estimate rather than choosing from a range of mine (the experimenter). For the purposes of my discussion I am not, at this juncture, particularly concerned with distinguishing between these methods. I simply want to recognize that there are numerous ways to enact estimation and methodological distinctions between the approaches we use to secure psychophysical assessments. Now suppose I ask you to estimate 2 metres, then 3, 4, 5, 6 meters, etc. What I would then have are data on one axis that are the actual meters (or rather what we all socially agree to call these numbers of metres and the standard definition of such a length). On the other axis, I would have your estimate of those differing lengths. Typically, the base axis is reserved for the actual lengths; subtly indicating the primacy and reality of the physics within psychophysics. The vertical axis represents your estimates, often now seen as the deviation of the psychological experience from the reality of the physical value. More colloquially your deviation from precise agreement is seen as *error*. This Cartesian representation shows the physics and the psychology in tandem; hence *psychophysics*. Now, with these estimates for differing lengths secured, I can plot a function which ostensibly relates reality to perception.

Typically, this is shown as a curve of some sort, even though such a curve is most often derived from connecting the discrete, individual point estimates. We have performed our thought experiment here using the dimension of length (distance) but, in principle, we could have used any physical parameter (e.g., light, sound, pressure, weight, etc.). In fact, any physical dimension that can be conceived of, if able to be manipulated in a reasonably controlled fashion, can be used in psychophysical testing. As I noted, there is an extended history of this form of research that I am not going to further elaborate upon here but which the interested reader can follow for themselves (see e.g., Fechner 1860; Luce 2002; Stevens 1957).

But how does psychophysics relate to the nature of time? Well, quite simply, we can conduct such experiments on the psychophysics of time (see e.g., Hancock and Block 2016). We can present ostensibly 'objective' durations and then ask people to estimate those intervals using either forced-choice (e.g., Lhamon and Goldstone 1956), or duration estimation techniques (Block, Hancock and Zakay 2010), or even employ alternate psychophysical approaches such as interval reproduction (Eisler 1976).[6] As these observations imply, such investigations have been frequently reported and summarized across the years (and see Hancock and Block 2012). The degree to which perception fails to match the physical values presented is given, in psychophysical parlance, by the exponent of the curve. This exponent value is most often derived from the summation of individual experiments (cf. Figure 6.1 and Figure 6.2). Some physical dimensions have exponent values that are less than unity i.e., < 1. Thus, the change in perception diminishes with increasing physical value (see Figure 6.1, line B). Others have exponents that are greater than unity. i.e., > 1; the percept increases disproportionately in comparison with the physical value; pain being an interesting example of this (see Figure 6.1, Line A). What is of concern for us here is that, on average, the psychophysical exponent for time is unity (Figure 6.1, Line C). More succinctly, the slope of the psychophysical function for time is 1. These three respective tendencies are each shown in Figure 6.1 (Lines A, B, and C). A comprehensive meta-analysis that have I previously conducted (Hancock 2011), has demonstrated that the average slope value that is derived (from male participants; of which more later), is 1. The slope for female participants is also linear, but the average slope value is 1.13 (see Figure 6.2).

---

6   I should note here that Eisler (1976) identifies the exponent value derived from a fairly comprehensive survey of the time reproduction literature as 0.9 (as opposed to 1.0). However, I should note here that he includes numerous studies which use drug manipulations as well as incorporating those studies involving mentally disturbed individuals. It may be such idiographic sources of variability that influence the 0.9 value as compared to the 1.0 that I have gone on to report as being more generally representative here.

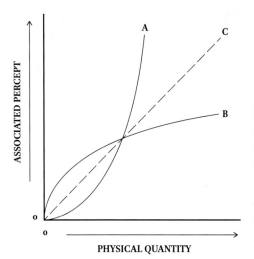

FIGURE 6.1
Exponential functions plotted in psychophysical space. Experiences such as pain increase disproportionately quickly with respect to the physical stimulus, thus the exponent value is great than 1 (curve A). Other experiences, such as loudness asymptote as the perceptual experience is saturated (as shown by curve B). In contrast to either of these, the perception of time follows closely to the line of unity (line C).

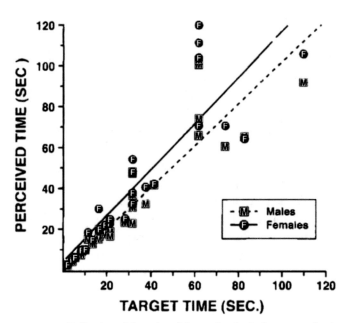

FIGURE 6.2 Indication of the unity of the psychophysical exponent for time. The illustration also shows that the 'creation' of the social reticulation of time is predominantly a male fabrication. Time-keeping devices naturally follow and perpetuate the narrative of time as a 'real' dimension. The illustration, taken from Hancock (2011) does not show individual participants but rather each point represents the average of a particular study involving up to hundreds of participants per study (and see Hancock and Rausch 2010).

But, is this observation in any way important? After all, does not this just mean that we humans are (on average) accurate at estimating time? Actually, as individuals, we humans prove to be quite variable at estimating time and the reasons for this wide individual variability have yet to be fully and satisfactorily determined (Doob 1971). But look again at the axes themselves. As I have noted, materialists take the base axis to be reality, i.e., the reification of the physical value (and the associated primacy of physics) over the psychological; (thus, the associated relegation of affect). The vertical axis is left as the perceptual variation (or tellingly here, the error) from such reality. Yet there is another way to interpret this graphic representation. Suppose that both axes actually represent created, designed, and fabricated social agreements.[7] This means what we are actually looking at is a correlational space between the estimates of a sample of modern human individuals (i.e., the present experimental participants) against an artificial dimension created by the estimates of an earlier sample of human individuals (i.e., the ancestral 'designers/fabricators' of the concept of time). When we let go of the reification of physics here, it is little wonder that on average, the expressed correlation (and the resultant exponent) is one. Further, this helps to explain why the slope value for males is 1.00 while the slope value for females is 1.13. Following the logic of the argument, this outcome derives because the original designers of such temporal measures were themselves male.[8] Thus, males have a greater opportunity to be accurate with respect to a past sample of their fellow males than do modern-day females. The implication here, the symptom of the crack in time, is that unlike other quantities with varying exponential values, the unity of the temporal slope can begin to show that time is a *fabricated* dimension and not one intrinsic to the world around us. We can, from such initial observations, begin tentatively to suggest that the dimension of time is a iatrogenic and not a fundamental one. At this stage, this is a highly contentious and admittedly tenuous postulate from this psychophysically-based argument. However, there

---

7 By this I mean that not only are the ways in which we *measure* time common social agreements, i.e., the standard metrics of seconds, minutes, hours, etc., but also the fundamental dimension of time is also a common social construction. It is actually the case that time proves to be an exceptionally useful and even indispensable human tool. What we may be seeing here is the confirmation of this prosthetic efficacy. And see below.

8 The primacy of 'male' time may help explain why some female members of society feel disjointed with respect to what is offered as intrinsic reality. In the same way that left-handers have to adapt to a right-handed world (Coren 2012), females (on average) are constrained to adapt to a male temporal world. However, with such large variations across individuals, such general statements cannot be considered idiographically specific (Pos 2006).

are more clues to the puzzle of time, so let us see whether the contention that time is a pure fabrication can be supported from other lines of evidence?

3   In 480 BC …

For those who seek to resolve the mysteries of time, 480 BC must have been a wonderful year in which to be alive. For, it is then that Wikipedia assures us, three of the most influential fathers of temporal thought were all alive at once. Respectively, these were *Protagoras, Parmenides*, and *Heraclitus*. It is surely a sadness that we possess only a limited sample of their respective thoughts and writings. However, what we do possess is instructive, even if it is brief. *Heraclitus* asks us to examine our world and to consider the proposition that all is change. In contrast, *Parmenides* challenged us to consider that the Universe is without change. The key to resolving these purportedly, diametrically opposed perspectives lies, I believe, in the observation of *Protagoras* who asserted that "man is the measure of all things." For completeness and symmetry, and as explicitly noted in Figure 6.3, there should be a fourth proposition which might be termed "contra-Protagoras," or, as I would assert: "man is the measure of nothing." This combination leaves us with an interesting 2 × 2 matrix which is illustrated in Figure 6.3 below.

Since Protagoras points the way forward here, let us consider his provocative proposition in more detail. We can, I think, assert that in general Protagoras's view has held sway, at least for a number of millennia and especially in Western thought.[9] I would argue that this dominance of human-centeredness is almost necessarily the case since, from birth, we humans are rather saddled with the narrative that we are the hero of our own story (Campbell 1949). This inherent narrative occurs even within more collectivist cultures where the emphasis is placed more on the group rather than on the individual. But further, there are good evolutionary and neuropsychological reasons to retain the primacy and centrality of personal consciousness at the forefront of existence for each and every individual. Quite simply, humans may often act in social groups but it is our tragedy that we each live and die alone. In essence, Protagoras formalized and generalized this ultimate ego-centric positioning in order to claim that we humans are the primordial 'frame of reference' through which all of reality

---

9 It is important to note here that much of our Western thought on time has been shaped by those who have, on occasion, asserted the vacuity of time themselves but it is within the Eastern mystic tradition that the fallacy of time has been most often and most thoroughly explored, although not codified in the manner necessarily coherent with the Western worldview.

Man is the Measure of

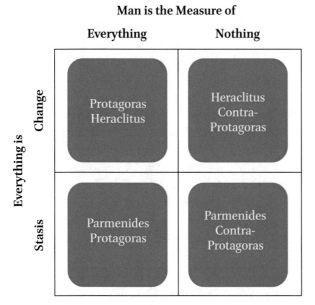

FIGURE 6.3   The Quadrants of Time. Our standard, and largely
             unquestioned model of reality occupies the top left
             of these quadrant. It is enlightening to consider
             what the nature of time is in the other quadrants.
             One can ask whether duration persists in a state of
             stasis, as well as what time looks like in the absence
             of all living entities (where the term 'man' here acts
             as a proxy for all such living systems).

has to be scaled. I seek to dispute this proposition. Although we are cognitively
aware of the insignificance of our physical presence and physical location in the
Universe, humans have never abandoned their essential cognitive hubris which
puts self above all. This stance proves to be a categorical error. The insightful illustrations of the visual images in the text "Powers of Ten" by Eames and Eames
(1998; see also Morrison and Morrison 1990) reinforces this centrality in an involving and interesting visual manner. Crucially, and despite the wonderful effort at 'perspective,' the 'Powers of Ten' project still places 'humanscale' at the
$10°$ location.[10] Such is the intellectual seduction of this scaling reference frame
that, in other work, I have also defaulted to this use of "humanscale" to represent the growth of perception and action in the Universe (Hancock 2009).
As I noted, this emphasis on the intrinsic centrality of human beings may even

10   If you have not yet had the opportunity, I strongly recommend viewing the associated film
      on-line. At the time of writing, the URL: https://www.youtube.com/watch?v=ofKBhvDjuyo
      was still working.

rise close to the level of necessity in the very strange and bizarre case of the nominally exceptionalist species *homo sapiens* (and see Hancock 2015a).

As we see in illuminating works such as that of Koestler (1990), the history of our human existence can be characterized as a progressive recognition of our displacement from the physical center of the Universe (i.e., the terracentric to heliocentric step; the step from galactic center to galactic periphery, etc.). In science, we now readily admit to this physical displacement and recognize that there is no necessarily privileged observer or observational point in the Universe; *for it is now the standard scientific narrative* (Hancock 2005). However, as religion (narrowly writ) and morality (more broadly writ) show, we humans have never even seriously considered displacing ourselves from the center of our own cognitive Universe. We remain very much bound by this intrinsic dogma; even in science. By championing an anti-Protagoran stance here, I want to explore what such a cognitive displacement would represent; and especially what it means for our understanding of time. Since I have already claimed that time is a fabricated dimension that emerges from the strange occurrence of living systems, and since I also argue that such living systems occupy no privileged position in existence (except that which they award themselves), then logically I doubt and actually dismiss time itself: although I fully comprehend that on its head, this appears to be a nonsensical assertion (but see Barbour, 1999; and of course, McTaggart 1908).

Since nothing impresses so much as the appearance of mathematical and computational attribution, I have expressed the four propositions of Figure 6.3 in somewhat quasi-Boolean logical terms. Thus:

> *IF Protagoras (Man is the Measure of All Things), THEN Heraclitus (All in Change),*
> *BUT*
> *IF NOT Protagoras (Man is Not the Measure of All Things), THEN Parmenides (No Change),*

Here, I think we can, with a little acceptable intellectual consideration, substitute the world "time" for the word "change." Thus:

> *IF Man is the Measure of All Things, THEN we have TIME,*
> *BUT*
> *IF Man is the NOT the Measure of All Things THEN TIME is delusion.*[11]

---

11   I am perfectly aware that many will argue that there is no logical foundation for assuming that if we do not measure all things by 'man' then time evaporates. However, for 'man,' we need to be more catholic in our interpretation and read instead 'living systems,' or life.

The step of ultimate cognitive dissonance is to ablate all that we have thought to understand of the idea of self, which is to starkly discard all of our auto-biographical treasures and consider the following, admittedly arid and un-palatable alternative. That is, living systems possess no exceptionalism. Their intrinsic emergent properties (e.g., mind) are only the result of the haphazard happenstance of the possible re-arrangement of matter in the Universe. Time then is an emergent sequela of such a nominally 'statistical' property. Time is not, nor necessarily can it be, an inherent dimension of, nor *sine qua non* of existence.[12]

4 The Nature of the Narrative

To this point, the assertions as to the non-existence of time come essentially from an external perspective; from the "outside-in" as it were; from a position that explicitly and a priori rejects our standard interpretation (e.g., the upper left quadrant of Figure 6.3). However, in order to understand the true vacu-ity of time from the "inside-out," we have to explicitly attack this standard narrative of human reality which appears, necessarily, to progress from the past through the present to the future. Thus, to disturb, and even move this standard perspective (if indeed any movement is morally advisable) from the top-left quadrant of Figure 6.3, we have to generate a 'push,' force as opposed to some attractive 'pulling' force from any of the other three quadrants.[13] In respect of this 'inside-out' exposition, examples from observations of athletic competitions help. As I write this work, the Rio Olympic Games of 2016 are in full swing. We are regaled, especially on American television, with a series of "up close and personal" stories in which we learn of each athlete's individu-al travails. Each of these stories feature how the cited individual has battled against great odds, great adversity, and great heartbreak in their lives along their own personal road to triumph and redemption. Proud parents, or even prouder foster parents, are shown beaming down upon their respective chil-dren as the standard "heroic" narrative is again recounted and the anthemic pride of each nation is played out over and over again. Of course, the Olympics are only one instructive but very public case among all such human narratives. But let us examine the instructional aspects of such events in detail. First,

---

12   Interestingly here, the word statistical actually reinforces the notion of time. That is, the uncertain distribution suggests change and variation. If the Universe is as I will propose, then such change itself is completely obviated.

13   As I shall reiterate extensively, none of these other quadrants actually represent an attrac-tive option. Indeed, this may be why the human narrative asymptotes to this particular form.

barring significant disruption (as for example in the case of World Wars) the
Olympics will take place and someone is going to win each Olympic event.
Olympic medals are not contingent upon absolute performance level. Rather
they are based on relative competition, i.e., you only have to beat those around
you and not the whole world in order to win. Such competitions are not the
same as setting a world record in which one has to beat all previous athletic
efforts; or alternatively invent one's own new sport. This means that in essence,
of all competitors present, someone will win.[14] And remember each of these
respective winners will *necessarily* have their own personal story which, inevitably, partakes of the prototypical heroic narrative (Campbell 1988).

Such myths are also promulgated, for many and diverse reasons, that such
success is open to any human being with sufficient moral fortitude, personal
motivation and physical endeavor.[15] Of course, this is patently false, absurd,
and even insulting. Almost all athletic competitions prove to be statistical sort-
ing mechanisms. If one wished to win a gold medal in a particular sport, one
would be well advised to choose one's forebears very carefully; even unto the
seventh generation. For even with extensive personal training one purportedly
"cannot put in what God has left out," (where the latter might more appealingly
be viewed scientifically as what genetic advantage confers). Thus, what we are
actually viewing in Olympic competition is a game of statistical dice in which
the vagaries of individual genes are exposed to the haphazard nuances of environmental variation in order to produce each specific event winner. These
winners (and their facilitators and promoters, in the form of television com-
mentators, product advertisers and the like) either implicitly or, most often,
explicitly collude with each other to promulgate an unquestioned hindsight
bias in which the standard delusion (of time) is sustained and reinforced (see
Hancock, 2015b). Of course, the Olympics only prove to be one small, but very
globally prominent, example of what happens in everyday human life anyway.
But if, with some small modicum of courage, we can let go of such iatrogenic
delusion, (and of course, in so doing also ablate the importance and centrality
of our own personal narrative), then what we are evidently viewing are simple 'spatio-temporal' distributions of such happenings and activities. Albeit
that this observation of varied matter distribution remains, at this juncture,
couched within the framework of time we have traditionally accepted.

Now, if we can accept the happenstance of such distributed events, we can
begin to look at patterns which are apparently involved in such distributions.

---

14   Of course, there are exceptional cases where all competitors are disqualified, but such
        instances do not negate the point I am making.
15   After all, the motto of the Olympics is: citius, altius, fortius.

Here, I want to suggest that recognition of any such patterns are inherently *apophenic*. That is, it is we humans who *impose* pattern on what we experience. Thus, there is no such thing as external 'order.' At this point I also want to assert that the know-ability of the distributions of such events and activities are essentially symmetrical about any arbitrary "now" or point of the perceptual present adopted by any living observer, (Hancock 2013, and see also Figure 6.4). Thus, the degree of one's personal and social knowledge of events which occurred some thousand years ago is formally, statistically equivalent to what the same individual can know of what will happen in one thousand years' time. Like the unreality of time, this proposition also grates upon the intellect which has (often unquestioningly) accepted the tenet that the past is known but the future is yet to come. However, I believe formal probability falls off symmetrically about any 'specious' present (which necessarily requires the presence of a living observer to underwrite it; James 1890). History, which acts as the slavish handmaid of time, appears to give the lie to this assertion. However, cogitation upon this proposition begins to reveal the self-evidence of its truth. Like a visual "magic eye," it eventually makes itself manifest. As my earlier allusions to the vacuity of time should now make clear, without any single privileged temporal observer (i.e., no one's "now" has any preference over any other; as no one's "here" does either), the observed events themselves are inherently timeless. They are only endowed with time through their discovery by a living observer.[16] We can now add in the prior observation concerning the false reification of living systems to understand that living observers can necessarily, never confer any privilege in observation; contra our own indomitable cognitive hubris.

The outcome of this line of thought is, of course, rather bleak. Hero worship proves to be without value; effort and perseverance are myths, social approbation proves vacuous. "Good" is not triumphant; and morality is totally obviated. Even cause and effect and their indemnifier "free-will" go by the board. Without time's delusion, the very fabric of social civilization, cultural evolution, personal responsibility – in fact all such myths are rendered naught. *Even meaning is without meaning.* This is not simply a nihilistic position, but it does deny the nature of the reality we have created and renders *iatro-telesis* the only remaining cognitive pabulum available.[17] While I shall hold out one small hope (for after all what writer can not), the simple fact is that the omnipresent

---

16   Since life, as we know it, is only one form of emergent property, we cannot be certain that other forms of 'observe' do not also exist.

17   Iatrotelesis can be defined as self-developed purposes for specifying meaning. This underlies a notion I have termed the *cognitive anthropic principle*. Although, it proves to be a false idol, I shall seek to present why it might prove to be a socially acceptable pabulum at this juncture in our collective comprehension.

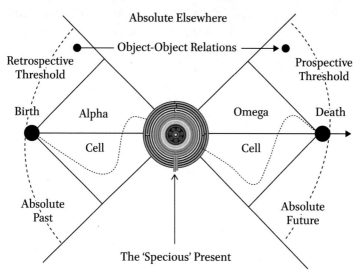

FIGURE 6.4 A space-time representation of the transience of any living
entity. It is bound by the genesis of its coherent order (birth) to
its subsequent cohesional dissolution (death). The necessary
presence of a specious present (now) gives rise to the boundary
between alpha and omega cells (and see Shaw and Kinsella-Shaw,
1988). The ratio of the perception of these cells is empirically
explorable (and see Figure 6.6). The detail and meaning of the
central boss is expanded in Figure 6.5.

but haphazard emergence of putative consciousness from specific arrangements of material matter serves to create time. From an *a priori* construct to
the constant renewal of living tissue, time has proved to be a wonderful, and
insight rendering tool (Hancock 2018), and we cannot wean ourselves off of
its addictive qualities. I doubt human beings ever could, ever can, ever will, or
would ever actually want to. As the structure of the previous sentence serves
to demonstrate, the tool of language is itself unhesitatingly contingent upon
the *a priori* tool of time. The Delphic Oracle, having inveighed us to "know ourselves," gave no accompanying guarantee we would enjoy such ultimate self-
knowledge. And when Locke protested his loyalty to the truth, he observed
that it is "worth the seeking," but not necessarily that it was worth the finding.[18]

---

18   The memorial plate in Christchurch Cathedral, Oxford to the great John Locke reads: "I
know there is truth opposite to falsehood that it may be found if people will and is worth
the seeking."

FIGURE 6.5 The present description details the central boss of Figure 6.4. It is representative
of the 'specious present' or now in Figure 6.4 and provides a leitmotif for ZTT.
Since there is no possibility of a direct percept of time, all representations of it
are constrained to be spatial in form. The present spatial image seeks to convey
a number of dualities embedded in the traditional concept of time. Thus, the
primary image is one showing a circular clock while the numerical values
emphasize the more modern digital allusion. Respectively, these reflect time's
arrow and time's cycle (Gould 1988). At the same visual level I have emphasized a
different duality. Here the contrast lies between the natural and the technological.
The featured line drawings show a classic 'flower clock.' The opening times of
these respective floral representatives act to parse time across the traditional
twelve hour differentiation. The circular representation again recapitulates the
cyclic nature of time. Superimposed upon these is my own representation of
the 'Centon' clock dividing the day into ten 'hour' segments with one-hundred
'minutes' per hour. Again, this features the conflict between the vestigial
sexagesimal system and the more modern SI based notation; so this duality
pits past against present in the human struggle with time. History is featured
in the background of the image, which shows the outline of the floor maze in
Chartres Cathedral (Hancock 2016). Sometimes described as a surrogate for the
path of life, there are intriguing suggestions as to how the path of any individual
life (pilgrim) turns back on itself in time. I have color coded the stages of the
labyrinth in order to generate a mandala and its associations with the visual
representation of the Universe in Eastern cultures. Unlike the classic four-entry
mandala, I have provided a single entry to feature the unique nature of time. I
have also hidden other rebuses within the image so that individuals can explore
deeper as they wish.

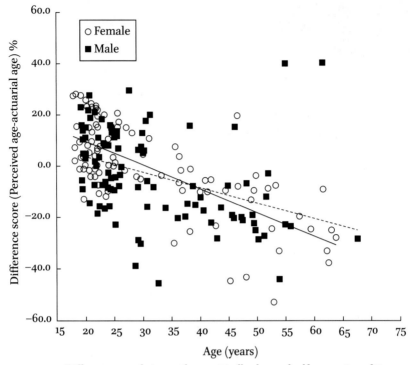

FIGURE 6.6 Difference score between the empirically observed self-perception of time
in life versus the actuarial projection of the same individual's lifespan
plotted as a function of the person's sex and age in years. Note that the 'life
indifference interval,' the age at which estimators prove to be maximally
accurate (i.e., on average a zero perceived difference) equals the allometric
estimate of 'natural' lifespan – being 26 years of age. (Data from Hancock
and Hancock 2014; for allometric scaling see also Thompson 1917.)

5   Explicit and Implicit Observers

Most in the time research community will already be aware of McTaggart's
(1908) now classic work: 'The Unreality of Time.' Such is its influence that ob-
ligatorily, I started this present discourse with reference to it. Thus, I want to
consider what the observations I have made mean for this central conceptual
position. In his work, McTaggart argued for two fundamental perspectives
on time which he labeled 'Series A' and 'Series B,' respectively. It is actually
important to note that he also identified a 'Series C' which is much less frequently cited in time's literature. The work stimulated the subsequent Nobel

prize winner,[19] Bertrand Russell (1915) to write in response. Both works provide sophisticated arguments about the possible nature of time and should be required reading for anyone looking to wrestle with time's *chronundra* (Hancock 2015c). In many ways, these two discourses are reminiscent of Abbot's (1884), classic "Flatland" in that they require us to think exceptionally and tangentially about our basic assumptions of reality (see also Haldane 1928). Their great gift is to loose us, even momentarily, from the dogmatism of the standard narrative (model) (i.e., upper left quadrant of Figure 6.3). Yet, these works are now more than a century old and I want to question their assumptions and assertions on two levels. The first ground is a philosophical one, which addresses and attacks their expressed assumptions upon the grounds upon which they were each formulated. The second attack is founded upon modern advances in neuroscience (and see also Montemayor 2010; 2013), a lacuna that, of course, cannot be laid at the doors of these two original authors.

The present essay is necessarily limited in length and scope and so a full critique of these positions is not tenable within the space available. However, one critical aspect concerns the observational basis upon which the absolute and relative nature of the Series A and Series B concepts are founded. Series A represents our standard, human-centric concern for time as progressing from past through present to future. Series B represents 'before-after' relations which putatively remain consistent, independent of any observer's temporal location (e.g., the Roman Empire occurs before the British Empire independent of my observational location). It is from comparisons across these series that McTaggart's subsequent argument concerning the unreality of time emerges. Russell looked to answer arising issues by explicating that Series A necessarily derives from subject-object relations, while Series B results from object-object relations. That is, Series A derives from the necessary presence of a living observer who *may* be seen to be obsolete with respect to Series B. Further, any such living observer has an inherent temporal privilege of access to knowledge of a unique now. More formally, the anchor point of the derived ratio scale of time has a zero point in the observations of the present moment.[20] This Series A conception of time requires an *explicit* observer. As Russell's position makes

---

19   It is of interest to note that Russell won the prize for Literature, perhaps because his many, fundamental contributions failed to fall in to the discipline-based categories that then existed.

20   McTaggart (1908) spends a little time on what is needed to coalesce and create such a 'now' in using terms such as 'event' and 'moment.'

clear, the presence of such an observer is an exceptional case; that is life (living systems) are themselves provided a privilege which extends beyond any mere specific arrangement of matter.[21] Series B seems initially to be set in direct contrast to Series A. At first blush we see no privileged position of observation; however, that appearance represents one fundamental shortcoming of both McTaggart's and Russell's positions. For, of course, the assertion of any 'before-after' relation actually mandates the preference of an *implicit* observer. This is not merely the trite assertion that some individual must be present to read the current passage of words upon the page, but rather it is a much more fundamental flaw such that before-after relations always imply the requirement for an observation of that order; and so before-after relations cannot be the sole realm of object relations. Thus, in the weeds of McTaggart's dichotomy lurks a hidden observer of the B Series; the same form of observer who stands proud and prominent in Series A. Here then, I already disagree, firstly with Russell, that life possesses any privileged position whatsoever. That is, fundamentally, subject-object relations are merely object-object relations; and nothing more. Essentially, the arrogance of life is chimerical. Second, we can now expose the assertion of McTaggart, who in essence promises us a zero point on a ratio scale of time somewhere at the beginning of B Series. The antiseptic, observerless Universe proves untrue. Series B is as contingent upon an *implicit* observation, as the Series A is on an *explicit* observer. Thus time, even from these basic conceptions, is a way of viewing and not what is to be viewed.[22]

But there is more to understand here, in particular the etiology of time and why our present standard model is so readily accepting of it and so continually reinforces it. The essential basis of this common acceptance derives from the fact that time represents our most effective and useful tool for survival – par excellence. To understand why time is such an effective tool, we have to briefly examine and elaborate upon modern-day advances in the neurosciences. It is truly 'anachronistic' to criticize either of the former, erudite and laudable individuals on the basis of their lacuna of knowledge of the modern neurosciences. Quite properly, we should not and cannot attribute this failure to them. However, much of the argument of McTaggart (1908) especially, relies upon an

---

21 While obviously inter-linked with mind-body issues, I wish to reserve such arguments for the following section on advances in neuroscience and our modern understanding of what such a dichotomy means and how it can be resolved in light of our present knowledge.

22 I do not mean to imply that this is an exhaustive or even comprehensive critique of the 'Series'-based view of time- it is not. However, it is sufficient here to point to yet another crack in time; albeit that I ultimately agree with McTaggart's general pronouncement on the unreality of time.

observers' recall capacity embodied in memory. For it is memory that permits us to distinguish events constituting the present moment from those no longer directly perceivable, either retrospectively in the past or intriguingly, prospectively in the future (Hancock 2015a). However, memory is itself, in actuality, an artifact of the apparent challenge of survival. For, memory is truly only for the future (Hancock 2005).

A growing cadre of memory researchers (see e.g., Hancock and Shahnami, 2010; Nairne and Pandeirada 2010; Schacter, Addis and Buckner 2007), under the influence of timesmiths, are coming to the view that memory is a capacity which has evolved primarily, and perhaps exclusively to support *prospection*. As certain learning capacities have invaded explicit, conscious processing, so memory has the by-product capacity to recall moments and situations no longer present. Memory also sustains skill assimilation and the momentary exercise of those skills when the human (organism) needs to distill an effective course of action (survive). Often it is better to anticipate threat rather than react to it, and humans (among other life forms) take extensive advantage of the facility that such memory capacity provides. However, explicit recall is a chance (or perhaps oblique) by-product of its utility, not the reason for it. *The fundamental purpose of memory is to anticipate the future, not recall the past.* Thus, memory supports and sustains future survival for which it has been differentially selected. We may choose to believe that "we are our memories," but our past is of no pragmatic meaning except that it serves our future. Modern neuroscience tells us that the brain is a prospection machine. It deals with the nuances of persistence, momentary response, and prospective planning with three rather disparate but interactive neural systems, each of which is necessarily erected, one upon the other (and see Hancock 2015a). These respective capacities have proved so successful that they help to vaunt us humans to the peak of the pyramid of living things upon this planet. But that is no great cause for jubilation. What has helped put us there is essentially underwritten by the iatrogenic invention of the tool of time. By this I do not mean just timing and time-keeping conventions (Cipolla 2003) but rather the fundamental concept itself. The tool of time has proved so effective, it is entrenched into our very nature. To seek to extract it is to seek to change what humans beings have been, what they presently are, and what they imagine they will become in the future (assuming our species has a future). Revelation of the iatrogenic nature of time is also a moral question; a question of purpose and our own perceived control over that purpose. The negation of time is thus central to our future iatro-telesis. To the 'present,' we have been seduced by time. It is perhaps our most successful invention ever. Its manifest utility continues. It is the framework and binding that holds together the fabric of our civilized society. To reveal its vacuity is to threaten all that we are and perhaps all that we can be. To wake the sleeper is indeed a dangerous act.

6   Out of Time?

The foregoing observations, albeit necessarily constrained, begin to show some of the cracks in the contemporary notion of the centrality and reality of time. In order to reinforce the present postulations, I would like to conclude by reiterating some of the central points but from a more technical, psychological perspective. Thus, the fundamental notion of past, present, and future (i.e., hindsight, eyesight, and foresight) derive from the way that the *apophenia* of space-time is conceived as a social reticulation (Figure 6.4). While spatial attributes can be 'directly' perceived, temporal attributes are necessarily indirectly constructed (Gibson 1975). It has often been argued that our unique character as human beings derives from our cognitive recognition of the prospective threshold of our non-existence (Figure 6.4). This pre-occupation with self has meant that ('subject') is almost ubiquitously and unquestionably contrasted with all other non-living entities ('objects'). This division begs Schrodinger's (1944) threshold question; "What is Life?" Simply considering these outer boundaries of the demesne of life versus non-life, logically and rationally pushes us toward the more crucial question; "Why is Life?" The latter question may be a manifestly teleological one, but the answer I have provided here is a distinctly non-teleologic and indeed pessimistic one. We can further inquire whether the degree to which living systems explicitly recognize 'time' actually provides us with a marker on the phylogenetic scale of their development. It may be that the designation 'subject' is not indicative of, nor embraces all living things equally but rather is indicative of a very restricted subset (i.e., human beings only). Indeed, to the best of my current understanding, I would argue that the concept of time derives from the human egocentric elevation of life alone. As I have noted, while Russell fought to resolve challenges posed by McTaggart, he did so by extracting and reifying sentience beyond insensate material. The problems of time, indeed the actual phenomenon of time begins to dissolve when we engage in an intentional act of cognitive negation. Paradoxically, this very particular act of ratiocination, contra Descartes, serves to depose completely, the tyrannical monarchy of consciousness (i.e., the I in the 'cogito' is obviated).

One way of taking this ultimate step of complete temporal negation is to consider super-symmetricality in space-time (Figure 6.4). This helps illuminate yet another expression of time's delusion. Independent of the observer's own idiographic history, enacted in their autobiographical memory, such symmetry recapitulates the notion that all living systems necessarily exist between the extremes of immalleable determinism (i.e., Absolute Zero; perhaps reflected in Parmenides view of an unchanging Universe) and random chaos of extremely high absolute temperatures. As Schrodinger (1944) insisted, the conditions for life as we have come to know it are, in their physical profile, highly constrained. Such cracks in time are telling and disturbing, yet for most individuals they seem to have no practical choice than to partake, ab initio, in the collective delusion. Pragmatically, this means that we live in a world epitomized by cognitive anthropic revelation.[23] Necessarily then, our circumstances inculcate within us a narrative that we 'discover' a future unto our own individual selves, bound within the social grouping of society. Sequellae of this referential frame are expressed in iatrogenic conceptions such as 'free-will,' 'causation,' etc. As I have noted, rejection of these social bulwarks can, would, and perhaps will, lead to vastly different forms of sociality and accompanying expressions of the human narrative. Therefore, let me bring this brief essay to a close with some concluding remarks.

7   No Time Left?

When we question the reality of time, we are engaged in a parlous enterprise indeed. The indistinguishable comedies and tragedies of the never-exhausted fountain of ignorance continue to rain down upon our public stage. Even the anodyne offered here of *cognitive anthropic revelation* will ultimately prove to be manifestly insufficient for social cohesion in the face of undeluded time. The conclusion of the vacuity of free will, of causation, and thus of personal and collective responsibility would threaten to burst society asunder. Humans look to avoid such disillusionment at all costs; even the cost of living in a continuing delusion. Perhaps a final and inevitable recognition of time's delusion implies a solution to Drake's conundrum (and see Scharf and Cronin

---

23   Cognitive anthropic revelation means that as limited entities that cannot exist, qua Schrödinger, in the determinism of absolute zero or the randomized chaos of excessive temperature, we necessarily live in a state of partial uncertainty/certainty. In consequence, what occurs to us can be partially envisaged and partially not. The brain evolves to seek the regularities of the former and minimize the potential damage of the latter. Thus, we will always be subject to a degree of informational 'revelation' which connotes this necessary uncertainty. Our traditional approach implicitly embraces this intrinsic uncertainty. What I am recommending here is an explicit embrace of this fundamental condition of (known) living systems.

2016). Drake has trenchantly observed that the number of opportunities for life in our Universe are so great that we cannot, statistically speaking, be alone. However, the antithesis of this optimistic pronouncement is to ask why we have not yet been contacted. Indeed, why have we not been bombarded with such communication ever since the origins of our contemporary consciousness would permit such recognition (Jaynes 2000). One potential answer is that eventually all living beings must recognize their un-exceptionalism from matter, subsequently recognize the vacuity of time (if they have created it in the first place) and then react accordingly. Like some of our own previously stultified cultures across the globe, but magnified to the nth degree, this recognition would bring all hope of 'progress' to an immediate and shuddering halt. Perhaps Pandora's leavings are all we have, and it may well be better to travel on in hope then arrive in misery? Is there life out there in the Universe – maybe so; but is there Time – no.

References

Abbott, E. (1884). *Flatland: A romance in multiple dimensions*. New York: New American Library.

Barbour, J. (1999). *The end of time: The next revolution in our understanding of the universe*. New York: Oxford University Press.

Block, R. A., Hancock, P. A., and Zakay, D. (2010). "Cognitive load affects duration judgments: A meta-analytic review." *Acta Psychologica* 134: 330–43.

Campbell, J. (1949). *The hero with a thousand faces*. Princeton: Princeton University Press.

Campbell, J. (1988). *The power of myth*. New York: Doubleday.

Coren, S. (2012). *The left-hander syndrome: The causes and consequences of left-handedness*. New York: Simon and Schuster.

Cipolla, C. M. (2003). *Clocks and culture*. New York: W.W. Norton.

Doob, L. W. (1971). *Patterning of time*. New Haven, CT: Yale University Press.

Eames, C., and Eames, R. (1998). *Powers of ten: A flipbook*. New York: W.H. Freeman & Co.

Eisler, H. (1976). "Experiments on subjective duration 1868–1975: A collection of power function exponents." *Psychological Bulletin* 83 (6): 1154–71.

Fechner, G. (1860/1965). *Elements of psychophysics*. (*Vorschule der aesthetik*. Breitkopf & Härtel.) New York: Holt, Rinehart & Winston.

Fraser, J. T. (1987). *Time, the familiar stranger*. Amherst, MA: University of Massachusetts Press.

Gibson, J. J. (1975). "Events are perceivable but time is not." In: J. T. Fraser and N. Lawrence. *The Study of Time II*. (pp. 295–301). Berlin: Springer-Verlag.

Gould, S. J. (1988). *Time's arrow, time's cycle*. Cambridge, MA: Harvard University Press.

Haldane, J. B. (1926). "On being the right size." *Harper's Magazine* 152: 424–27.

Hancock, P. A. (2002). "The time of your life." *Kronoscope* 2 (2): 135–65.

Hancock, P. A. (2005). "Time and the privileged observer." *Kronoscope* 5 (2): 177–91.

Hancock, P. A. (2009). *Mind, machine and morality*. Ashgate: Chichester, England.

Hancock, P. A. (2011a). "On the left hand of time." *American Journal of Psychology* 124 (2): 177–88.

Hancock, P. A. (2011b). *Cognitive differences in the ways men and women perceive the dimension and duration of time: Contrasting Gaia and Chronos*. Lewiston, NY: Edwin Mellen Press.

Hancock, P. A. (2013). *On the symmetricality of formal knowability in space-time*. Paper presented at the Triennial Conference of the International Society for the Study of Time, Orthodox Academy of Crete, Kolymbari, Crete, July.

Hancock, P. A. (2015a). "The royal road to time: How understanding of the evolution of time in the brain addresses memory, dreaming, flow and other psychological phenomena." *American Journal of Psychology* 128 (1): 1–14.

Hancock, P. A. (2015b). *Hoax springs eternal: The psychology of cognitive deception*. Cambridge: Cambridge University Press.

Hancock, P. A. (2015c). "Chronundra." *Time's News*, No. 46, April.

Hancock, P. A. (2017a). *Transports of delight*. Cham, Switzerland: Springer.

Hancock, P. A. (2018). "The design of time." *Ergonomics in Design* 26 (2): 4–9.

Hancock, P. A., and Block, R. A. (2012). "The psychology of time: A view backward and forward." *American Journal of Psychology* 125 (3): 267–74.

Hancock, P. A., and Block, R. (2016). "A new law for time perception?" *American Journal of Psychology* 129 (2): 111–24.

Hancock, P. A., and Hancock, G. M. (2014). "The effects of age, sex, body temperature, heart rate, and time of day on the perception of time in life." *Time & Society* 23 (2): 195–211.

Hancock, P. A., and Rausch, R. (2010). "The effect of sex, age and interval duration on the perception of time." *Acta Psychologica* 133: 170–79.

Hancock, P. A., and Shahnami, N. (2010). "Memory as a string of pearls." *Kronoscope* 10: 77–82.

James, W. (1890). *Principles of psychology*. Holt, New York.

Jaynes, J. (2000). *The origin of consciousness in the breakdown of the bicameral mind*. Itsaca, IL: Houghton Mifflin Harcourt.

Koestler, A. (1990). *The sleepwalkers: A history of man's changing vision of the universe*. London: Penguin.

Lhamon, W. T., and Goldstone, S. (1956). "The time sense." *Archives of Neurology and Psychiatry* 76 (6): 625–29.

Luce, R. D. (2002). "A psychophysical theory of intensity proportions, joint presentations, and matches." *Psychological Review* 109 (3): 520–32.

McTaggart, J. E. (1908). "The unreality of time." *Mind* 17 (68): 457–74.

Montemayor, C. (2010). "Time: Biological, intentional and cultural." In J. A. Parker, P. Harris, and C. Steineck (eds.), *Time: Limits and constraints.* (pp. 39–63). Leiden: Brill.

Montemayor, C. (2013). *Minding time: A philosophical and theoretical approach to the psychology of time.* Leiden: Brill.

Moray, N., and Hancock, P. A. (2009). "Minkowski spaces as models of human – machine communication." *Theoretical Issues in Ergonomics Science* 10 (4): 315–34.

Morrison, P., and Morrison, P. (1990). *Powers of ten: About the relative size of things in the Universe.* Scientific American Books.

Nairne, J. S., and Pandeirada, J. N. S. (2010). "Adaptive memory: Ancestral priorities and the mnemonic value of survival processing." *Cognitive Psychology* 61: 1–22.

Pos, R. (2006). *The gender beyond sex: Two distinct ways of living in time.* Bloomington, IN: Trafford on Demand Pub.

Russell, B. (1915). "On the experience of time." *Monist* 25 (2): 212–33.

Schacter, D. L., Addis, D. R., and Buckner, R. L. (2007). "Remembering the past to imagine the future: The prospective brain." *Nature Reviews Neuroscience* 8: 657–61.

Scharf, C., and Cronin, L. (2016). "Quantifying the origins of life on a planetary scale." *Proceedings of the National Academy of Sciences* 113 (29): 8127–32.

Schrödinger, E. (1944). *What is life?* Cambridge, England: Cambridge University Press.

Shaw, R., and Kinsella Shaw, J. (1988). "Ecological mechanics: A physical geometry for intentional constraints." *Human Movement Science* 7 (2): 155–200.

Stevens, S. S. (1957). "On the psychophysical law." *Psychological Review* 64 (3): 153–81.

Thompson, D. W. (1992). *On growth and form.* Dover reprint of 1942 2nd ed. (1st ed., Cambridge University Press, 1917).

Wissner-Gross, A. D., and Freer, C. E. (2013). "Causal entropic forces." *Physical Review Letters* 110 (16), 168702.

CHAPTER 7

Deconstructing the Zero Time Theory

Frederick Turner

Popular misconceptions are usually right

G. K. Chesterton

∴

1   Definitions of "Truth"

"We must each write our own truth. The veracity of that truth is only ostensibly tested by time. However, here, I seek to write the truth of time itself." So declare the first two sentences of the abstract of the paper under discussion. The abstract does well to introduce the concept of truth at once, though in the paper itself more indirect terminology is used and the assertion is not so much about the truth of the paper's contention as about the illusion perpetrated by the view it attacks. But the essential distinction drawn in the paper is clearly that between truth and illusion.

So we are entitled to ask how those terms are defined by the author, especially considering the éclat with which its professedly novel and contrarian assertion – that time does not exist – is presented. Obligingly, the author gives us immediately not just one but two definitions of "truth":

a.) "We must each write our own truth." Here truth is defined as subjective – each of us has our own truth. Thus for any individual, a view of the world that is his or her own view is truth; another's view, if it is different, is untrue, an illusion.

b.) "The veracity of that truth is only ostensibly tested by time." Perhaps because the author senses the potential solipsism of the first definition, "truth" is here intersubjectively "cashed" in the Wittgenstinian sense ("the meaning of a word is its use"[1]) by its being something "ostensibly tested by time." That is, one can point out to others (the act of ostension) the evidence for something's veracity, and so over time the truth is revealed.

---

[1] *Philosophical Investigations*, section 43.

The problem here is that if the test of truth is time, and if, as the paper maintains, time does not exist, there is no test for truth.

So since alternative b.) is empty, we are thrown back on definition a.), that something is true because "I" say so. A joke? Perhaps.

2   A Brief History of Time Denial

Nevertheless, perhaps it *is* true that time does not exist. Many people have believed that time is an illusion: it is one of the oldest and most comforting philosophical assertions in the human tradition. Let us look briefly at the history of the concept. For the classic Hinduism of the Vedas and the Upanishads time is literally an "illusion," in the sense of the Latin derivation of the word, from *ludus, ludere*, game, play. The Sanskrit word *maya* is a preferred term for the illusory and delusory play (*lila*) of the world.[2] Buddhism, especially in its Zen form, follows suit: the wheel of time is a wheel of illusion and suffering, and it is only when we escape that wheel that we live in the truth and escape the grief and anxiety of the world of time. Parmenides' block universe, as the paper acknowledges, also denies the reality of time, as does Plato in the *Timaeus*. Some of the more philosophical branches of Islam (though not, perhaps, the *Qur'an* itself) deny the reality of humanly-experienced history in the face of the eternity of heaven.[3] Medieval and Renaissance Platonic Christianity (though not the Hebraic strain of it, including that of Jesus himself) also denied the being of time.

Pierre-Simon Laplace's "demon" (1814), a universal calculator that contains all past information and can predict everything that will happen, is an Enlightenment example of the idea.[4] Nietzsche's notion of the eternal return, an idea introduced with the same Voltairean deflationary glee as in this paper, is at bottom the same concept.[5] Closer to our own era, as the paper points out, Einsteinian relativistic physics, with its geometry of spacetime, effectively reduces time to another spatial dimension. That dimension, perpendicular to the other three, as the imaginary number series is perpendicular to the natural number series, shares only the zero in common where they

---

2   See the Appendix to the *American Heritage Dictionary*, under "leid-."

3   For instance, Al-Hazen's *Book of Optics*, 1021 CE.

4   Laplace, Pierre Simon, *A Philosophical Essay on Probabilities*, translated into English from the original French 6th ed. by Truscott, F. W. and Emory, F. L., Dover Publications (New York, 1951) p. 4.

5   *The Gay Science: With a Prelude in Rhymes and an Appendix of Songs* by Friedrich Nietzsche; translated, with commentary, by Walter Kaufmann (Vintage Books, March 1974), §341.

cross. And Modernist existentialism, with its powerful concepts of the timeless moment where all reality is gathered, the epiphany or chairotic moment, explores the idea in poetic terms – the moment in and out of time.[6] Cubism, as in Duchamp's famous "Nude Descending a Staircase," attempts to include all the states of a world-line at once, achieving, as it were, the viewpoint of a divine being observing simultaneously all the steps of the lady as she goes downstairs. Even political theories hanker after a final state of changeless perfection, whether it succeeds the triumph of the master race or the dictatorship of the proletariat. Finality – whether the Last Judgement or the withering away of the State – is another version of the ideology of timeless perfection.

The point is that this assertion of the unreality of time is not an astounding new discovery, but a core principle of a whole series of major regimes of power and knowledge (as Foucault would have put it). It is the ideology of a temporal panopticon,[7] where the prison guards are able to see not only the present of their inmates, but their future and past as well, reducing both to an eternal present. It is this condition, presumably, that the last sentence of the abstract refers to: "I here present some keys to unlock the bounden hands of time in the hope of its final redemption." The guardians are indeed truly free; their hands are not bound by the sentimentalities of time.

3 Time Denial in Physics, and Its Discontents

The paper rather neglects what may be the most persuasive argument for the nonexistence of time, the aforementioned relativistic perspective in which all "time" is represented as an eternal present, a tesseract or four-dimensional cube, where the world-lines of observers such as animals, plants and humans snake about, expanding and contracting, between their two endpoints. Perhaps the author avoids an extended discussion of relativistic models of time, because if time is even recognized as a dimension, then it exists as such. Or perhaps the author relies on these arguments, but abbreviates them because they are well-stated elsewhere. However, let us dispose of these first to cover up any potential bolt-hole, before we go on to the main piece of evidence the author produces.

The relativistic physics arguments about the block universe have, of course, spectacularly shipwrecked upon two major uncomfortable facts. The first is the constitutive randomness, superposition, and "state of knowledge" conditions

---

6   T. S. Eliot: *Four Quartets*. Harcourt Brace, 1943. pp. 15, 44.
7   Michel Foucault: *Discipline and Punish: The Birth of the Prison*, Random House, 1975.

of quantum mechanics, and their profound difference from the "real world" of classical mechanics, a difference nicely illustrated by Schrödinger's cat. The moment in time of the emission of a charged particle from an excited atom cannot be predicted, and is unpredictable by its nature. Nature plays dice. Once the particle is emitted and takes its role in the classical-mechanical macrocosm, the rest of the universe is by some very minute amount changed forever. If it is emitted sooner, you get one universe; if later, another. A block universe would have to include all the different moments of emission and their different universes, involving impossibilities both of physics (for instance, violations of the Pauli exclusion principle) and of logic, as with the dead/alive cat: A is not A.

Even if that irrational universe is accepted as the timeless reality, there is still the problem of the lawfulness of classical mechanics. The absolute indeterminacy of the ideal quantum world must coexist with the absolute determinacy of the ideal classical world, and time is the only solution. Either quantum mechanics or classical mechanics could be interpreted as the way a previously deluded temporal observer would come to see the timeless reality: the first as a block probabilistic state, and the second as a block universe crisscrossed with unchangeable world lines. Each system (like, for instance, classical supply and demand economic theory) is a useful timeless idealization, even if corrupted always by reality. But the coexistence of the two systems in strict experimental terms cannot do without time. As the graffito goes, time is nature's way of making sure everything does not happen at once. Or, better, making sure that things do not both happen *and* not happen at once. Another way of putting this is the CPT theorem: To define an elementary particle it is not enough to identify its charge and its parity (the chiral direction of its spin); its time must also be specified.[8]

The second shipwreck of the relativistic block universe is the whole science of thermodynamics. Thermal disorder, strictly defined, increases in any closed system over time.[9] If the universe is, as the paper suggests, like the foreordained print of a movie, the movie is changed by being shown. A movie of particles or massive bodies in a vacuum flying about individually can be run backwards without violation of the laws of physics (if it were not for that pesky "T" in the CPT theorem), providing the past/future symmetry the paper proposes. But when those particles are affecting each other in a nonlinear system of mutual

8   Lüders, G. (1954). "On the Equivalence of Invariance under Time Reversal and under Particle-Antiparticle Conjugation for Relativistic Field Theories." *Kongelige Danske Videnskabernes Selskab, Matematisk-Fysiske Meddelelser* 28 (5): 1–17.
9   Halliwell, J. J. et al. *Physical Origins of Time Asymmetry*. Cambridge, 1994.

cause and effect, new properties such as temperature, entropy, enthalpy, pressure, free energy, and work emerge. And those properties are entropic by definition: local variations in them sooner or later decay with time. Some new waste heat is created whenever work is done. Information theory, which interprets entropy in a different way, maintains that information can be created but not destroyed (even in Black Holes, as Hawking showed: they leak), providing an inherent direction for time. Even the apparently predictable, and thus reversible, movement of massive bodies in a vacuum is immediately compromised in the case of the three-body problem – not to speak of the infinitely deep, various and unrepeating strange attractor of a mutually revolving star cluster.[10] Novelty is inherent in chaos; and novelty cannot be detached from time.

Perhaps we can get a better grasp of the problem of the block universe if we specify what it would need to contain. For a start, not only all the consequences of the random emissions of quantum particles, but also all the alternative universes postulated by big bang cosmology. Contemporary many-worlds theory, anxious to avoid any unexplained suggestion of design or uniqueness in the initial conditions of this universe, postulates that the original quantum foam can and does coalesce into an infinity of universes, most of which cannot continue or do continue lifelessly without change, ours being a lucky survivor; we perceive an ordered yet changeable universe because only such a universe would produce observers, and no other explanation is necessary.

The alternative universes can contain different possible mathematical bases, such as classical and Riemannian topologies, and those with commutative or noncommutative mathematics, and those with both, like this one. The timeless block universe would have to include them all, including ones with lots of weird kinds of (illusory) time and weird observers all convinced wrongly that time exists (since the author identifies consciousness, change and time as necessarily interdependent). One consequence of this view is that there must be a tiny fraction of those universes that by sheer chance emerge from the quantum foam completely organized already as a vast, benevolent, endlessly subsisting, all-knowing, locally omnipotent conscious observer, i.e., God. Tiny though the fraction would be, it would still be infinite. Here we encounter the multitudinous divinity of the *Bhagavadgita*. *Our* universe, though, is somewhere in between the short-lived chaotic and the boringly unchanging ones, and between the ones containing no consciousness at all and those that are instantly divine. We live in an (illusorily) evolving universe, where the

---

10   James Gleick. *Chaos: Making a New Science*. Penguin, 1987, 2008, pp. 44–50.

possibility of a God (illusorily) emerging is up for grabs. All this too goes into the big timeless bundle.

There does now exist a hypothesis that may go some way to satisfying the author's desire for a different model of what we call time. It is known as the "growing block universe" theory. This theory postulates that the past does exist as a subsisting spacetime block, and the present is real as its edge or boundary. The past is all the paths that were actually taken, rather than all the paths that could have been taken, thus the concept is pruned of all the quantum and probabilistic alternatives that were not realized. Those complications are reserved for the present, and constitute the field of the present's freedom. In this view the future does not exist, but many futures can come into existence, as long as they remain within the generous limits of past natural laws mitigated by the constitutive wiggle room offered by quantum or thermodynamic or chaotic indeterminacy. Even new natural laws can come into being. The present is the growing point of the universe where new reality is continually generated, a place of emergences that create new truths in the interstices left by the old.[11]

If the various levels of organization in the universe – the periodic tables of the elementary particles and the chemical elements, the expressive codes of DNA, RNA and epigenetic regulation, the orders of social species and ecosystems, and the realm of human communication – are all construed as languages, the present is generated by the illocutionary acts within them.[12] If the paper's author is willing to give up two thirds of his claim against the reality of time, and be content with the nonexistence of the future, perhaps some of his hypothesis can be saved.

To use terms like "emergence" and "sequelae" as the paper does – to avoid the implication of temporal order in explaining how the illusion of time might have arisen – does not get it off the hook. They already imply a logical precedence and consequential order. It would be much more accurate to interpret the meaning of these words as an actual proof of the existence of time. Since the relation between the prior state of enlightenment and the secondary state of illusion is a one-way connection (otherwise illusion could be given the ontological status of ground or founding), even logical inference requires at least a very primitive kind of time. You cannot have an illusion without a reality about

---

11   See: C. D. Broad. *Scientific Thought*. Harcourt Brace, 1923.
        Frederick Turner. "Poiesis: Time and Artistic Discourse." *The Study of Time III*. Ed. J. T. Fraser, N. Lawrence, D. Park. Springer Verlag, 1978.
        Michael Tooley. *Time, Tense, and Causation*. Oxford University Press, 1997.
        Peter Forrest. (2004) "The Real but Dead Past." *Analysis* 64: 284.
12   J. T. Fraser, *passim*.

which one has the illusion. See my later remarks about real illusions and commutative vs. noncommutative kinds of arithmetic.

4 Time Skepticism's Reduction to Nothingness

If the only things that are real are timeless – i.e., the block universe or some theological version of it – then the following must be true.

There are no such things as color and sound, since both have vibrational frequencies that involve time. Color and sound are illusions held by animals. There are no such things as a living organisms, since organisms are defined by terms such as metabolism, reproduction, etc., which involve a time dimension: life is an illusion of matter.

The author of the paper takes the process of deconstructing the temporal universe this far, while using some quiet legerdemain in sometimes blaming our folly on human hubris, sometimes on the sensoria of living organisms in general. Living organisms presumably include not only the impressionable mammals and birds, but also the much more stolid plants and bacteria, which can hardly be accused of hubris. All living organisms by definition possess metabolism and reproduce themselves, both undeniably temporal characteristics.

But why stop with living organisms? As we have seen, inanimate matter itself, composed as it is of vibrational particles subject to the CPT theorem, is irreducibly temporal in nature. So, for consistency, the paper should also argue that not only are there no living organisms, there is no such thing as matter, since matter is defined as such by the relative temporal endurance of its component particles: matter is an illusion of pure energy.

Ah, but there is no such thing as energy either, since as essentially vibratory it is indefinable without a t-axis, as we have already agreed in the case of color and sound: energy is an illusion of timeless light, defined as Einstein defines it, existing in a timeless moment. A halted light wave or sound wave is not a something that is not moving, but a nothing at all.

So there is no such thing as light, since light is defined as an ordered distortion of spacetime, and if there is no time, there cannot be a spacetime. If light is an ordered distortion of space alone, how is space to be defined? Certainly not by any measure involving time, such as how long it takes to get from one point in space to another. And what other measure would there be? Mathematics?

If space be defined mathematically, even here very timelike properties appear, especially the mathematical property of algorithmic difficulty, known as "hardness," measurable by computation science in terms of how long it would take a calculator to achieve the solution to very difficult NP complete problems (completable only in "*Non-P*olynomial *time*").[13] Likewise, arithmetic comes in two kinds: commutative and non-commutative. The familiar commutative arithmetic defines 2+ 3 as being the same as 3+ 2; non-commutative arithmetic does not hold that postulate, so that 2+ 3 might be very slightly different from 3 + 2; after all, once the initial "2" has been established, the "3" enters the situation with the "2" already in place, as opposed to the other way around. In this case "evenness" has got to the head of the queue before "oddness." The Polish mathematician Michael Heller argues that the basic universal unit of time, the smallest chronon, is this difference made by the sequence itself.[14] Now non-commutative arithmetic is as much a legitimate mathematics as is the commutative kind; we live in a universe with problems accurately solvable by either one or the other. A very primitive temporality even emerges in mathematics. So maybe mathematics, too, must go into the grand bonfire of time.

Many religions, at their more mystical frontiers, have in their own ways pursued such a train of reasoning, braving its *reductio ad absurdum*, and postulated that the universe is basically nothing. It is but a dream of Brahman, a *lila*, whose positives are perfectly balanced by its negatives. The universe is made of nothing. But it *is* made, say the religions, needing to explain why the illusion of the world appears so real, and it takes time for its unmaking, until presumably all positives randomly end up meeting with their negatives and are annihilated. So here again time has reared its ugly head, as a sort of divinely-permitted bubble in the eternal moment.

5   The False Tinsel of Hollywood: Real and Unreal Illusion

So much for the paper's version of "truth." What about "illusion"? So, let us concede that time is an illusion. The problem now becomes the definition of "illusion."

If the only stuff that is really there is illusion, then does not illusion get promoted to the status of reality, absent any other contender? If time does not exist, and there is a conscious knower that is under the illusion that it does exist, then *illusory* time certainly exists. And if knowing, itself, is a timeless state, then the only knower who knows time does not exist would have to be unconscious, since consciousness itself is essentially sequential, involving at

---

13   See the excellent summary in Wikipedia under "NP-completeness."
14   Michael Heller: *Creative Tension: Essays on Science and Religion*. Templeton Foundation Press, 2003. See also my review of this book in *Kronoscope* 11: 1–2. (Summer/Fall 2011).

least the order of logical inference and, in practice of course, the awareness and gradual forgetfulness of what one was thinking earlier, an awareness of its earlierness.

There is an old joke about the false tinsel of Hollywood, to the effect that it is either better or worse than the real tinsel of Hollywood. Is the illusion a real illusion? If it is not a real illusion, then we do not have it and the problem disappears: we see things as they are. If time is a real illusion – *real* tinsel! – and if, as we have shown, the only alternative is nothingness, the real illusion – or better, the whole hierarchy of illusion: space, energy, matter, life, and people – is the only reality there is. If the people in Plato's cave are right, and there is no reality outside after all, at least they have the very real play of shadows on the wall – or movies in the theater (Fraser 1980).[15]

And here we have the real value of the paper, which forces us to imagine the necessity of a deeply flawed, constitutively subjective and intersubjective, continually transforming and fairly insubstantial universe that is creative in its essence. It may not be much good as *being* – there is no Dasein there, so we can stop yearning after *that* illusion – but it is the only one we have got. And it seems to have its own regularities and perceivable consistencies, such as the laws of mathematics, physics, and chemistry and the great law of evolution; and it is quite interesting, and it is home to a lot of beautiful things. The author actually concedes this, but seems to want to debunk it anyway in favor of the block universe. Another view might be possible: that time, this whole insubstantial pageant, as Shakespeare puts it in *The Tempest*, is an unbelievably courageous effort (if admittedly a foolhardy one, given the suffering involved) by all of us – atoms, plants, animals, humans in solidarity – to invent a reality where there was none before.

6   The Evidence and Argument of the Paper

Here we must turn to the only concrete evidence presented in the paper for the author's contention that time is an illusion: the different estimates of time given by males and females. This is interesting and provocative material, and much might be made of it on a more modest scale. Perhaps the evidence can be explained in a different way than by the apocalyptic hypothesis that all temporal reality is an illusion. Possible explanations of the small divergences observed between male and female estimates of time run into the dozens. Here are a

---

15   J. T. Fraser. "Out of Plato's Cave: The Natural History of Time." *The Kenyon Review* New Series, Vol. 2, No. 1 (Winter, 1980), pp. 143–62.

few less interesting explanations, involving the notorious unreproduceability of social psychology studies:

> Sampling error: the diagrams presented showed a very small sample.
> Unconscious bias in the self-selection of subjects.
> Reporting bias by interviewers.
> Reporting bias by interviewees.
> Etcetera.

> More interesting explanations include:
> The social psychology of gender – proactive vs reactive gender roles.
> The life expectancy of female humans is about 4–5 years longer than that of males, providing a different subjective timescale of the same order of magnitude as the difference in the experiment.
> Gender role distinctions in the future discount rate.
> Gender role based risk strategies.
> Attention span differences.
> Etcetera.

To borrow what Horatio says to Hamlet, there needs no ghost come from the grave to tell us that humans estimate all sorts of things differently from each other and from what a strictly rational assessment would require. Indeed, there are often good evolutionary adaptive reasons for unwarranted optimism, unreasonable risk aversion, conformity to group opinions, high self-esteem, even depression, not to speak of all the biases associated with the seven deadly sins.

7   Evolution Denial

But evolution itself is one of the things the paper denies. In the timeless monoblock the fact that there are many forms of horse from Eohippus to the modern quarter horse, or humans from Homo erectus to Homo sapiens, is a pure coincidence. Genetic and selective cause and effect, in this view, is itself an illusion. An odd kind of evolution denial, apparently without the usual theistic ulterior motives! The paper might as well deny climate change and the Holocaust together with evolution, while it is at it, since all three are profoundly temporal events, involving radical and constitutive differences among past, present, and future. If time does not exist – and the author implicitly concedes this – all these events are illusory and thus morally insignificant.

The paper's rather heated and unaccountable attack on the achievements of Olympic athletes may be construed as part of the "vanity of vanities: all is vanity" theme that seems to underlie the paper. But it is rather inconsistent in this context to base a condemnation of athletic glory on the idea that winners are fated to win because of their genetic history and evolutionary advantages, when temporal concepts like history and evolution are among the targets of the paper's polemic.

The paper takes a stand against the argument that memory is a proof of the reality of time. It invokes good recent research on the use of memory as a means of anticipating future events that might affect the organism's survival; but in this it is likewise a victim of its own premise. If memory actually works in figuring out the future, then cause and effect must work, since memory, as part of a "prospection machine," infers future effects from past causes. But cause and effect are explicitly rejected in the paper as a temporal illusion. One cannot use one theory that one is debunking to support the debunking of another, unless one has one's tongue in one's cheek or is satirically demonstrating the fallacy of the very argument one professes to advance.

An evolutionist would argue that if memory does not actually work at predicting the future, because cause and effect do not operate, and the world is completely random, then surely evolutionary adaptation would have long ago culled out the metabolically expensive genetic, neural and immune-system hardware for observing, storing, and retrieving past information. If, as the author claims, future and past are symmetrical as regards knowledge – we know as little about the past as about the future – why should the past be privileged as a source of information about the future? Why not just anticipate twice as hard? For that matter, why know the future anyway, if knowledge is illusory?

Any argument about the usefulness of anything is itself meaningless if time does not exist: something that is useful is something that can be used. "Can" is a word that denotes potential, the difference between what is and what could be in the future. In a timeless universe there are only states that are there and others that are not, and since there is no transition between such states, usefulness is neither here nor there.

8   The Intentions of the Paper

The "growing block universe" might allow for some of what the paper apparently wants, without getting into much of the trouble outlined here. If it is only the reality of the future that must be jettisoned, many of the logical problems disappear. But I am not sure that the author would make this concession, for he or she seems to have other fish to fry, that require a more wholesale consignment of time to illusion and demolition. In order to understand the author's intentions, we must ask a question: How are we to take the paper? Is it:

a.   An intervention designed to elicit this sort of response, a "gadfly" piece like *Erewhon*, that challenges, for the sake of deeper understanding, the contemporary received wisdom?

b.   An experiment with the media, like the famous Sokal paper, designed as a satirical hoax and implied critique of the position it professedly espouses?

c.   A case of *Monocausotaxophilia*, a term coined by Ernst Pöppel, a former member of the ISST: the love of a single cause that explains everything? (In this instance, the single cause would be the block universe.)

d.   An act of cognitive terrorism, like Samson's bringing down of the Philistine temple?

e.   The enactment of a (perhaps misplaced) desire for absolute moral clarity and intellectual honesty: an attempt at radical purification?

f.   A religious statement, clearing the way for a turn toward an eternal God?

My own guess is that it is a mixture of all of these. Many of the contradictions and inconsistencies of the paper might be explained by the competition of these motives. The paper does erect a mountain of dark and terrifying inferences, including the despairing extinction of all intelligent species, upon a relatively small experiment in social psychology – it is almost as if that experiment were a useful occasion to launch a jeremiad. But is the jeremiad launched, like Jonathan Swift's *A Modest Proposal*, against the holders of the view it professedly espouses, or against that view's opponents? If the former, then the contradictions and inconsistencies would be the whole point: the fictional time skeptic would be refuted out of his own mouth while the real author smiles at his misadventures. If the latter, the case becomes more complex, and perhaps we should conclude this interrogation of a very interesting paper with a few speculations about its intent.

One clue lies in the discussion of the changing view of the Earth's position in the universe, the so-called "displacement" of humanity from the center to the outskirts. This is of course an old chestnut of the Enlightenment debate between the Ancients and the Moderns, the Moderns' triumphant putdown of ancient knowledge, and was always cited as a corrective to human pride, hubris, exceptionalism, and egocentricity.

But closer examination of the Judeo-Christian interpretation of Ptolemaic/ Aristotelian cosmology reveals the fallacy of the Moderns' slur. As we see in Dante's *Inferno*, for the Ancient cosmology the Earth is the nearest place to Hell in the whole of the cosmos. It is at the bottom. Spiritual things are weightless and good, and fly upwards. Bad material things are heavy, and thus fall downwards. The fallen earth has fallen as far as it can without actually descending into Hell, which is in the way of its going even further. The centralness of the Earth is not a cause for pride, but for utter humility and abasement. Thus Copernican and Galilean cosmology at least promoted us upward, and made us no worse than any other heavenly body. The Renaissance and Enlightenment, though I for one honor them, are not known exactly as models of humility and self-doubt.

But what is interesting about the paper's reminder of the Moderns' old reproach is the resentment it apparently displays about the fools and true believers and inflated humanists who believe in the reality of time. In the context, read at a triennial meeting of the International Society for the Study of Time, it is clearly making a statement. One detail in the paper as it came to me – perhaps it has been edited since – was a parenthetical remark introducing a diagram representing the views of the Greek philosophers:

> Since nothing impresses so much as the appearance of mathematical and computational competence I have expressed the four propositions of Figure 3 in quasi-Boolean logical terms.

Such a remark presupposes an audience that is likely to be impressed by such terms, unlike the author, who is not swayed by superficial displays of knowledge. The audience is a common herd that needs to be led along by scientific magic tricks until it is ready for the great grim truths that only the far-sighted can endure.

The terminology of the paper and much of its rhetoric strikes the same pose. The author's journey is a lonely odyssey; the adept, his hair streaming (so to speak) in the cold wind of the "stark" truth, speaks to us like Zarathustra from a lonely place of "bleak" reality; his audience of time addicts needs to be "weaned" away from its "pabulum." We are commanded to be humble and abase ourselves, and enjoined to "sit and cogitate" like good students, while the master tells us of his titanic moral struggle, over whether the dreadful illusion should be revealed as what it is – lest, as it inevitably must, the unveiling should destroy all morality, value, and civilization itself. But even these sacrifices are justified by the heroic integrity of knowing the truth.

I believe that this language is at least partly tongue in cheek, more to entertain a sophisticated audience than to insult a stupid and naïve one. I choose to take the paper as a complex and not entirely successful attempt to open up further thought on the nature of time. And there may even be a further suggestion.

We live in an era when "truth" and truth-claims are increasingly problematic. Fake news, false facts, denial of evolution, climate change, the Holocaust, etc, are constant, and our present public space seems more a matter of who can best spin the facts and discredit the fact-checkers than of the normal order of business. So the author's insistence on "truth" may be well-taken. Perhaps the paper is a sort of test, like the one I proposed thirty years ago in my poem *Genesis: An Epic Poem*. Here the character Ruhollah, the leader of a future nihilistic movement, attempts to recruit an acolyte, the troubled and anguished Garrison, in the presence of another potential convert, Tripitaka:

> "Why," asks Ruhollah, "My devout young friends,
> Should Purity reside even in Nature?
> Is not the world of nature one of eating?
> Is it not vile that one being's life should flow
> From the appropriation of another's?
> Nature is but a pit of mouths, a nest
> Of bellies brewing acids into gas;
> That sin we stand condemned of was committed
> At the accursed beginning of the world.
> We are the fragments of the Enemy,
> The demiurge who would rebel at God,
> And take kinetic form from the potential,
> The sea of sightless light in the beginning;
> And he, the Angel, fell into the shape
> Of sensual energy and grossest matter,
> Away from that perfection he enjoyed
> At first in those transparent fields of light.
> The ground state of the world is all potential,
> And since the cause is fuller of that essence
> That gives effect its being, than the effect,
> All action is a fall and a declension
> Into the brute adultery of time.
> The Ground is pure and indeterminate;
> Is the white joy of unenactedness;
> The big bang where the laws of time all fail;
> The place we must return to, to be saved."
> "But what," asks Garrison, dismayed, "about
> The innocence of nature? I can see
> How human beings are to blame; I know
> Myself the gulfs of consciousness,

The slimy things that grow inside the skin
Of candor and apparent honesty;
But as for nature, she is pure, she must be.
Matter has no awareness what it is."
Now Tripitaka listens silently;
His eyes glitter and his kendo hand
Wanders to touch the lupus of his cheek.
But still Ruhollah speaks again, and smiles.
"Is not the nature that you so admire
That which your senses in their corruption
Pick out to find delectable and pure?
And if your senses, calibrated as
They are by that depraved and self-subomed
Selfconsciousness you loathe, were given you
As your inheritance by nature too,
Are they not prone to partiality,
To hanker for the mother of their form?
And is not consciousness itself the last
Of many gross conceptions on itself
That nature practiced long before the birth
Of humankind, its monstrous masterpiece?"
"What then is pure? asks Garrison despairing,
And Tripitaka breathes the same breath through.
"The chiffre," says Ruhollah; "Nothing is
Pure; seek out that nothing and you will see."[16]

---

16   Frederick Turner. *Genesis, An Epic Poem.* Saybrook Publishing Co., 1988. (Revised and
republished, with a new preface by the author: Ilium Press 2011), II.i.291–345.

CHAPTER 8

Time's up: Clarifying Misunderstandings of Zero-Time Theory

Peter A. Hancock

1 A Preamble and the Rules of Engagement

Firstly, I would like to take the 'time' here to thank Professor Frederick Turner for the response that he has made with respect to my chapter entitled "Zero Time Theory," herein after (ZTT). This cannot have been an easy exercise and thus, I am indebted to him for the effort that he has taken in this matter. It will become apparent that we diverge rather radically in our positions and opinions. However, such theoretical and philosophical disputes in no way alleviate me from the debt of gratitude that I owe for his interesting and polemic response. There is at least one thing I am hopeful that we can definitely agree upon. That is, that such interchanges between rationale and well-intentioned beings is one of the few avenues through which our species can at least hope to make logical and consistent advance. That being said, then let's go to battle ...

The choice that one faces in creating any response to a disputational tract or commentary is twofold. First, one can amass all of the critical points together and respond via one global rejoinder. Almost inevitably, this approach leads to a reiteration of the original proposition and consequently, one in which the debate devolves to an impasse and stand-off with the two perspectives now simply re-juxtaposed; if perhaps now expressed in a little more vehement a manner. The second strategy is to deal with the critical observations seriatim and thereby aspire to a more targeted argument and so bound the search for either resolution, or compromise, and if that is not possible, then at least to achieve a clearer statement of difference and disagreement. There is no guarantee, of course, that the latter strategy will result in any substantive progress, but it is the one I have chosen to adopt here. The following sections then refer back specifically to the headings in the cited commentary (Turner 2019) and should be read as specific responses to each of these observations.

1.1 *Definitions of Truth*

Turner starts us off down an unhappy path indeed by resurrecting and reiterating Pontius Pilot's centuries old cri-de-coeur *'what is truth?'* Turner presents an unnecessary dichotomy here in order to attack my assertions but the dichotomy that he offers concerning truth is itself a false one. Thus, the putatively solipsist conclusion that he then seeks to attribute to and even force upon me, is itself fallacious. Let me seek to resolve this issue of 'truth' in this present instance. To accomplish this we must first begin by acknowledging the fundamental limitations of all human beings. As the British empiricists understood and articulated so endearingly, we can each individually only observe the Universe around us via a very limited sensory and cognitive window. For example, human sensory systems are narrowly confined. Thus, for example, we do not directly perceive much of the possible electro-magnetic spectrum to which we are exposed. Infra-red and ultra-violet displays can be created for us which we can experience when they are translated back to the range of the light spectrum we can perceive. But these are just simple exemplars that feature the limited human sensory capacities, connoting what the psychologist James Gibson termed the perceivable 'affordance' structure of the environment around us (Gibson 1966, 1979; and see Lewin 1936). More recently, innovative thinkers such as Nagel (1974) have challenged us to empathize with the perceptual reality of bats and other creatures. Inspired thinkers, such as Wise and Fey (1981) have asked what our world of design would be like if human beings were shaped and acted like centaurs! It is certain that if we were shaped like centaurs, then the technologically-mediated world which surrounds us now would be very different and not merely in physical composition alone. These modern works remind us of Abbott's (1885) classic 'Flatland' and his reverie of multiple dimensions which inveighed us to consider the reality of a two-dimensional world.[1] These are each respective exercises in fundamental epistemology, one of the threateningly arid highways of philosophy that asks us to empathize with that which we are not. The point here is simple, we all individually look through these limited key-holes of experience and thus, any truth expressed by any such limited entity is itself necessarily limited. Necessarily, such expression can only be our own.[2] This is what I mean by the statement that we must each write our own truth. This and nothing more.

Thus, I can never be you and you cannot be me. So, to the present, experience remains an individually specific phenomenon, although what will be revealed

---

1   Even in this wonderful little text, Abbot did not consider the ablation of time, although a comparative text with a title such as 'Timeland' is surely feasible to conceive and produce.

2   All experience, however nomothetic we would like to believe it to be, is necessarily idiographic (individual-specific) in its final expression. I may learn from others, and this will bias my perspective on the world, yet that perspective remains uniquely my own.

to us also necessarily remains uncertain to our limited cognition.[3] However, it is the most colossal height of hubris to believe that any such limited vision can contain all of truth. There are craters on the far sides of planets we have yet to discover that have their own structure, their own form, and thus their own truth of which we know absolutely nothing at all. And such physical lacunae most clearly extend also to the many possible worlds of cognition. In this protestation, I am with John Locke. For like him I believe that; "... *there is a truth opposite to falsehood that it may be found if people will and is worth the seeking.*" The nominal dichotomy then between the relativistic and absolute nature of truth is exposed as simply a function of the continual and overbearing lack of human humility. We are not all that there is, in fact we are a staggeringly, some might say insignificant and even diminishingly, small part of the truth of reality. Our knowledge is also sadly miniscule in proportion. Notwithstanding such recognition, some humans remain strongly and strangely compelled to write of what we may understand, and that is the purpose of my original chapter (Hancock, this volume). There is no discord between the statements I have made. It is only a misunderstanding of the frame of reference which leads to any such appearance. I am gratified that Turner has rendered me the opportunity to resolve this confusion. We now move to Turner's next objection.

1.2   *A Brief History of Time Denial*[4]

Turner first concedes and then elaborates upon certain historical considerations of time as illusion. But first, just a brief, yet essential point of clarification. The term illusion implies a contrasting reality. Thus, for example, it is illusions in the environment which fool us (and see Hancock 2015a). So rather than the use of the term illusion that can implicitly, and even explicitly, indicate the presence of some alternative, non-illusory reality. I prefer to use the term delusion, or better self-delusion since the reality of time only derives from such a iatrogenic process. In consequence, in such instances of delusion, time

---

3 Interesting work has been reported recently, for example, on the use of virtual reality to induce out-of-body experiences. This line of effort might prove fruitful in terms of mutually shared conscious experience. And see: Lenggenhager et al., (2007).

4 There is a brief but important objection I must raise to this particular heading. The problem here is the connotation. As is evident later in the response, this attribution of 'denying' is intentional and troubling. I think we are engaged in an important discourse as to one of the perceived fundamental dimension of human existence. The problem is that the denier label awakes in many of us certain disquieting associations. We talk, often with collective disapprobation about 'climate-change deniers;' we reserve even greater social excoriation for 'Holocaust deniers.' I do not believe we should use such rhetorical devices in our debate on time. The appellation and especially the association demeans us. I hope that the strategy was an unintentional one.

is certainly not unique (Hancock 2013). There are many psychological states to which we have attributed a 'reality,' but in what manner these attributes match to an externality remains a matter of great contention and debate in both experimental psychology and in neuroscience (see Hancock 2017). It is important to observe that human beings continue to fool themselves about many aspects of reality which we can, for example, observe in a growing propensity in social media as well as other social contexts (Hancock 2000; Waltzman 2017). But arguably, time is the most powerful of these delusions. Here, Turner's synopsis of these prior historical propositions is fair, and to the degree that it can be in a brief commentary, comprehensive. However, to posit that this is *"one of the oldest and most comforting philosophical assertions ..."* is diametrically opposed to my own experience in battling with the monster of absolute determinism. Indeed, I can state with painful certainty that I have found no 'glee' involved in this process at all; Voltairean or otherwise. In actuality, I wrestle constantly to find any degree of positivity or salvation in the position I have promulgated. To date, and sadly, I still find little, if any comfort; either pragmatic or intellectual.[5] Turner then adroitly invokes the immeasurable and unjudgeable in forms such as the poetry of Eliot and references to art, where any drawn equivalences are so fundamentally subjective that employed analogies and metaphors become grist to never ending discourse and dispute.[6] I am happy, as most who pursue understanding are, to engage in these delightful intellectual chautauquas, but I know they will not provide us resolution even were we granted an almost endless cavalcade of 'time.'[7] Turner then asserts that the notion of the illusion of time is not a new one. Indeed not. But as I noted, the idea that time is an effective, iatrogenic organizational tool derived from the imperative placed on

---

5   It is one of the hallmarks of the overall human narrative that we embrace heroic redemption and fulfillment in the stories that we tell ourselves. That such tales are underwritten by the fundamental social fabric, which we have generated, and then accepted as the mirrored reflections of that reality is a delusion that it is dangerous to challenge and perhaps even existentially threatening to embrace. I have no personal grounds for rejoicing or glee in this matter.

6   Any observer of the current media will immediately recognize this 'appeal to authority' in making such subjective judgments generally socially acceptable. One need only peruse any cooking show contest or beauty pageant, to realize this form of reification of 'authority.' As the foundational epithet of the Royal Society indicates, 'nullius in verba' is the hallmark of science; opinionated but unsupported demagogues, however apparently persuasive, are its antithesis.

7   It is one of those quirks of life that I actually grew up in the English village adjacent to Burnt Norton. I am sure all timesmiths will be familiar with Eliot's opening lines: "Time present and time past, Are both perhaps present in time future, And time future contained in time past. If all time is eternally present, All time is unredeemable...."

the purportedly evolving brain is, I believe, a differing and original perspective from those who have, in previous works, simply asserted time's illusion. As with Montemayor's (2013) imperative, I derive my position from an understanding of neuroscience, not simply from interesting and well-expressed philosophical tracts and ratiocinations (see Hancock 2015b). However, I am happy to let the reader decide on the originality of my contribution and perspective for themselves for we, in our society, have an informed membership.

1.3   *Time Denial in Physics and Its Discontents*

Turner next evaluates the evidence for ZTT in physics. Before I engage with his specific observations, I would like here, with respect to physics, to make two foundational points of my own. The first is a rather tangential one, the second more immediately relevant. I hope, however, that the importance of the first of these points will become more evident as this response proceeds. Progress at the forefront of both theoretical and practical physics has its greatest exposure at the scale of the very large and the very small. Innovative understanding in astrophysics and quantum mechanics fill our scientific literatures and, on occasion, exert their impact sufficient even to move the mass media (see Musser 2015). Two instruments lie at the forefront of these fields. On the large scale, we presently rely much on the information derived from the Hubble Space Telescope. These wonderful, false color, images do much to fire the public and professional imagination concerning the Universe that we occupy. I fully accept that there are large projects concerning radio telescopic exploration and their efforts to penetrate to the edges of the known Universe, but it is the visuals from Hubble that affect the public the most. At the level of the very small, the Large Hadron Collider (LHC) serves us in the same fashion. Looking into inner space, the LHC was able to find a particle in the mass region around 126 GeV consistent with predictions from the standard model for the existence of the Higgs boson. These are wonderful instruments at the very forefront of physics; but do they fully represent science? These multi-million, and even billion dollar facilities (Hubble and LHC) are each unique. However, their observations, at present, cannot be independently replicated. By this I am not casting aspersions upon any of the scientists involved, nor their integrity or research. I am simply stating that independent replication, one of the hallmarks of science, is problematic when we seek to explore the very edges of our understanding. At the relative signal to noise ratios at which such instruments operate, there exist certain inherent problems as to what it is possible to be observed (and see arguments for example in e.g., Hancock, Masalonis and Parasuraman 2000; Parasuraman, Hancock and Olofinboba 1997).

The second point that I have to make is that all physics is strained through one final common pathway: the brain. This means that all such information is interpreted via the processes of perception and the processes of cognition. As is very evident from our current state of understanding, the brain itself is subject to many inherent biases and much implicit processing of which we still only have a meagre degree of direct knowledge (and see Kahneman 2011). In this context, take for example the original observations under-pinning the concept of the 'arrow of time.' In those observations, Edington (1929), derived from his presentations made in the 1927 Gifford Lectures, provided three fundamental characteristics concerning the arrow. He specified first, that the arrow is: "vividly recognized in consciousness" (i.e., immediate, direct perception). Second, he observes that it is "equally insisted upon by our reasoning faculty, which tells us that a reversal of the arrow would render the external world nonsensical" (i.e., subsequent acts of cognition and then decision-making). Finally, he observes that; "It makes no appearance in physical science except in the study of organization of a number of individuals" (in which what connotes 'organization' itself is also subject to the pattern-recognition capacities of the brain). Please note, I only use these observations as limited *exemplars*. I do not base my whole case concerning time on these few phrases. However, it is very clear that Eddington, at that juncture in history, largely specified the directionality of time primarily through appeal to human perception and cognition. If we take Eddington's observations as being rank-ordered, then his most important specifications are perceptual and cognitive in nature. Finally, the latter criterion refers to order (the more modern appeal being to entropy). But pattern and order are also cognitive identifications. Encoding one's deeply held beliefs in a mathematical form may be wonderfully appealing and comforting, but in the final analysis it remains a translation of thought to another communicative medium. I make these two points prior to examining Turner's observations on physics only to illustrate the nature of the ground upon which we are engaging in this segment of debate. Such grounds are, I believe, far less stable than the standard account or narrative of reality attests.

In his now emergingly characteristic manner, Turner seeks to tie my arguments to those of previous authors, thereby tagging me with their assumptions and frailties. Later this enables him to dismiss my own observations via reference to certain standard objections to those previously stated propositions. This he refers to as a sealing up of 'bolt-holes.' However, I require no such subterranean havens. In part, both of the assumptions he makes about my original reasoning are themselves, on the surface, reasonable. First, I did not want to burden my readers with a re-capitulation of a series of frequently made observations on the physics of time, and also I do believe that, with Hume, giving something a name (and in the case of time, a virtually ubiquitous and uncontestable name) provides it with a power and a subsequent reality that is hard to dissolve. Thus, talk of tesseracts (directly founded on the employment of and belief in the utility of Cartesian coordinate systems), and space-time diagrams, imputes to the temporal dimension the same immediacy and reality as to that of the spatial dimensions (and see Gibson 1975, for arguments contrary to this). But, despite my use of these stepping stones as forms of 'stations of the cross' to reach my eventual goal, it is the use of time in these representations as if it were simply another dimension of space which specifically I contest. Thus, orthogonal, four-dimensional block interpretations are anathema to the ZTT position I have presented. Hence, this is one major reason why Turner's standard denials are inappropriate to my, different, case. But just as an aside, Turner's analysis is simply flawed even with respect to these standard objections. For what Turner uses to deny the delusion of time are temporal propositions. Thus, "if it (a charged particle) is emitted sooner, you get one universe; if later, another. A block universe would have to include all of the different moments of emission and the different universes, involving impossibilities ..." (parentheses mine). So here, like in so many parts, Turner assumes time (i.e., emitted *sooner* ... if *later*), and then extends this proposition to argue for the impossibility of a block universe of all potential possibilities (another temporal referent). As I noted above, these contentions are themselves predicated upon the reality of a single temporal dimension, in order then to subsequently argue for the disproof of ZTT. *It is one of the greatest challenges of time research to seek to comprehend a universe without it.* I think the flaws of a self-referential nature embedded in this argument should hopefully, now be clear. However, we must be grateful to Turner for exemplifying these flaws and the way in which my own conception differs from the standard position in physics on the negation of time that he wishes to dismiss (and see Barbour 2001). Turner then goes on to conduct comparisons which do not apply to what I have offered and I shall pass over his contentions about 'growing block universes' as irrelevant to the present response.

Where we pick Turner up next is with his foray into thermodynamics. Again, he very unfortunately prejudices the discourse with epithets such as 'shipwreck,' but again I must respond that he is erecting his own windmills against which to tilt. It is again important to reiterate that, in essence, he is searching to fit ZTT into one of any number of previous propositions so that he can then use the trails of past arguments to contradict it. But he has missed my entire point (and I partially take that shortfall upon myself), since fully letting go of time is a step of over-whelming cognitive effort. This being one reason why I believe ZTT is liable to make little immediate impact.[8] For example, Turner's next analogy is with a film (which is changed by being shown). But this is a completely misleading equivalency. It means that Turner is using this analogical relationship(s) in circumstances that are not analogous. Thus, a film must move and when he references it being 'run backwards,' he provides it with a temporality that is denied in ZTT. So again, the objection fails. I will not get involved with the disputes about what constitutes the boundary conditions of a 'system' (either open or closed), since this too is fundamentally an arbitrary human decision (but for a related discussion of this issue, see Hancock 2018a). Nor will I wander down some of the sad, tried, and tired routes that have led us to the current impasse in our greater understanding of the Universe. Suffice it to say that terms such as 'pattern,' 'decay,' and 'novelty' are redolent of cognitive apophenia. However much one wishes to strain these appellations through the orthotics of physics (either theoretically or empirically expressed), or mathematics (in all its various incarnations), it remains a question of psychological attribution. That the standard model(s) in physics proves pragmatically useful, I do not deny. After all, I believe that time is one of, if not perhaps, the greatest invention and tool that human beings have ever created (Hancock 2018b). But the tool is now proving to be our unchallenged master as opposed to our most useful servant. As I shall conclude, time is not ubiquitously benign, nor will it ever give up its grip on the human mind easily.

Turner then uses the rhetorical device of implicit (and sometimes an explicit) appeal to authority, and also he references, mostly in passing, some of the more interesting elements of the academy (e.g., strange attractors, etc.). I do not blame him for this strategy, for I too have been seduced by such excitements myself across the years (Hancock and Pierce 1989). But at heart, Turner sees my work as a re-iteration of the block-universe conception and it is against this latter proposition that he continues to battle. For example, he then delves into the many 'universes' conjecture and mentions the anthropic principle in essence, but not by name. But here again we encounter the unbound hubris of human beings. Note that the anthropic principle reifies 'observers.' It is the same reification that places a notional point in time, a 'now,' in between a past and a future which divides McTaggart's (1908) differing Series, and parenthetically

---

8   It is, as we know, hard enough trying to change people's minds about things that they *think* that they *know*, but changing minds over things that they *know* that they know is almost impossible. Such social epiphanies only occur after many decades and even centuries of maturation, and then not even in any guaranteed manner.

appears to differentiate them, as specified in Russell (1915) (cf. my response in the original chapter, Hancock 2019). It is indeed remarkable to see the diverse invocations of these explicit and/or implicit observers pop up in physics without any particular recognition of the cognitive nature and dimensions of any such 'observers.' Turner's unbound speculations upon various incarnations of possible universes emanating from quantum foam are, at one level, entertaining and fodder for science fiction and more serious science speculation, but again, these statements in probability and statistical possibilities are not germane to the resolution of the concept of timelessness and therefore beyond the immediate purview of ZTT. The fact that the evident cognitive abilities that we do possess can provide such forms of speculation is interesting to us but not necessarily to anything beyond us. For, as I have previously observed, *the Universe is not interested in us, but we are not disinterred in it.* Turner's reference to the 'growing block universe' conception again harps on the geometric and probabilistic aspects of theories advanced elsewhere and how they match with what is known empirically by the highly-constrained specific observer we label homo-sapiens. Even as a member of that self-same species, my interest in such propositions is a passing and diminishing one (both terms here being necessarily laden with temporal import).

Turner's final point in this particular section is well-made. Earlier, I opined that time is the most powerful tool that humans have ever created (and see Hancock 2018b, 2018c). However, not far behind time we must also place language (and if we are willing to stretch the language category a little, mathematics also). These latter, one might say, dominant human influences are themselves each founded upon time. Thus, when we speak, when we write, compute, or when we communicate in general, the process is itself predicated upon temporal assumptions, most frequently implicitly embedded, but sometimes explicitly stated. Turner is correct that in seeking to communicate the delusion of time, I am constrained to use the tools of the 'enemy,' as it were. But not simply the words or symbols. For example I am, perhaps even constitutionally, constrained to believe that the present moment (the 'now' when I am creating this text) will be followed by some later moment when the text will reach the press and then subsequently the reader. Earlier, I abnegated the charge of 'glee' with respect to my original discourse and also referred to the continual cognitive 'effort' required to hold the delusion of time to the forefront of consciousness and at bay. I have dealt with the difference between 'illusion' and 'delusion' before and so do not here repeat the point. However, the fact that the tool of language is erected upon the tool of time does serve to make my whole notion of timelessness much more difficult to communicate. This apparently necessary constraint is recognized and accepted.

1.4   Time's Skepticism's Reduction to Nothingnesss

Turner now turns to life as proof of time. He argues that since life exhibits characteristics such as 'metabolism,' 'reproduction' and the like, it must be temporal in very essence. Here again we hear the siren call of the reified observer.
Turner's appeal is nicely couched since he references plants and bacteria as
now being forced into this role of observer; or rather hubris-free observers that
cannot possibly possess the shattering lack of humility that I have attributed
to humans! He even extends his argument to inanimate matter, which he observes itself 'resonates' in time, so thereby 'proving' the existence of time and
the vacuity of ZTT. He again erects a number of straw-men, familiar to experimental psychologists, concerning a necessary illusion for the existence of color
and sound, but these are, like his central point, to pre-suppose the existence of
time in order to prove as incorrect the non-existence of time. I am, of course,
sympathetic to this appeal to life (Hancock 1984), in the same way that I am
sympathetic with the hierarchy of time championed by Fraser (see e.g., Fraser
1987). In some ways, Turner moves us here from the realms of physics-type issues of the *proto-temporal* and the *eo-temporal*, to the biological levels of the
bio-temporal and the *noo-temporal*. The comparison is not exact of course, but
the sentiment is equivalent. One of the problems here is the epistemology of
bacteria, or indeed inanimate matter. Earlier, I appealed to human readers to
take the step of imagining differing forms of existence (humans being centaurs,
Flatland, and the like). Here, I ask again. In limited written responses such as
the present one, it is often the case of replying to specific observations, so here
I can reference my position with respect to this issue which is given in an earlier article (Hancock 2015b). This argues that we humans have and continue
to draw an immutable barrier between life and non-life (see also Schrodinger
1944). And we continue to attribute autobiographical sentience to some orders
of life, but not to others (few, for example, perhaps with the exception of empathic micro-biologists consider bacteria as having a personal life story). As far
as non-living systems go, we humans rarely if ever consider awarding such an
entity with a 'living' story. Yet they do. Turner's objections here are elaborated
into an apparent negation of light and indeed all phenomena which are only
interpretable as temporal processes. But the problem of an inappropriate circularity in the employment of a concept of time is again clear. One assumes an
initial foundation (time), but it is the foundation which is at issue – here the
logic fails, and is perhaps one of the reasons why time has seemingly always
proved such an intractable study across centuries of formal exploration. We
then take a peripatetic intellectual excursion into mathematics in which the
role of the observer is replaced by the role of the 'calculator.' As I noted above,
the presence of some (to a degree) sentient entity is often implicitly expressed

and so my objection remains the same. Therefore, I do not have to consign anything to the *"grand bonfire of time,"* as Turner would seek to force me to do. The
transition to religious considerations is one of critical import. As with philosophy, religions over the centuries have often engaged in unbound speculations.
One of these recurring themes in religion(s) is indeed the 'illusion of time.'
As with many similar philosophical speculations, they remain without logical constraint and thus essentially unresolvable.[9] They appeal to our intellect
but cannot be decided upon by unequivocal rational disquisition, contingent
upon empirical observation. It is to this topic of unbounded speculation and
the emerging role of the neurosciences in acting as such a potentially bound-
ing force that I return at the end of this response.

1.5   *The False Tinsel of Hollywood: Real and Unreal Illusions*

The section on false tinsel is well named as it presents yet another series of
false dichotomies. The rhetorical device of creating an either/or situation and
then rejecting the one which purportedly represents your interlocutor's position is proving a somewhat tired device by this juncture. Here again Turner is
using the term illusion and its implication that behind the illusion necessarily
resides an extant reality of time (e.g., that which serves to produce the dynamic shadows on Plato's cave) which thus now 'really' exists. If the nomencla-
ture and the assumptions are wrong, it is little wonder that such an argument
leads us down inappropriate paths. For example, Turner asserts that the only
alternative to ZTT is nothingness. This is an incorrect supposition. He is again
caught up in the differentiation between illusion and delusion and thus even
framing his argument in terms of Plato's cave does not resolve this misunderstanding. And if I did not make this difference sufficiently clear in the original
work, then this is, at least in part, my shortcoming. His insistence on casting
my own position as exactly equivalent to the notional 'block universe' often
induces these incommensurate comparisons. Indeed, the emergent property
of mind can derive many self-intriguing patterns and then apply its own defi-
nitional terms to them (i.e., unbound apophenia). In this discourse, those that
have been mentioned include beauty, uncertainty, mathematics, etc. These
phenomenological realities are not negated by ZTT, but rather re-cast by recognition of their origins. Finally, I am not a fabulist. I do believe that humans,
as a result of their inherent perceptual and cognitive capacities, impose many
(if not all) patterns upon the environments in which they are found. However,

---

9   One can, of course, argue that this concern applies to all human knowledge. However, the
     point I make here echoes Montemayor's appeal to constrain philosophical speculation with
     neuroscientific understanding.

it is yet another of the faces of hubris to believe that the observer, and specially the human observer, actually creates all of that which underlies the possibility of those patterned observations.

1.6   The Evidence and Argument of the Paper

Turner now considers some of the experimental information I provided that can be interpreted as suggestive first, of the iatrogenic nature of our time-keeping systems. It is critical here to differentiate between the framework(s) we have created in order to keep track of time (see e.g., Cipolla 1967), as opposed to the iatrogenic nature of the concept of time itself. In concatenating the two, Turner makes a misstep and it is not one I would have expected him to commit. The data that I presented, amongst many others (see Hancock 2011a), demonstrates a fundamental coherence between the way in which the average male perceives duration and the way in which certain common temporal accounting mechanisms has been structured. This is expressed in its essence by the psycho-physical exponent being very close to, or at unity (but see also Eisler 1976). Either this can mean that men are (on average) extremely accurate, and thus females less so. Or, the alternative hypothesis is that the scale itself was invented by males (and so male estimates correlate more closely with the 'metrical' scale than do female estimates). Clearly, I have chosen this latter interpretation of creation. The stronger argument, and contrary to what I have asserted, is to object that time-keeping might well have been a unique invention by a singular individual; assumedly male also? And thus, why would estimates by a grouping of males accord with the perception of this one single individual? I would have responded to this objection that time-keeping is fundamentally a social imperative, and thus one not necessarily needed by totally isolated individuals. So time-keeping mechanisms are socially-mediated tools. However, such an observation and objection was not made. I also provided a further example, that such differential perceptions (of nominally 'objective' time versus perceived time) persist across the lifespan and were not confined to short intervals of time up to two minutes in duration as shown in the first, summated Figure. In exemplifying how these effects exist in time-keeping strategies, I was endeavoring to provide a provocative example through which to approach the greater issue associated with the iatrogenesis of the actual dimension of time itself. Few would argue that such social time-keeping formalizations are *not* human-created and human-designed conventions. Who, for example, protests that the unit of one minute adheres anywhere in nature directly; who would posit that that a unit of one hour necessarily emerges directly from the orbits of the planets? We can certainly see how the idea of one day derives from our own terrestrial revolution with respect to the sun and

this shows how social conventions often compromise with, or are partly constrained by natural imperatives. However, the choice of seven days to compose
one week is clearly and solely an arbitrary human decision. I doubt whether
our aforementioned bacteria deal in units like a fortnight, or are glad of an approaching weekend?[10] Thus, most timesmiths can readily agree that certain of
the formalities that we use to index the passage of time are arbitrary human
creations. What I am asking here is to take that line of thought one giant step
further forward and consider the whole concept of time as an exercise in
iatrogenesis.

Turner proceeds to articulate some of the potential sources of extraneous
variance which apply essentially to all observational analyses. But these are
the fodder of standard statistics courses and the appeal smacks of a rather des-
perate grasping for reasons why the nomothetic patterns observed may not
be veridical ones. I could proceed, on a point by point basis to refute each of
these objections, but in reality it is simply not worth it, given the current limits on space and the greater concerns of the present discourse. And so I have
relegated such responses to an explanatory footnote.[11] Hamlet now makes his
obligatory appearance, but the greater issue that could have been raised concerns of individual differences in such estimates; an issue that has plagued
the research psychologists of time now for an extended interval (and see
Doob 1971). The observation offered, i.e., that statistical variation within any

---

10   We have, of course, already discussed the great problem of steps of empathy with other
      forms of life; and indeed non-life. I must, logically admit that bacteria just might organize
      according to a fortnightly schedule, or that French bacteria yearn for 'le weekend.' But
      levity aside, oscillation (as we humans view it), represents one essential characteristic of
      such 'living' organisms. But as I have protested, emergent properties (such as life) possess
      no privileged position. The latter I here take a chance to re-iterate simply because of its
      fundamental importance to my position and zero time theory.

11   *Sampling Error:* Turner fails to understand that the point in Figure 1 each represent indi-
      vidual experiments, many of which compose tens, if not hundreds of participants each;
      but such was clearly stated in the Figure Caption. *Selection Bias:* As the summation is
      across multiple studies. His objection must posit that each disparate group of researchers
      had exactly the same bias, across more than a century of data collection – unlikely. The
      time estimation studies are conducted in a Laboratory setting where the data are col-
      lected (most often) by electronic timers. These methods do not 'interview' the participant
      in any linguistic sense. *Reporting Bias:* Likewise a button press allows little opportunity
      to express any such bias. *Gender Roles:* Have and do indeed change, yet the data are con-
      sistent across more than a century of such social change. *Differences in Life Expectancy:*
      The axes in Figure 2 (for example) are scaled to national life-expectancy figures and thus
      are adjusted for same. We then devolve to a series of undefined speculations, culminating
      in the etcetera. I could continue with more detailed responses but the interested reader
      is directed to the original papers. The relationship between variance and entropy should
      here also be emphasized.

particular population provides the best defense against uncertain variation in the environment, most probably contains a kernel of truth. But again the implications as to the non-importance of each individual in that population is not then explored – more is the pity.

1.7   Evolution Denial

Turner then proceeds to recast my rejection of time as a rejection of evolution. But of course, this latter assertion is simply a reflection of the former. I think that it is important to first reference the fact that I have already written on this topic and have sought to understand how Darwin himself derived his insight when he was able to break free, intellectually speaking, of the unrelenting stranglehold that time has on human cognition (see Hancock 2007). Turner then makes explicit his earlier and unfortunate rhetorical device, now directly accusing me of denying the reality of events; one element of which action is even labelled a 'crime' in some nations of the world. This is surely beneath him.[12] I make no such claim. That human beings phenomenologically experience a continuous stream of consciousness (as I do myself), has led that consciousness to imprint a past, present, and future; a hindsight, eyesight and foresight structure on existence. This gives a necessary phenomenal 'reality' to that passage of experience. I am no nihilist. I do not believe that there is no such thing as reality. Nor do I believe that reality is simply the construction of the individual human mind, nor for that matter any celestial mind. Let me be very clear. I explicitly reject Turner's innuendos here, and am saddened by his way of expressing them. That such forms of consciousness can weave together an autobiographical trace, as it were, and thus impose a dimension of time derived from that perceived continuity has proved seductive in the extreme. But *emergent properties*, (minds included) *possess no privileged position*; other than the one it arrogates to itself. Memory is the cord upon which the emergent mind strings its pearls of recollection (Hancock and Shahnami 2010). ZTT does not deny reality, it just obliterates our traditional view of time in the experience of such a reality.

In respect of the observations on Olympic athletes, Turner simply misses the point. Had he explored his earlier, largely ad hoc, observations about

---

12   I must reassert here that the central postulate of ZTT does not deny reality. Thus, there is no 'denying' in the same fashion that others have sought to deny the reality of the events of the holocaust, global-warming, and evolution. As a different order of argument there is no essential parallel and indeed the group of 'deniers' of the latter realities are most certainly temporally-bound, and on occasion deniers of reality itself. It is disturbing for all of us if by the implied parallel, we were here to stifle discourse. I would be alarmed by such an imprecation, if this is true.

individual variation, he might have better understood my observation. Here again, we witness in summated human communication, that the 'story,' the 'narrative' of each individual proves overwhelmingly seductive, most especially to the person themself. And in affirming this narrative, as evident in the stories associated with large social events, the media especially reinforces the collective social agreement that humans are each individually unique and valuable. Morally, we as humans can convince ourselves, and even socially agree, that this is so (and see Hancock 2009). However, any non-human oriented level of analysis can attest to the vacuity of such self-aggrandizement. I am asking here that we progress beyond our blinkers of human exceptionalism to see other views of reality and one especially in which time is not the central arbiter.

The next paragraph of Turner's commentary beginning "The paper takes a stand ..." now rightly takes me to task over the problem that one cannot use one set of observations, consistent with the existence of time, to postulate the non-existence of time. Such an objection is, as I have endeavored to show with respect to Turner's own comment, valid. But this is largely a stations of the cross issue. ZTT is so counter-intuitive that one cannot expect individuals to jump to embrace it in one cognitive leap. Rather, like the earlier observation concerning the use of temporally-redolent language to attack temporality, this present concern lies in the same vein. Here, I have attempted to explicate ZTT through the use of contemporary advances in psychological science. For more than a century, studies of human memory have been focused on the past. That is, memory has been, by definition, the recall of prior events. But more recently, memory researchers have re-oriented their perspective to consider memory's prospective utility as much more important than traditional retrospective recall (at least explicit recall; since implicit recall may be a different issue). In essence, in standard terminology, memory is now being considered more as a future survival mechanism. Like other similar exemplars (e.g., Darwin's own fundamental insight), this is a step of re-conceptualization. I wanted to use this particular example to show how, consistent, logical and insightful steps along the road toward ZTT are actually being made. Turner however, rightly observes that these initial steps remain couched in the standard temporal narrative (e.g., the evident use of the word future) and so remain temporal in their orientation. It was inconsistent here to use the two in conjunction as I freely acknowledge. However, this was not, nor is, a tongue-in-cheek exercise.

Turner then proceeds to yet another assumption that I would certainly not make. He asserts that: "(as) cause and effect do not operate, and the world is completely random ..." [please examine the whole observation so that I do not quote out of context here] (parentheses mine). Yet, I would make no such assertion. The absence of cause and effect to me certainly does not imply the necessity of complete randomization. Rather, I would suggest that any sentient being could find sufficient regularity in the distribution of energy observed to 'find' and/or 'impose' patterns of their own utilitarian creation. Subsequently, I am asked by Turner why evolution, long ago has not culled out certain mechanisms. Here again, Turner wants to play in the traditional play-yard of temporality to justify his opposition to ZTT. It is evident that this is his central objection but again to reiterate, it is a misplaced one. However, the observation itself is telling. The temporal perspective employed again reinforces the standard notion that humans are close to, if not at the 'peak' of evolution. It implies a teleology for evolution and again represents hubris epitomized. Even within the standard narrative, humans (if we can use this as a general term) have only been around approximately four million years. In light of traditional perspectives on time, it can be seen as an infinitesimally small portion of existence that this interval represents. Where then do we derive this ultimate lack of humility? I shall, at the end, conclude my response with brief observations on this question.

The issue concerning meaning is much more vital to our present considerations. If, as I have posited, ZTT proves to be a veridical description of our fundamental reality then, as Turner notes, meaning is obviated. Indeed, there are only states that are, and then states that are not – that is a fair assessment. Where then does 'meaning' come from? As should by now be evident, I argue that meaning is one of the outcome derivations of the explicit or implicit acceptance and imposition of time. That is, for any emergent property of matter that can be described as 'living,' the transience of that life which is involved renders the creation of time a useful solution. But then, for the 'higher' levels of that form of emergence, meaning is a temporal extension derived from conscious consideration of possible future states. In my own previous terms, *time is the punishment for consciousness* (Hancock 2011b). However, if ZTT holds sway, it acts to negate meaning in the mind of the perceiver. This is dangerous ground indeed, and no cause for glee in any form. To this juncture, the only human anodyne that I can offer is the alternative of cognitive anthropic revelation that I have explained in the initial text. Such revelation demands that we explicitly acknowledge our limited window of perception and proceed from the necessary indeterminacy of those limited perceptual and cognitive abilities. While the nominal future remains 'set,' it is our perception of that apparent future that remains unknown to us. How we integrate any utility associated with the remnant concept of free-will in this framework is obviously problematic. However, we are talking here of a species that can praise an all-powerful god as the fountainhead of ultimate good, while still continuing to observe an ever-growing litany of human disasters on our planet. Hence, si-multaneously held concepts that are cognitively dissonant do not at all appear to be beyond us.

1.8   The Intention of the Paper

In his final set of observations, Turner begins now with his standard strategy. He attributes to me certain perspectives that he then looks to negate. He again uses terms such as 'illusion' and 'demolition' in the hope of retaining the certi-tude that time is real by implying that the reality lies beyond the illusion and that to be demolished, something must have existed in the first place. These are incorrect attributions and I do not accept them. I am then cast as a 'gadfly,' but I can assure all readers that I do not goad cattle; at least not intentionally. I am next cast as satirical hoaxer, but I am no hoaxer, satirical or otherwise (and see Hancock 2015a). Next, I am told I possess a bad and even terminal case of 'monocausitaxophilia.'[13] I have no such disease; although I agree with Locke and am a believer in truth. I am next termed a 'cognitive terrorist.' While my observations can indeed be disturbing, at no point have I sought to induce ter-ror. In the penultimate accusation, I am deemed to have a 'misplaced' desire for absolute moral clarity and intellectual honesty. As I noted earlier, morality is a human invention and thus I remain doubtful whether absolute moral clarity can ever be achieved. I hope that I do aspire to intellectual honesty and do not ever resort to ad hominin attacks to achieve it. Finally, and perhaps from my perspective most problematic, I am alleged to have produced a 'religious state-ment' to clear the way for a turn toward an eternal God. I can assure my readers that this accusation supersedes only the vacuity of the others which precede it. Having accused me of this series of motivations, Turner then believes that I suffer from an agglomeration of all of his alleged motivations. I find no value in this. Turner accuses me of erecting my whole thesis upon one 'relatively small experiment in social psychology' but this simply mischaracterizes my text and my case. Like the 'Discourse Between Two Worlds,' Turner proposes that I set up

---

13   I have used the work of the cited Professor Poppel on many occasions. I find his work interesting and informative. Although the invention of a linguistic term tends to provide a form of power to the concept so expressed, I do not find the search for the pattern of meaning amongst the meaning of pattern necessarily at odds with the search for unified explanation. As to whether any organism so constituted as ourselves must necessarily search for and believe in some form of monolithic mono-causality, the question depends upon the degree to which one believes that pattern adheres in reality, as opposed to the degree that pattern is imposed by the perceiver. Despite my effort to emphasize the latter in relation to the organizing principle of time, I do not believe that all regularities are the creation of an emergent mind. However, the interpretation of such patterns (and even their basic observation in the first place) must necessarily be so.

the present postulate only to knock it down to prove its antithesis. A well-used device of past historical disputes, it is out of favor at the present, and I do not resurrect it here. Turner then blithely concatenates physical disposition with theological assertion to seek to negate my observations about human hubris. He happily employs the Copernican cosmology alongside the Alighierian allegory to conclude that those of the Renaissance and Enlightenment were not of a humble disposition anyway. My concern here is with the physical realms, the subtleties and nuances of spiritual geography I leave, pro tem, to others more adept in them (Rusca 1621). Rather piquantly, Turner then accuses me of employing terms (and methods) that are likely to impress. However, Boole's approach is clearly useful in this context and I employ it for no other reason than its present utility. Its impact and impression I leave to the reader.

In his final ad hominin foray, Turner creates what can only be seen as some sort of Wagnerian intellectual hero (hair literally flying in the wind), a sage sat upon a mountain-top dispensing wisdom to underserving masses; a Gandalf locked in battle with the Balrog of Time. I found it all rather touching. In some ways, I would love to believe many of those things about myself – how wonderful it would be to be called "The Wizard of Time!" Alas, I remain human, all too human. I cannot fulfill this role that Turner wishes to believe of me. While I may aspire to such mythic status, I am not too far gone in self-delusion to ever believe it even partly true. Perhaps time will tell? But one point is well-taken. I am indeed concerned about the morality associated with ZTT. Words have power, as we see in our current, dysfunctional political discourse (and see Waltzman 2017). What we say has ramifications, (Hoffman and Hancock 2014) and, as I have written before on such effects, I am aware of some of the responsibilities involved (Hancock 2009, 2012). Of course, as with all scientific discourse, the overwhelming probability is that the modal number of non-obligated readers of the present response will be zero. The value of Turner's poetic observations, I leave to others who self-identify as more qualified to judge them than myself.

2   Final Observations

2.1   On Philosophy and Neuroscience

I realize that the present response has been reactive and rather scattered in nature as I have looked to respond to each of the points made. Thus, I think it is important to finish here on a positive note. After all, disputations, however well-intentioned can often dissolve into polemic contrarianism, even if we do not intend them to do so. Thus, my penultimate comments concern philosophy and neuroscience. The observant reader will have noticed my earlier piquant phrase "seemingly arid highways of philosophy." But how can this be so, one might ask? After all, has philosophy not constituted much of what has pushed knowledge forward over the whole epoch of recorded human history? How can they be arid highways; are they not actually our fertile routes of progress? To a degree, this is so. Unfortunately, philosophical reverie and conceptualizations have, up to almost the present time, been insufficiently bounded by what we have learned of the human brain. In essence, philosophy presents a virtually unbounded journey of meta-cognition about itself and the world around it. And, as I have noted, the vast majority of the human species find themselves wonderfully, even endlessly, interested in themselves and so the better equipped members of such a species can come up with a seemingly endless series of speculations and postulates. Until the recent revolution in neuroscience we have not been able to sufficiently prune the tree of philosophy. But now we have the beginnings of such an effective tool. This point has been made by Montemayor (2013) who asked us to consider what such an emerging understanding of these neuroscience-based constraints mean to traditional philosophical theories and stances; especially metaphysics related to the mind. In short, such constraints mean much (and see Montemayor 2010). In my own work, I have looked to synthesize such evidence to provide a coherent picture of human interaction in 'dynamic' worlds. Especially, I have sought to understand why the human 'invention' of time has proved such a seductive proposition and why time has proved such an over-dominant tool in the human ascendancy across the living ecosystem (and see Hancock 2015b). It may be through the capacities of modeling and simulation that we can begin to formalize and enact such neurophysiological and neuropsychological constraints on philosophical theorizing. If so, the present era will continue to be a phenomenologically exciting one to live in. Although it appears *that I have no choice but to believe in free will*, I believe that I can actually exercise such cognitive anthropic revelation even as the ZTT world is revealed to our necessarily impoverished understanding.

2.2   *The Humility of Humankind*
Throughout this interchange I have continued to reference that ascendancy of hubris and the remarkable lack of humility of the human species; this Turner dismisses with the Biblical "vanity of vanities; all is vanity." I would, however, suggest that good things happen in human existence when such vanity is logically recognized and rationally acknowledged. Sadly, due to the very temporary nature of the emergent property of individual human existence, we rarely pass on the empathic advantage of recognizing such humility directly. Yet we do appreciate some echo of it in our collective passage of knowledge.

Some millennia, or at least centuries ago, human beings saw themselves at the center of the physical universe. The sun, the moon, the wand'ring stars and the fixed constellations all revolved around the eye-point of each single human being (Koestler 1959). Such a perspective is certainly understandable from psychological theories such as 'direct perception.' It is also internally logical since individually; we are cognitively, all that we possess. Why would not the Universe revolve around us? It becomes evident with social interaction that we, as individuals, are not alone. So, do the heavens now revolve around the eyes of our companions also? This realization, had in later childhood, is one small push away from personal exceptionalism. Then empirical observation begins to encourage us away first from an ego-centric and then from the earth-centric perspective. In essence, in terms of our physical location in the Universe, humility has exerted a small victory over hubris. Now the sun becomes the center of the 'solar' system, and Earth, our home, only one of a number of planets within such a system. Observations are consistent with that picture and the standard narrative of human existence changes accordingly.[14] Further observations by Brahe underlie Kepler and then inform Newton to derive what is still, quite frequently used today, as our computational basis of celestial mechanics. Each of these steps, from Brahe's chilly nights on early telescopes, through Kepler's search for celestial unity, to Newton's "non-fingo'ed hypothesis," each small step reduces the geographical exceptionality of the human species and places it ever more distant from the physical center of existence. With our contemporary observations, we now accept this narrative and appreciate our non-central physical location and thus the ramifications of such a displacement. ZTT requires such a step; but here displacing ourselves from the cognitive, the psychological center of existence. It is, most likely, a much harder step to take. I am not sanguine about the chances of ZTT. Indeed, my own personal skepticism about its acceptance is expressed in proportion to the degree of self-absorption in each human being.[15] Let me conclude simply by saying once again that time is perhaps one of the best of servants that humans have ever

---

14 The usual accounts of 'flat-Earth' beliefs are frequently distorted and anachronistic in their specification and assertions (This, of course, retains the standard temporal based narrative approach to history).

15 Let us, in bidding goodbye to the present dispute, consider the epithets that have been arrayed against me in this exchange. I have been labelled a time-denier, a Holocaust denier, a global-warming denier, a cognitive terrorist, a gadfly, a jeremiad launcher, a magician, a perverter of the 'common herd,' an adept – my 'hair streaming' in the cold wind of the stark truth, a Zarathustra, a purveyor of 'old chestnuts,' a pabulum weaner, in addition to a number of others. When someone chooses to use these labels, I can only ask the reader to consider why such a strategy should be involved?

created; but equally if unquestioned, time has the capacity to be perhaps the most oppressive of masters. Hopefully, a little revolution can go a long way.

References

Abbott, E. (1884). *Flatland: A romance in multiple dimensions.* New York: New American Library.

Barbour, I. G., and Barbour, I. G. (1997). *Religion and science: Historical and contemporary issues.* San Francisco: Harper.

Barbour, J. (2001). *The end of time: The next revolution in physics.* Oxford: Oxford University Press.

Cipolla, C. M. (1967). *Clocks and culture: 1300–1700.* Boston: W.W. Norton.

Doob, L. W. (1971). *The patterning of time.* Yale: Yale University Press.

Eddington, A. S. (1929). *The nature of the physical world.* New York: MacMillan.

Eisler, H. (1976). "Experiments on subjective duration 1868–1975: A collection of power function exponents." *Psychological Bulletin* 83 (6): 1154–71.

Fraser, J. T. (1987). *Time, the familiar stranger.* Amherst: University of Massachusetts Press, 1987.

Gibson, J. J. (1966). *The senses considered as perceptual systems.* Boston: Houghton Mifflin.

Gibson, J. J. (1979). *The ecological approach to visual perception.* Boston: Houghton Mifflin.

Gibson, J. J. (1975). "Events are perceivable but time is not." In: Fraser, J. T. and Lawrence, N. (Eds.). *The study of time II.* Berlin: Springer.

Hancock, P. A. (1984). "An endogenous metric for the control of perception of brief temporal intervals." *Annals of the New York Academy of Sciences* 423: 594–96.

Hancock, P. A. (2000). "Is truth soluble in politics." *Human Factors and Ergonomics Society Bulletin* 43 (4): 1–4.

Hancock, P. A. (2007). "On time and the origin of the theory of evolution." *Kronoscope,* 6 (2): 192–203.

Hancock, P. A. (2009). *Mind, machine, and morality.* Aldershot, England: Ashgate Publishing.

Hancock, P. A. (2011a). *Cognitive differences in the ways men and women perceive the dimension and duration of time: Contrasting Gaia and Chronos.* Lewiston, NY: Edwin Mellen Press.

Hancock, P. A. (2011b). "On the left hand of time." *American Journal of Psychology* 124 (2): 177–88.

Hancock, P. A. (2012). "Notre trahison des clercs: Implicit aspiration, explicit exploitation." In: Proctor, R. W. and Capaldi, E. J. (Eds.), *Psychology of science: Implicit and explicit reasoning.* (pp. 479–95), New York: Oxford University Press.

Hancock, P. A. (2013). "In search of vigilance: The problem of iatrogenically created psychological phenomena." *American Psychologist* 68 (2): 97–109.

Hancock, P. A. (2015a). *Hoax springs eternal: The psychology of cognitive deception.* Cambridge: Cambridge University Press.

Hancock, P. A. (2015b). "The royal road to time: How understanding of the evolution of time in the brain addresses memory, dreaming, flow and other psychological phenomena." *American Journal of Psychology* 128 (1): 1–14.

Hancock, P. A. (2017). "Whither workload? Mapping a path for its future development." In: Longo, L. (Ed.). *Human mental workload: Models and applications.* Cham, Switzerland: Springer.

Hancock, P. A. (2018a). "The humane use of human beings." *Applied Ergonomics*, in press.

Hancock, P. A. (2018b). "On the design of time." *Ergonomics in Design* 26: 4–9.

Hancock, P. A. (2018c). "Can time really be designed? Some observations on the comments and critiques of my interlocutors." *Ergonomics in Design* 26 (2): 19–22.

Hancock, P. A. (2019). "Zero-time theory." *This Volume.*

Hancock, P. A., Masalonis, A. J., and Parasuraman, R. (2000). "On the theory of fuzzy signal detection: Theoretical and practical considerations and extensions." *Theoretical Issues in Ergonomic Science* 1 (3): 207–30.

Hancock, P. A., and Pierce, J. O. (1989). *Integrating signal detection theory and catastrophe theory as an approach to the quantification of human error.* Paper presented at the Annual Meeting of the American Industrial Hygiene Association, St. Louis, MO: May.

Hancock, P. A., and Shahnami, N. (2010). "Memory as a string of pearls." *Kronoscope* 10: 77–82.

Hoffman, R. R., and Hancock, P. A. (2014). "Words matter." *Human Factors and Ergonomics Society Bulletin* 57 (8): 3–7.

Kahneman, D. (2011). *Thinking fast and slow.* New York: Farrar, Strauss & Giroux.

Koestler, A. (1959). *The sleepwalkers: A history of man's changing vision of the universe.* London: Hutchinson.

Lenggenhager, B., Tadi, T., Metzinger, T., and Blanke, O. (2007). "Video ergo sum: Manipulating bodily self-consciousness." *Science* 317: 1096–99.

Lewin, K. (1936). *Principles of topological psychology.* New York: McGraw-Hill.

McTaggart, J. E. (1908). "The unreality of time." *Mind* 17 (68): 457–74.

Montemayor, C. (2010). "Time: Biological, intentional and cultural." In Parker, J. A., Harris, P., and Steineck, C. (Eds.), *Time: Limits and constraints.* (pp. 39–63). Brill: Leiden.

Montemayor, C. (2013). *Minding time: A philosophical and theoretical approach to the psychology of time.* Leiden: Brill.

Musser, G. (2015). "Where is here." *Scientific American* 313 (5): 70–73.

Nagel, T. (1974). "What is it like to be a bat?" *The Philosophical Review* 83: 435–50.

Parasuraman, R., Hancock, P. A., and Olofinboba, O. (1997). "Alarm effectiveness in driver-centered collision-warning systems." *Ergonomics* 40: 390–99.

Rusca, A. (1621). *De Inferno, et statu Dæmonum ante mundi exitium, libri quinque. In quibus Tartarea cavitas, parata ibi cruciamentorum genera, ethnicorum etiam de his opiniones, daemonumque conditio, usque ad magnum judicii diem, varia eruditione describuntur.* Milan: Ambrosian College.

Russell, B. (1915). "On the experience of time." *Monist* 25 (2): 212–33.

Schrodinger, E. (1944). *What is life?* (Based on Lectures to Trinity College Dublin). Cambridge: Cambridge University Press.

Turner, F. (2019). *Deconstructing the zero time theory. This Volume.*

Waltzman, R. (2017). *The weaponization of information.* Santa Monica, CA: Rand Corporation.

Wise, J. A., and Fey, D. (1981). "Principles of centaurian design." *Proceedings of the Human Factors and Ergonomics Society* 25 (1): 245–49.

PART 2

Urgency and Time Scales

∴

CHAPTER 9

Eternal Recursion, the Emergence of Metaconsciousness, and the Imperative for Closure

Jo Alyson Parker and Thomas Weissert

Abstract

In the following paper, we discuss a sub-genre of fiction that we call time-loop fiction, texts that are predicated upon a situation of seemingly eternal recursion. Drawing upon examples taken from literature (Kate Atkinson's *Life after Life*, Ken Grimwood's *Replay*, Richard Lupoff's "12:01," and Hiroshi Sakurazaka's *All You Need is Kill*) and from film/telefilm (*Edge of Tomorrow, Groundhog Day, Source Code, 12:01 PM*, and "Cause and Effect"), we explore the "time as conflict" between endless looping and narrative closure for both the protagonists and the readers/viewers who follow their plights. After describing the forking-paths structure of time-loop narratives, we examine a key feature of them – the emerging metaconsciousness, a character's ability to transcend the loop in which they currently exists and to recall past loops in order to bring about a better outcome with each passage through the loop. We conclude by discussing three different types of time-loop narratives and their implications for our understanding of time.

Keywords

time-loop narratives – metaconsciouness – eternal recurrence – recursion – closure

> In practice, of course, we don't get caught in infinite loops – life is
> simply too short for that.
> MICHAEL C. CORBALLIS, *The Recursive Mind*

•••

> All these different worlds – none of them gets it right.
> THE TENTH DOCTOR

•••

> We cannot, of course, be denied an end; it is one of the great charms
> of books that they have to end.
>> FRANK KERMODE, *The Sense of an Ending*

∴

1   Introduction

In life, we do not get do-overs; we cannot save and reset the game. In our narrative fictions, however, we regularly return to the motif of the do-over, in modes
ranging from the ridiculous to the sublime – from the absurdist comedy film
Groundhog Day to Kate Atkinson's provocative novel *Life after Life*. Constituting
a sub-genre in fiction, time-loop narratives are predicated upon a situation of
seemingly eternal recursion, whereby a character (or in some instances a collection of characters) is thrust back into an earlier point in time in order to
relive that time.[1] In the following paper, we examine key aspects of this sub-
genre, focusing on both the above two narratives and also the following: the
telefilm "Cause and Effect," the short story "12:01" and its film version *12:01 PM*,
the film *Source Code*, Ken Grimwood's novel *Replay*, and Hiroshi Sakurazaka's
novel *All You Need is Kill* and its film incarnation *Edge of Tomorrow*.[2]

As Mark Currie has remarked, "fiction is capable of temporal distortion that
cannot be reproduced as lived experience" (2007, 85). Although we cannot actually loop backwards in time, a hallmark of our humanity, in the words of
Michael Corballis, is the "mental time travel" (2014, 100) in which we engage,
inserting previous experience into present awareness. Moreover, we often consider our previous experience in light of alternative possibilities, the roads not
taken. Our human predilection to consider alternative paths is prompted by
our wish that, if we could relive some specific period of our life, we could "finally [...] get it right" (Atkinson 2013b, 446).

Although widely diverse with regard to content, structure, and style, the
narratives that we examine here all share certain characteristics: (1) each time
loop begins at a particular temporal juncture; (2) the loop generally lasts a

---

1   Indeed, we are dealing here with what we might call a sub-sub-genre. In *Time Travel: The
    Popular Philosophy of Narrative*, David Wittenberg (2013) makes a distinction between the
    paradoxical "'time loop' or 'closed causal loop'" story (31), whereby a character returns to an
    earlier point in their life and meets an earlier version of the self, and "later loop fictions [that]
    explore directions away from purely closed plots" (81). We deal here with this latter.
2   There are many other examples of time-loop fiction. For our purposes, however, the texts
    cited above offer the most provocative or representative examples of the genre.

fixed amount of time – eight minutes, 24 hours, et cetera – although the span
is variable in some cases; (3) a particular event – sleep, death, or even a specific
timespan – precipitates another do-over; (4) within each do-over, the protagonist makes different choices; and (5) the different choices are influenced by
what we call a metaconsciousness, a consciousness capable of surviving into
the next passage through the loop and, integrating knowledge obtained in the
previous passages, of affecting the outcome within the current passage or even
of effecting the ultimate dissolution of the loop itself.

   We think that it is important to distinguish between the topological entity
that is the loop and the character's passage within the loop. One might think
of the loop itself as a racecourse and the character's passage as a particular
race on that course. To push the analogy – although the specific topology of
the racecourse persists throughout a time-loop narrative, with each re-do the
key character begins a new race wherein different conditions can lead to new
outcomes.

   Richard Lupoff's classic 1973 science-fiction short story "12:01 PM" serves as a
succinct exemplar of time-loop narratives. As the title indicates, the story highlights the crucial temporal juncture that initiates each loop; it begins with the
protagonist, Myron Castleman, hearing a "loud sound" and then noting that
"the clock on the Grand Central Tower said 12:01, as it always did at resumption
time" (Lupoff [1973] 2013, 144). The word "always" and the fact that he has given
a name to the phenomenon ("resumptions") indicate that Castleman has been
caught in the loop for awhile, although we never learn precisely how many iterations he has undergone. Castleman's passage through the loop always lasts
a fixed amount of time – precisely one hour, the ending of this passage through
the loop and the beginning of a new one heralded by "a single loud sound resembling the sound made by the implosion of air into a shattered vacuum tube
or the report of a small-caliber firearm" (148). Castleman then finds himself
once again standing before the clock tower. In the story, we go through five
iterations of this process with him. Within these five iterations, Castleman
makes different choices during the hour (for example, chatting with a woman
at a diner or trying to contact a physicist about the "time bounce" he is experiencing), and these choices are prompted by the fact that he alone has a metaconsciousness that allows him to know that he is looping: "Everybody bounces
back and forgets everything, but I don't. I don't!" (154). At the end of the fifth
iteration (in our reading experience), Castleman apparently dies, but he finds
himself once again back at 12:01.[3]

---

3 The 30-minute film *12:01 PM*, although varying some of the action, follows essentially the
   same storyline. There is also a feature-length version of *12:01*, which was released in 1993, but
   it is only loosely based on the story. Lupoff wrote two sequels to "12:01 PM"–"12:02 PM" (2011)

The overall narrative arc of "12:01 PM" deals with Castleman's attempt to break the hour-long loop. In many of the texts we examine, the desire to "get it right" drives the narratives, and the character's sense of the urgency often intensifies with each successive passage through the loop. We first explore how various alternative choices made by the protagonists as they experience a new passage along the loop map out a veritable garden of forking paths that may threaten to proliferate endlessly (within the physical constraints of the narrative). We next discuss how the temporal, even topological, conflict between endlessly passing through the loop and narrative closure is often mediated by the emergent metaconsciousness.[4] We then examine various permutations of the time-loop motif, from the sentimental to the disillusioned. Finally, we attempt to draw some conclusions about what time-loop narratives have to tell us about the human response to time.

2   The Road Not Taken and the Garden of Forking Paths

We are aware that we cannot step back in time – that we are traveling down a single path, one that leads inevitably to death. We do, however, continually engage in mental time travel, a recursive process whereby "one has projected one's self into the past to re-experience some earlier episode" (Coraballis 2014, 85), embedding in present consciousness the memories of the past. Moreover, in mentally time-traveling into the past, we often imagine the forks that we encountered and the roads that we could have taken, ones that might have led us to different outcomes. As J. T. Fraser has noted, "the idea of a freedom of choice with limitations" may have "originated in the recognition that the paths of action one could have taken in the past were, usually, more numerous than the single path one actually did take" (2007, 237). Michael Coraballis argues that our fictions "allow us to go beyond personal experience to what might have been, or what might be in the future" (2014, 124), and time-loop fiction epitomizes this circumstance. Its very estrangement from our lived experience is what makes time-loop fiction poignantly human, for, in exploring what cannot occur in fact – "to do it again and again" (Atkinson 2013, 446) –, it highlights

---

and "12:03 PM" (2012). In the former, Castleman finally breaks the loop, and, in the latter, he continues to have strange temporal experiences.

4   In "Crossing the Junctureless Backloop," a paper given at the 15th Triennial Conference of the International Society for the Study of Time: Time and Trace," Weissert (2013) described this process as the Temporal Topological Trope – that is, the topological change from a line to a loop or a loop to a line.

what occurs regularly in our mental landscapes as we envision choosing a different fork that culminates in a different outcome.

In his short story of the same name, Jorge Luis Borges eloquently describes *The Garden of Forking Paths*, the mythical "chaotic novel" created by the monk Ts'ui Pên:

> In all fictional works, each time a man is confronted with several alternatives, he chooses one and eliminates the others; in the fiction of Ts'ui Pên, he chooses – simultaneously – all of them. *He creates*, in this way, diverse futures, diverse times which themselves also proliferate and fork.
>
> BORGES [1962] 1983, 26

One way that we could conceive of generating the garden of forking paths aligns with the multiple-worlds hypothesis in which each bifurcation creates new realities splitting away from one another; however, time-loop narratives provide an alternate mechanism wherein each unique path through the garden is generated by the changes from passage to passage within the loop. (See figure 9.1.) In this way, time-loop narratives provide us with a dynamic mechanism for generating the proliferating forks. Although a trapped character lives a linear sequential progression, the garden of forking paths maps the history of the character's passages through the loop.[5]

Indeed, particularly in regard to the more sustained narratives such as *Life After Life* or *Replay*, our reading or viewing experience, although linear, involves our mapping previous passages through the loop onto current ones. We are thus reading or viewing both backwards and forwards as we move to a new passage through the loop: that is, as in a typical reading or viewing experience, we are drawing on our memory of what we have read or viewed earlier as a way

---

5 Although it may seem as if we are dealing with two different topological structures (a racecourse and a garden of forking paths), we would have readers consider that each new race along the course is essentially a different path through the garden, leading to a different outcome. In "Hollywood Goes Computer Game: Narrative Remediation in the Time-Loop Quests *Groundhog Day* and *12:01*" (2011), Martin Hermann differentiates between time-loop narratives and forking-path narratives, citing as examples of the latter such films as *Przypadek (Blind Chance)*, *Sliding Doors*, and *Lola rennt (Run, Lola, Run)*. He points out that time-loop narratives, unlike forking-path narratives, "do not present parallel worlds centering on the question of 'what if' but instead follow the linear logic of trial-and-error" (147). Although we agree with this distinction in part, what we argue is that, in laying out the various paths that a character follows sequentially, time-loop narratives can be mapped as a garden of forking paths, as figure 9.1 demonstrates. A key distinguishing feature between forking-path narratives, such as those mentioned by Hermann, and time-loop narratives, is that there is no metaconsciousness in the former.

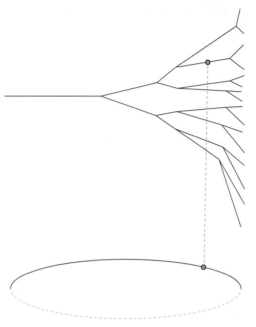

FIGURE 9.1
Each distinct path through the
garden is a variation of decisions
made in one of the passages through
the loop.

of making sense of what is happening in the present of the text, but we are
also anticipating a repetition of events that have occurred in an earlier passage
through the loop. In effect, we are "remembering forward" and thus creating
a deeper context. We are attuned to the fact that what might seem to be an
insignificant event marks a key decision point for a character.[6] We may, in fact,
feel a sense of urgency as the character nears the crucial point, particularly if
our awareness is greater than that of the character.

When talking to the psychiatrist enlisted to treat her problems with time,
Ursula Todd, the heroine of Kate Atkinson's *Life After Life*, provides an apt figure for the temporal dimension in time-loop narratives: "Time isn't circular,"

---

6   In his discussion of foreshadowing, Gary Morson points out that, in life, "we do not remem-
    ber forward" (1994, 46). In some sense, time-loop fiction is predicated on foreshadowing, but
    foreshadowing with a difference. As we move into a new passage through the loop, we know
    what is supposed to happen next, but we know, too, that, once a character reaches a fork,
    some change will occur. Although Hermann argues, "A forking point [...] does not exist in
    time-loop narratives" (149), we would argue that such instances do occur, but that occasion-
    ally such forks may be elided. So, for example, readers or viewers may discover the protago-
    nist in different circumstances during a second passage through the loop without necessarily
    being presented with the fork or decisive choice that has led to these different circumstances,
    yet it is clear that the protagonist has at some point chosen an alternate fork.

she said to Dr. Kellet. "It's like a … palimpsest" (Atkinson 2013, 506). We might even add "a dynamic palimpsest." Traces of a character's previous lives (or passages through the loop) underlie the current life (or passage), enabling the reader to build up a composite portrait. Each passage through the loop is in some way a self-contained plot, having a beginning, middle, and end in good Aristotelian fashion, but the overall plot of each time-loop narrative involves all of the character's passages through the loop working together toward a coherent end – although some narratives problematize our sense of an ending.[7] Significantly, it is often not only the reader who can visualize all the passages through the loop but a character (or in some cases characters). Thus the metaconsciousness.

3   The Emerging Metaconsciousness and the Desire for Closure

In *Life after Life*, Ursula's beloved brother Teddy asks, "What if we had the chance to do it again and again […] until we finally did get it right? Wouldn't that be wonderful?" Ursula responds that "it would be exhausting" and that she "would quote Nietzsche to you but you would probably thump me" (446). In one of the epigraphs for the novel, Atkinson does indeed quote Nietzsche, specifically the well-known passage from *The Gay Science* dealing with Eternal Recurrence, in which a demon proposes, "This life as you live it and have lived it, you will have to live once more and innumerable times more" (Nietzsche, qtd. in Atkinson 2013b).[8] Significantly, however, Atkinson elides the words that follow: "and there will be nothing new in it, but every pain and every joy and every thought and sigh and everything unspeakably small or great in your life must return to you, all in the same succession and sequence" (Nietzsche [1882] 2008, 194). Ursula, like the other time-loopers we examine, is not subject to eternal recurrence but to *quasi-eternal recursion*, whereby each life varies in some way from the lives prior to it and the

---

7   We might consider Hilary Dannenberg's point in *Coincidence and Counterfactuality* that a reader's cognitive desire is "the final configuration achieved in narrative closure when (the reader hopes) a coherent and definitive constellation of events have been achieved" (2008, 13). In "Try Again: The Time Loop as a Problem-Solving Process in *Source Code* and *Save the Date*" (2015), Victor Navarro-Remesal and Shaila García-Catalán note, "Even though the protagonist and the viewer want to believe that the 'happy ending' is the only real ending, both works seem to consider each loop as a valid telling of their stories" (214).

8   Atkinson also playfully uses Teddy's question as an epigraph, ascribing it to "Edward Beresford Todd."

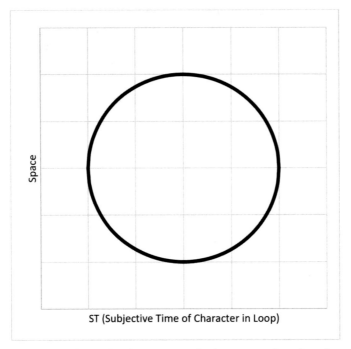

FIGURE 9.2 The simple subjective time of a character caught within a loop.

potential for getting it right pertains. The term "recursion" suggests that
what has gone before feeds into what will come next. But that can only be
true if there is someone aware of what went before and thus has a perspective outside the events of the current passage through the loop – a metaconsciousness. Just as the time frame of the reader is an extra dimension of
time, beyond the experiential time of the characters in the loop, the metaconsciousness, too, must occupy an extra dimension of time. (See figures 9.2
and 9.3.)

The variation during each passage through the loop occurs when the result
of a previous passage influences what happens in a subsequent one – a situation that is enabled by the metaconsciousness. The metaconsciousness manifests in a Nietzschean Über-character with an extra perspective from beyond
the confines of a single loop. With each iteration, variations in actions accumulate and serve to enrich the Über-character for the reader, just as the successive
iterations of a trajectory on a strange attractor tend to enrich our understanding of its infinitely complex structure. It can be an uncanny experience for the
character, as the following description from *Life After Life* indicates, "The past
seemed to *leak* into the present, as if there were a fault somewhere. Or was it

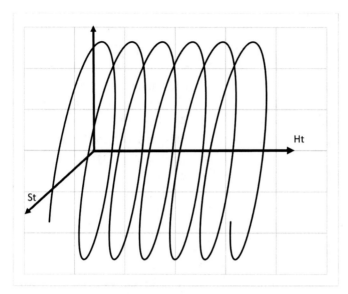

FIGURE 9.3   The same looping as in the previous figure, now with a new
dimension of providing a perspective that allows a character to
distinguish among the passages through the loop.
Note: St = subjective time; Ht = historical (objective) time.

the future spilling into the past? Either way it was nightmarish, as if her inner
dark landscape had become manifest" (505).[9] Depending on the narrative contract and because of its extra perspective, the Über-character may possess the
power knowingly to take actions within the loop that affect the course of the
iteration process itself, as well as actions that shape the future nature of the
underlying character.

For we readers/viewers, variations in each passage through the loop are, of
course, necessary for keeping our interest, but our interest might flag if we are
simply given variation after variation without a purpose. It is perhaps no sur-
prise that time-loop narratives and video games share certain features. Hiroshi
Sakurazaka, in fact, discusses this connection in his afterword to *All You Need
Is Kill*, wherein he discusses why he loves video games: "I reset the game hundreds of times until my special attack finally went off perfectly. Victory was
inevitable" (2004, 199).[10] The game would seem purposeless if there was not

---

9   Weissert (2013) identifies this leakage of information from one passage in the loop to a
      future passage through it as the "Crossing the Junctureless Backloop" Trope.
10   Both Hermann and Navarro-Remesal/García-Catalán discuss time-loop narratives in
      light of computer games. Hermann notes that, "whereas computer games turn users into

ultimately a way of winning. The narrative would lose our interest if there was not a resolution.

Ultimately, we expect and desire closure (and not simply because a text is finite). Teddy's point about doing it "again and again" until we finally "get it right" is germane. The metaconscious Über-characters in time-loop narratives are driven by the aim of getting it right or putting things *to* rights, which in most cases entails finding a way to break the loop and bring the story – and their story – to closure. The metaconsciousness both enables this aim and, in some cases, arises from it, effectually emerging from an emergency – a threat to one life, a threat to many lives, a threat to the world.

As we earlier discussed, a temporal conflict between endless passages through the loop and narrative closure arises in time-loop narratives, and this conflict is related to the metaconsciousness. Another temporal conflict, also related to the metaconsciousness, is that between the seemingly infinite time of recurring passages through the loop versus the finite time within the loop during which certain actions must be completed. In *All You Need Is Kill*, the protagonist's complaint, "I had a whole day to repeat, but I didn't have enough time when I needed it" (Sakurazaka 2014, 39) sharply sums up this second conflict, which we explore in more detail below.

4   Emergency, Recursion, and Getting It Right

A character's (or characters') awareness about being stuck in a loop varies – from a nascent impression, hardly an awareness at all, to one that is fully aware and in place at the outset – as does the character's success in getting it right and bringing the story to a satisfying close. As we will see, however, success does not necessarily depend on how heightened a character's awareness is, but the writer's or filmmakers' approach to the implications of eternal recursion.

4.1   *Saving the Day: The Sentimental Conclusion*
Many time-loop narratives, particularly those that have a greater popular-culture appeal, feature plots wherein the accumulating knowledge of the metaconscious character or characters leads to a traditional happy ending. The notion is that, by passing enough times through the loop, the metaconscious character can begin to figure out exactly when and how mistakes were made and what exactly they can do to fix them and thus, it is hoped, break the

---

characters, filmic narrative remediations of computer games turn characters into users" (2011, 152) – in effect, gamers who get to keep resetting the game.

loop.[11] For the reader/viewer, these plots offer the reassuring notion that there is somehow a right path, one that leads to fulfillment, if one can only manage to figure out which it might be.

The *Star Trek: The Next Generation* episode "Cause and Effect" provides a good example of this plot structure. It features an example of an emerging, but never fully realized, metaconciousness – in this case, on the part of several characters. A collision with another spacecraft causes an explosion that has propelled the starship *Enterprise* into a time loop, thus destroying the future-going *Enterprise*. Within the loop, which seems to last less than a day, select members of the crew begin to experience odd instances of what seem to be déja vu while they are playing poker – for instance, one character predicts accurately when another is bluffing, and, during a later iteration, predicts the exact order of the cards. After several iterations, with their typical *Star Trek* acumen, the crewmembers figure out that they are in a loop and determine the cause, but they realize that they are unable to carry over their knowledge into the next passage through the loop. However, the android Data's "positronic brain" serves as a fleeting metaconsciousness, for he is able to send a brief, ambiguous message to himself in the future iteration, one that enables him to take the measures that will allow the *Enterprise* to avoid the collision and prevent it from entering the loop in the first place.[12]

The film *Source Code* also enables the protagonist to put things to rights and break the loop.[13] Interestingly, it features the briefest fixed interval in the narratives that we examine – just eight minutes. Severely wounded veteran Colter Stevens has been charged by a secret government organization with discovering the identity of a bomber who has blown up a commuter train in order for the organization to stop the bomber from detonating bombs in the future. His mind embedded in the consciousness of Sean, one of the now-dead commuters, through the Hollywood miracle of "parabolic calculus" and "quantum physics," Colter is sent back to the last eight minutes of Sean's life, eight minutes being the interval in which things can be stored in short-term memory. Reliving the same interval over and over, Colter draws upon his previous passages through the loop in order to come closer to identifying the bomber, who is presumably still on the train during the crucial time period.

---

11 As Navarro-Remesal/García-Catalán note, "Diegetically, the time loop becomes a problem-solving process" (2015, 206) – at least for the "save-the-day" narratives that we discuss here.

12 After the crewmembers are sprung from the loop, they estimate that they have been in it for approximately two weeks. Writer Brannon Braga accounts for the instances of déja vu as caused by distortions in the "dekyon field" that surrounds the ship.

13 Navarro-Remesal and García-Catalán (2015) provide a detailed discussion of the plot.

Furthermore, recursive interchanges with the young woman sitting beside him give him an additional goal – saving her life – because (through the miracle of almost instant attraction) he has come to care for her. Thus two imperatives drive the plot, and, in both cases, because the reality outside of Sean/Colter's memory continues to progress in time, Colter's task becomes increasingly urgent: he must find the bomber before the ever-multiplying eight-minute passages converge with the time of the next explosion in the "real" world, and disable the bomb on the commuter train itself, which is technically only in Colter's mind. Happily, drawing on his metaconscious awareness of previous passages through the loop, Colter is eventually able to identify the bomber and so prevent future bombings *and* stop the original bomb on the commuter train from detonating. But does Colter thus change history? No. Through the miracle of "the source code," he brings about an alternate reality à la many-worlds theory.[14]

In *All You Need Is Kill* and its film incarnation *Edge of Tomorrow*, the imperative is to defeat the invading aliens and only then to end the loop. In the novel, Keiji Kirikya is caught in a time loop via a tachyon pulse that can travel back in time. He receives "a portent, or window into the future" (Sakurazaka 2004, 121) that enables him to modify his behavior with each time he seemingly repeats his day; the loop resets approximately every 30 hours (specifically, once he has been killed in battle by the Mimics, the extraterrestrial force he is fighting). As former time-looper Rita Vrataski (a.k.a. the Full Metal Bitch), tells him, "from the point of view of the person with the memory, there was no difference between having had an actual experience and only having the memory of it" – a point that Keiji thinks is "philosophical bullshit" (142). Keiji ends up passing through the loop 160 times, and, although he realizes that he does not "have time to waste building muscle that didn't carry over into the next loop," his metaconsciousness enables him to spend part of each new passage through the loop "programming [his] mind for battle" (58), recursively drawing on his memory of the previous battle in order to make strides in the next passage and, eventually mimic Rita's battle prowess. Ultimately, attaining the goal becomes crucial for Keiji. The Mimics end up changing the playbook;

---

14   The time-loop motif is complicated by a transfer-of-consciousness motif, and there are more than a few gaping plot holes related to the alternate-reality scenario. Also, although the young woman feels a reciprocal attraction for Colter, as far as she is concerned, she is attracted to the man with whom she has been sitting during presumably many commutes – enough, anyway, for her to have revealed quite a few personal details to him. For an entertaining discussion of the film by physics professor Jim Kakalios, see the following: http://www.preposterousuniverse.com/blog/2011/04/12/guest-post-jim-kakalios-on-the-quantum-mechanics-of-source-code/.

"You're not the only one that remembers what happened in each loop" (167), Rita tells him. Rita has also realized that, "If humanity is going to win, we need someone who can break the loop" (181) by killing the Mimic server. Whether it is she or Keiji, only one can survive, one who has vanquished the other – and it must happen during the battle at hand.[15] Keiji, who has "learned to fight the Mimics watching Rita" (186), manages to land the killing blow but, smitten, he considers dying into another passage through the loop, asking Rita if they can "just keep repeating this," staying together forever, "longer than a lifetime" (187). Pointing out that every morning he'd "wake up to a Rita Vrataski who doesn't know you exist," she persuades Keiji "not to sacrifice the human race for himself" (188), and he goes on to kill the Mimic server, win the battle, and break the loop, resolving that he will "not let the world lose" (195) and that he will win the war – still ongoing – for Rita.

Although both novel and film culminate in the loop being broken, the film ending is much more in the traditional happy-ending vein. The film changes the loop-generating mechanism in that the so-called Alpha Mimic has the ability to reset time, an ability transferred to William Cage, the film avatar of Keiji.[16] Winning the final battle takes on urgency because a blood transfusion has disabled Cage from resetting the day. Due to the extensive training he has received at the hands of Rita, Cage is able to kill the Omega Mimic, and, despite his warning to Rita that, if he does kill it, "you're dead forever," the filmmakers pull out a Hollywood ending that allows both Cage and Rita to survive and even meet cute in the final scene, wherein Cage, apparently bent on explaining what has actually transpired, approaches a Rita who does not recognize him – but, we are led to believe, will believe him and possibly fall for him, as she did in previous passages through the loop.[17] Notwithstanding their differences, the plots of both novel and film focus on the protagonists' putting things to rights – and being able to do so by virtue of the metaconsciousness.

The plot of the 1993 film *Groundhog Day*, perhaps one of the most well-known and beloved of time-loop narratives, also is predicated upon getting it right. It is, however, a long, strange trip that gets us to that point, and the clear cause-effect of metaconsciousness-accumulated knowledge leading to the

---

15   What we hope has become apparent by now is that explanations for the mechanisms behind time loops are often predicated upon scientific gobbledygook if they are not elided completely. Despite the somewhat elaborate explanations that occur in both novel and film, we may well agree with Keiji that "None of it made any sense" (181).

16   In a nice tip of the hat to the novel, "Cage" is the way in which American soldiers pronounce Keiji's name.

17   As with *Source Code*, the filmmakers finagle the ending, allowing for one more reset – one that takes Cage back to an earlier time than the original loop started.

desired outcome is lacking. Nevertheless, when the acerbic television weather-man Phil Connors finds himself in a loop on, of course, Groundhog Day, we viewers cannot help but make a post-hoc assumption that somehow his bad behavior has led to this plight. What follows involves a variety of widely different trajectories: in some passages through the loop, Phil uses accumulated knowledge for suspect ends (maneuvering a gullible woman into bed, for example); in some, he is beset with despair over his seemingly eternal plight, even to the point of futilely attempting suicide; and in some he begins to improve himself, including taking piano lessons.[18] Ultimately, the attempts at self-improvement become paramount, as Phil becomes increasingly attracted to his producer Rita, and he sets out to win her love. Despite various missteps, he finally becomes the man she can admire, and Phil wakes up to find her in bed beside him on February 3, the loop finally broken. Once again, we viewers supply the cause-effect, seeing Phil's gradual development into a mensch as enabling him to break out of the loop.[19]

Despite their differences, what the narratives above have in common is the reassuring message that, with a metaconscious perspective, we could choose the right path and reach the desired destination. In the texts that follow, we enter more murky textual waters.

4.2   *Killing Hitler: The Ambiguous Conclusion*

In *Life after Life*, Ursula's metaconsciousness gradually emerges. Like the characters in "Cause and Effect," she has inklings that certain events are repeating.

---

18   Phil's inability to die and thus break the loop has affinities with Castleman's in "12:01 PM" and its film version. Interestingly, Lupoff and Jonathan Heap, the director of the short film, claimed that *Groundhog Day* had been based on their idea, but they eventually dropped the idea of filing a lawsuit.

19   Hermann discusses *Groundhog Day* in terms of an "internal quest" structure whereby "Phil's internal search ... leads him to the insight of what matters in life and how to become a content human being" (2011, 158). How long Phil stays in the loop has been a matter for much fan conjecture. Although in a DVD commentary director Harold Ramis estimated ten years, he later revised the number upwards. See http://birthmoviesdeath.com/2011/02/02/how-many-days-was-phil-connors-stuck-in-a-groundhog-day-loop.
Screenwriter Danny Rubin has made the following comment: "I know that I have been quoted as having originally intended for Phil to have lived 'ten thousand years,' a time-frame with Buddhist overtones. I find that so incredibly cool that I put no effort into disputing it. But it's not true. For me, any lifetime for Phil longer than one would have sufficed, and even so, that statistic never had to leave my head. As long as the audience understood it to be a very, very long time, it never had to become specific." See http://www.dannyrubin.com/blogusgroundhogus/2008/01/29/the-magic-of-friendship/. Of course the possibly 10,000 years of looping is brief in relation to the looping done by the Twelfth Doctor in the "Heaven Sent" episode of *Doctor Who* – four and a half billion years.

Like them, too, her incipient sense of past lives arises at crucial moments. Yet, unlike the narratives discussed in the above section, the forking paths in *Life After Life* at times seem simply that, and an attempt at getting it right occurs only when Ursula has achieved total metaconsciousness near the end of the novel.[20] The novel thus suggests that, although choosing a different path may lead to a different outcome, there may be trade-offs, and that no one path can bring about the most desirable conclusion.

Ursula Todd lives her life again and again (always beginning at birth), although, rather than living through a fixed period, she lives until death catches her, signaled by the oft-repeated phrase "Darkness fell." Her first life stops before it begins as she is strangled at birth by the umbilical cord around her neck, her longest life lasts until she dies of a stroke at age 57, and there are all sorts of time periods in between, from four years to 27, including one awful yet mordantly funny sequence when, after the Armistice is declared, Ursula dies of influenza at age at age eight again and again.

Whether Ursula lives or dies at a certain point depends on the random fluctuations that occur in all our lives. At her first birth, for example, she dies because the doctor has not arrived in time; at her second, he is there to snip the cord strangling her. A doctor's presence or a pair of scissors can mean the difference between death and life, as we see when, in the penultimate birth scene Ursula's mother herself snips the cord with a pair of surgical scissors – an indication perhaps that others besides Ursula may have their metaconscious moments.

Throughout the novel, Ursula is confronted with forking paths that lead to wildly different outcomes. When embedded in a particular passage through the loop, she generally has no conscious idea that there is indeed a fork in the path, but we readers, by virtue of our own metaconscious perspective (our ability to oversee all the passages that she has made), know when Ursula has reached one, recognizing in a later iteration what made the difference in an earlier one and feeling some urgency with regard to what her decision will be. As Ursula reaches a key bifurcation point a second (or third or fourth) time in her life, an incipient metaconsciousness does seem to guide her so that she chooses a different path than she had earlier. In one of the first examples, Ursula dies at age five when she slips off the roof while attempting to rescue the doll her brother had tossed out the window; in the next life, she hesitates before plunging out of the window, and her sister helps her rescue the doll with a net. Later, in the first

---

20   In some sense, *Life After Life* is somewhat of a hybrid between a time-loop narrative and what Hermann classifies as a forking-path narrative like *Sliding Doors*, but the metaconsciousness that Ursula eventually obtains shifts it into the time-loop camp.

sequence entitled "Armistice," Ursula dies of the Spanish flu after she contracts
it from the maid Bridget. In the next iteration, something stops her from going
downstairs to "interrogate Bridget about her high jinks":

> As she lay listening to the dark, a wave of something horrible washed
> over her, a great dread as if something truly treacherous were about to
> happen. [...]
>   [...] They mustn't see Bridget. Ursula didn't know why this was so,
> where this awful sense of dread came from, but she pulled the blankets
> over her head to hide from whatever was out there.
>   ATKINSON 2013b, 91

Despite her attempt to save herself, Ursula ends up dying of the flu once
again. It takes three more iterations before Ursula finally pushes Bridget
down the stairs so that the maid breaks her leg and cannot mingle with the
disease-ridden Armistice Day crowds. "Practice makes perfect," Ursula thinks
to herself, but she is also consumed with guilt at the unaccountable "wicked
thing" she has done: "All she knew was that she *had* to do it. The great sense
of dread had come over her and she had to do it" (124). Ursula's aunt speaks of
these experiences as déja vu, and the psychiatrist she is made to visit spouts
Buddhist philosophy. Yet Ursula remains puzzled by the "obscure memories
of elation, of falling into darkness" that belong "to that world of shadows and
dreams that was ever present and yet almost impossible to pin down" (156).

Perhaps the most poignant illustration of the way in which an unmarked
fork can make a difference in Ursula's life occurs in the section "Like a Fox
in a Hole," which deals mainly with what we might call life 9. On her sixteenth birthday, Ursula is kissed by her brother's brutish friend Howie, and,
a few months later, he cavalierly rapes her. The rape leads to a pregnancy and
an abortion that almost kills her, with Ursula barely comprehending what is
happening to her body, and she subsequently feels shame, culpability, and
worthlessness. She leaves school, attends a secretarial college run by a sexual
harasser, becomes a secret drinker, marries a man out of "relief that someone
wanted to look after her, someone who knew nothing of her shameful past"
(215), and endures physical and psychological abuse by him until he smashes
her head into a coffee table and kills her.

In the midst of this dreadful existence, Ursula has the sense that she
somehow made a wrong choice at some point: "She no longer recognized
herself, she thought. She had taken the wrong path, opened the wrong door,
and was unable to find her way back" (231). This line makes clear that, even
if there is no one right path, there can be wrong ones, and that the seemingly

insignificant effects of a choice can amplify into major consequences as one continues down "the wrong path." Happily for both Ursula and the reader, after the scene of her murder, Atkinson, bypassing the first sixteen years of Ursula's next life, returns to a revised encounter between Ursula and Howie. Rather than letting him kiss her, Ursula punches him, and, when he complains, "It wasn't like I was trying to *rape* you or anything," she thinks to herself that rape "was what boys like Howie did to girls like Ursula" and that "girls, especially those celebrating their sixteenth birthdays, had to be cautious when walking through the dark, wild wood" (241). Although she ultimately dies in the Blitz during this life (her tenth), she spends the intervening 14 years as a self-possessed and independent woman, leading a revolt against the harassing typing teacher, engaging in mutually satisfying relationships, and making a decent career for herself in the Home Office. The awfulness of life 9 is reinforced when Nancy, a young neighbor girl who is the "twin soul" (201) of Ursula's brother Teddy, is found raped and murdered. In life 10, however, on the day that Nancy was murdered in the previous life, a casual phrase triggers a "familiar but long-forgotten terror" (202) in Ursula, and, compelled by the feeling, she walks along the lane where Nancy was accosted during Ursula'a previous passage through the loop, meets Nancy, and scares away a strange man, presumably the killer.[21]

Not until what seems to be life 15 (out of a possible 17 detailed in the novel) does Ursula actually have some idea of what might be taking place in her life (or lives). In life 15, engaged in a tryst with a handsome young neighbor, Ursula misses Nancy in the lane on the fateful day, and, once again, Nancy's violated corpse is discovered. Ursula's sister tells the guilt-wracked Ursula that she is not responsible, but Ursula thinks "she was" and that "something was riven, broken, a lightning fork cutting open a swollen sky" (504). Ursula's accession to metaconciousness becomes fully realized when she encounters the man who was her murderous husband in life 9: "The past seem[s] to *leak* into the present" (505), and she flees in panic. Consigned to a sanitorium, she comes to that crucial realization that time is a "palimpsest" (506).

At this point, Ursula seems to have finally achieved metaconsciousness. She thinks to herself that "She had wasted so much time but she had a plan now" (507) – a complete plan for the first twenty years of her (next) life. During life 10, Ursula had mused to her lover Ralph about what might have happened "If just one small thing had been changed, in the past. [...] If Hitler had died

---

21   With Ursula's life 9 in particular, Atkinson explores the issue of male violence against women, a common motif in her novels. In addition to what happens to Ursula and Nancy, an unknown young girl is found sexually violated and strangled by a killer who is never caught (possibly the same one who later strangles Nancy).

at birth, or if someone had kidnapped him as a baby and brought him up in
[…] a Quaker household" (277). Ralph had responded that Hitler "might have
turned out the same, Quakers or no Quakers," and had then asked Ursula if she
could kill him "[i]n cold blood." Ursula had thought, "If I thought it would save
Teddy. […] Not just Teddy, of course, the rest of the world too." Her plan thus
harkens back to this earlier discussion, for it involves her getting a German
degree, doing target practice, taking a trip to Germany, and deliberately meeting Eva Braun so that she can eventually kill Hitler and presumably avert the
terrible world conflagration that looms ahead.[22] Recalling various people who
were doomed in previous lives, she seems filled with an awareness of all that
has transpired in her forking-paths existence and channels Nietzsche: *"Become
such as you are, having learned what this is"* (509; Atkinson's italics). She then
jumps to her death, thinking, "This is love" (509), aware that she will be reborn
to put her plan into action.

Ursula's plan thus has nothing to do with breaking the loop, but it does seem
to be an attempt at getting it right. In the subsequent scene, set in 1930, Ursula
manages to get off a shot at Hitler before being killed by his bodyguards, a fate
signaled by the ubiquitous phrase "Darkness fell" (515) However, this is not the
final scene in the novel, and, in fact, the novel actually opens with a variation
of the Hitler-killing scene that is apparently unanchored to any of the actual
lives that follow, thus suggesting that this attempt may be ongoing. In an interview, Atkinson herself suggests that Ursula does not achieve her goal within
the bounds of the text:

> My mantra is there's nothing beyond text, but I've not stuck to that, I
> think, in this book. That's in my head – *Life after Life* goes on and she's
> still out there somewhere in 1910 being born in a snowstorm and there
> is still a possibility that she will kill Hitler. And that to me was incredibly
> satisfying. (2013a)

So there is no actual end.

---

22   In the section "The Land of Begin Again," during what seems to be life 13, Ursula actually
       lives from 1933 to 1945 in Germany and spends a stultifying holiday in the Berghof with
       Eva and Hitler. Despite the fact that she has already died three times during the Blitz and
       that she finds Hitler a rather repulsive "megalomaniac," because her metaconsciousness
       is undeveloped at this point, she can only muse idly that "it would be easy enough to get
       hold of a Luger and shoot him through the heart or the head" (361).

Yet we would suggest that the text nevertheless achieves a provisional closure. In the penultimate section, "The Broad Sunlit Uplands,"[23] Ursula's brother Teddy, shot down in flames in other lifetimes of Ursula, and his soul-mate Nancy, violated and murdered in some of Ursula's passages through the loop, are reunited in a pub at the end of the war. Teddy gives Ursula "a smart salute" and then shouts something to her: "She thought it was 'Thank you,' but she might have been wrong" (525). The sequence implies that Ursula's agency has somehow brought about this desirable conclusion – but the very next section brings us back to the day of Ursula's birth, leaving us with the impression that this is not the last word on Ursula's lives.[24]

Despite the provisional closure effected by the penultimate scene, *Life After Life* seems more intent on detailing the various paths upon which an Englishwoman born in the early twentieth might embark than on exploring how multiple do-overs can lead to "getting it right."[25] Ursula herself may seem to get it right in one lifetime but actually get it wrong in a subsequent one, as Nancy's murder in life 15 indicates.

Interestingly, during the discussion with Teddy about getting it right, Ursula says, "We can never get it *right*, but we must *try*" (446). In the section that follows, we look at narratives that indeed confirm Ursula's point, that undermine the idea that one could ever really get it right, even with a metaconscious perspective.

4.3   (*Not*) *Saving Kennedy: The Disillusioned Conclusion*

Although *All You Need Is Kill* features a save-the-day time-loop, it highlights the existential plight of the looping protagonist. Keiji desires an ultimate release,

---

23 This section title and other section titles "A Long Hard War" and "Be Ye Men of Valor" are phrases from key speeches Winston Churchill made during World War II (The Churchill Centre: Speeches of Winston Churchill), reinforcement of Atkinson's point that "The Blitz may be the dark bleeding heart of the novel [...]" (2014, 6).

24 Atkinson herself has said, "And somewhere in that past, in the ethereal world of fiction, it is always a snowy night in February 1910 and Mrs. Haddock [the midwife] is always settling down to her third tot of rum" (2014, 7). In her subsequent novel, *A God in Ruins*, Atkinson returns to the story of Teddy and Nancy, seeming to confirm what happens in "The Broad Sunlit Uplands," but, unsurprisingly, she has a few surprises in store for the reader.

25 One wonders whether this different focus leads to Atkinson's work being regarded as "literary" fiction rather than fantasy/science fiction. Atkinson has said, "Seeing [Ursula] go through the war years multiple times – not only in her native England, but also in Germany – highlights the female experience of these events" (2013b, 139). Atkinson also has pointed out that the novel details "the making of a heroine, and it's a slow birth" (139).

but, when he shoots a 9mm bullet into his mouth, he wakes to find that his life has reset and bemoans his lack of choice:

> Day after day I go back and forth between the base and the battlefield, where I'm squashed like a bug crawling on the ground. So long as the wind blows, I'm born again, and I die. I can't take anything with me to my next life. The only things I get are my solitude, a fear that no one can understand, and the feel of the trigger against my finger.
>
> SAKURAZAKA 2004, 54[26]

The existential despair besetting Keiji is emphasized in narratives where getting it right shifts to simply "making it different."

Ken Grimwood's novel *Replay* features a character who finds himself caught in a loop wherein (initially) he keeps reliving the 25-year period between age 18 and age 43. In his first life (where we begin), Jeff Winston dies at age 43, childless, a dissatisfied man in a loveless marriage; he wakes up after death as his apparently 18-year-old self, accrues millions, marries a trophy wife, and fathers a daughter he adores, only to die again at age 43. Seven more passages through the loop follow, each shorter than the last because of the "skew" that, with each reset, places Jeff at successively later points in his 43-year life trajectory and, in each of these lives, different choices lead to different outcomes – although the ultimate outcome is always death.[27] At the end of his ninth life, a flurry of waking and dying besets Jeff, until he wakes a final time to discover that the clock has advanced from the moment when he has always died and that he has now resumed what seems to be the life that he was leading at the outset of the novel.

Although Jeff may wish to get it right, he finds instead that he just makes it different. Metaconscious from the outset, during his second passage through the loop, Jeff simply tries to make a viable new life for himself, aided by his (fore)knowledge of the outcome of sporting events on which he can place enormous bets. When he attempts to influence the past in one significant way, Jeff finds that his efforts fail; he alerts authorities to the threat that Lee Harvey

---

26   For some inexplicable reason, the markings Keiji scrawls on his hand to indicate how many times he has gone through the loop survive the loop.

27   Brian Richardson notes, "The fundamental nature of the difference between fiction and nonfiction is most prominent once death appears" (2012, 23). Grimwood's, Sakurazaka's, and Atkinson's protagonists all die to be born again, as do the characters in the various films and telefilms that we have examined. As Navarro-Remesal and García-Catalán point out, "Time and death are immovable, but the idea of a time loop dismantles and diminishes them" (2015, 212).

Oswald poses to Kennedy, but a different assassin simply takes the incarcerated Oswald's place and kills Kennedy.[28] Interestingly, as if making an implicit nod to Heraclitus's aphorism on stepping in the same river twice, the novel shows that Jeff cannot repeat his own past even if he would like to. For example, he engineers a meeting with Linda, his wife from his first life, on the day that they had originally met, and he attempts to revive the original attraction. However, discomposed by Jeff's boast of financial success and his uncanny knowledge about her, Linda ends up regarding him as "some kind of psycho" (75) and cuts off communication with him. When Jeff dies at the same moment that he had in his previous life and awakens from his second death to find himself 18 again and on a date, he is filled with a feeling of loss: "Everything he'd accomplished had been erased, his financial empire, the home in Dutchess County ... but most devastating of all, he had lost his child" (87). And in the next lifetime, devastated by the apparent erasure of his daughter from existence, he marries his college sweetheart, undergoes a vasectomy so as not to undergo the heartache of a child's negation, and adopts children who "already existed" and who would "still exist, though with a different life in store for them" (99) even if he died – which he once again does at age 43 despite having himself hooked up to electrodes and monitored during the crucial death-zone time period.

Jeff may be able to predict the moment of his death, but he cannot predict the impact of his actions, and, in fact, a concerted effort to get it right often entails the opposite effect, the metaconsciousness thus unable to guide him to a desired end. When his attempt to avert the reset in life 3 fails and he begins a fourth life, he determines that "[n]othing mattered" (104), and turns first to hedonism and then solitude before meeting fellow replayer Pamela and creating an idyllic, albeit short, life together with her. Although, after a somewhat rocky start, their next life together promises to be equally idyllic, their curiosity about whether there are others replayers and what is causing the "skew" leads them to discover an insane replayer who uses his replays to get away with mass murder, and this discovery makes "the difference between a lifetime of happiness and one of almost unrelieved anxiety" (218). When, in the following lifetime, the two attempt "with all good intentions" (253) to bring public attention to replaying by making accurate predictions about things to come, matters go horribly awry: the 1988 United States ends up in a state of martial law, the Golden Gate Bridge has been destroyed, and a massacre has occurred at

---

28   Speaking to his fellow replayer Pamela, Jeff opines, "it's impossible for us to use our fore-
      knowledge to effect *any* major change in history. There are limits to what we can do; I
      don't know what those limits are, or how they're imposed, but I think they're there" (144).
      Such are the rules of the game in this text.

the United Nations Building.[29] Jeff manages to make something of each of his lives, but he discovers that his attempts to control the direction they take generally backfire, and that, with each reset, he loses all that he has gained in the intervening years, retaining only the bittersweet consolation of his memories.

In the final pages of the novel, the skew accelerates to such an extent that Jeff dies and wakes "again, again, again: waking and dying, awareness and void, alternating almost faster than he could perceive" (305) – until the loop breaks. Why it breaks is no more explained than why it began in the first place, and Jeff's agency regarding both is nil, as is his agency in effecting a desired outcome. As Pamela tells him when they realize that, because of the skew, they may not meet again, "I got so used to the endless possibilities, the *time* ... never being bound by our mistakes, always knowing we could go back and change things, make them better. But we didn't, did we? We only made things different" (287). Despite this outcome, *Replay* does provide readers with the satisfaction of closure because the loop breaks. In an interesting twist, however, Jeff's realization that now "there was only *this* time, this sole finite time of whose direction and outcome Jeff knew absolutely nothing" (309–10) is balanced by his concurrent realization that "the possibilities [...] were endless" (310). All the forking paths now lie before him – but he will only be able to take one from this point on.[30]

For the most disillusioned – even bleak – conclusion of a time-loop narrative, we loop back to where we began: "12:01 PM." This story conveys the greatest sense of urgency because the shortness of the loop (one hour) makes it difficult, indeed ultimately impossible, for Castleman to effect the significant change of breaking the loop.[31] He may make changes each time through the loop, but when he attempts to contact Rosenbluth, the physicist

---

29   Particularly in the later parts of the novel, Jeff muses about whether, with each replay, a divergent time-line is created – so that, for example, the daughter from his second replay might still exist. Thus Grimwood makes a nod to many-worlds theory, but, keeping focalization on Jeff (and occasionally Pamela), he does not confirm or deny whether it pertains to what is occurring to the characters in the novel.

30   In an anticlimactic and unnecessary sequence, this provocative line is followed by a brief epilogue wherein a new character awakens from death to discover that he is replaying.

31   Castleman muses about the problem entailed by the shortness of the loop: "Castleman wished that the resumptions came further apart, really an hour wasn't long enough to do much. But then, he thought philosophically, it could be a lot worse. Hung up at a period of five minutes, he's never get *anything* done. And if it were *really* short – say, a second or less – it would be a living hell" (Lupoff [1973] 2013, 46). Interestingly, Colter in *Source Code* manages to effect change in the eight-minute loop to which he is subject, but his task is made easier because he is confined to a small space in which to maneuver.

who has predicted the time-bounce phenomenon, in order to find out if there is a way of averting it, Castleman discovers that he does not have the time to engage in a meaningful interaction: "What frustration [...] if he ever did succeed in making Rosenbluth realize that the strange phenomenon he had theorized was an actuality. [...] [a]t the end of the hour, the next resumption would find Rosenbluth as ignorant as ever [...]" (Lupoff 1973, 156–57). Despite this metaconscious knowledge, Castleman's time in each successive passage through the loop takes on increasing urgency, and, in the final iteration occurring in the story, he frantically rushes from New York City to the Long Island University in hopes that, by explaining to Rosenbluth in person, he can somehow bring the endless looping to an end. But his rushing seemingly brings on the heart attack that kills him. Apparently dying, Castleman is relieved that death will "bring him dissolution and release from the terrible form of immortality that fate had thrust upon him" (60), but, alas, he finds himself once again back at 12:01, his dying body returned to its previous form at the beginning of the loop yet his memory of previous passages through the loop remaining intact. Just as Castleman realizes that ultimately any choices he makes are futile, we readers realize that as well – which may indeed be the point of the story.

In some sense, the time-loop narratives involving disillusionment or futility may be the most realistic of the genre in that they make clear that, even if we had a metaconscious view, we might find that no path leads to a happy ending.

5   Conclusion

Although exploring situations impossible in the real world, time-loop narratives, with their focus on the metaconsciousness, mimic the human condition. In the present day, when we think back to our past actions and imagine alternative scenarios had we made different choices, we are employing a metaconsciousness that seemingly can stand outside our time-bound existence. As recent neuroscience studies have shown, such are the workings of memory:

> episodic memory supports the construction of future events by extracting and recombining stored information into a simulation of a novel event. Such a system is adaptive because it enables past information to be used flexibly in simulating alternative future scenarios without engaging in actual behavior.
>
> SCHACHTER AND ADDIS 2009, S111

Time-loop fictions presuppose that what we regard as a one-time-only linear path is equivalent to other journeys along different paths aided by our metaconsciousness.

Additionally, the existential plight faced by the looping characters is not so different from the plight that besets us all. Pamela, Jeff's fellow replayer in *Replay*, makes this point as she and Jeff ponder their powerlessness to stop looping:

> Our dilemma, extraordinary though it is, is essentially no different than that faced by everyone who's ever walked this earth: We're here, and we don't know why. We can philosophize all we want, pursue the key to that secret along a thousand different paths, and we'll never be closer to unlocking it. (158)

Discussing how the reset wipes all achievements away, Jeff notes, "All life includes loss. [...] But that doesn't mean we have to turn away from the world, or stop striving for the best that we can do and be" (159).[32] This notion of the best that we can do and be indeed seems a hallmark of many time-loop narratives.

In our own lives, we can only speculate whether a particular choice would have led or would lead to a better or worse outcome, and we cannot know which choices might be crucial or whether, in fact, every choice is crucial.[33] Although our episodic memory enables us to move backwards and forwards in time, we ultimately cannot move beyond our time-bound existence. There is no map of our own garden of forking paths; we cannot foresee the ramifications of each choice we make. Time-loop fiction, however, can satisfy our desire to explore the forking paths barred to us by our singular route along time's arrow yet still give us the fulfilling sense of an ending.[34]

---

32   The reader/viewer may be similarly beset by feelings of loss when we invest in a character's progress and then, like the character, must start over with a new scenario. More than one reader of *Life after Life* whom we have encountered has asked what the point was, what was at stake if Ursula would die and then just start again.

33   The classic example of the crucial nature of seemingly insignificant events (the so-called "butterfly effect") occurs in Ray Bradbury's short story "The Sound of Thunder" ([1952] 1962), wherein a time traveller accidentally kills a butterfly during a dinosaur hunt and returns to his present to discover drastic changes. As one character explains, "The stomp of your foot, on one mouse, could start an earthquake, the effects of which would shake our earth and destinies down through Time, to their very foundations."

34   Jo Parker would like to thank the Saint Joseph's University Board on Faculty Research and Development for a summer grant that enabled her to work on this project.

References

Atkinson, Kate. 2014. "Kate Atkinson on *Life After Life* and Researching World War II." In Reading Group Guide, 5–11, Appendix to *Life After Life*. New York: Back Bay Books.

Atkinson, Kate. 2013a. "Kate Atkinson – *Life After Life* – Costa Novel Award 2013 Winner." http://www.kateatkinson.co.uk/book_detail.php?b=Life_After_Life.

Atkinson, Kate. 2013b. *Life After Life*. New York: Little, Brown and Company.

Atkinson, Kate. 2013c. "*PW* Talks with Kate Atkinson: The Making of a Heroine." By Martha Shulman. *Publisher's Weekly*, February 25, 139.

Borges, Jorge-Luis. (1962, 1964) 1983. "The Garden of Forking Paths." In *Labyrinths: Selected Stories and Other Writings*, edited by Donald A. Yates and James E. Irby, 19–29. New York: Modern Library.

Bradbury, Ray. 1962. "A Sound of Thunder." In *R Is for Rocket*. New York: Doubleday. https://docs.google.com/document/d/1XFtrc-PgR8XPbKtU5j–HnzYNydHbub-Q9EEnomNO8CI/edit.

"Cause and Effect." 1992. *Star Trek: The Next Generation*. Teleplay by Brannon Braga. Dir. Jonathan Frakes. Perf. Patrick Stewart, Jonathan Frakes, Gates McFadden. Paramount.

The Churchill Centre. http://www.winstonchurchill.org/resources/speeches.

Coraballis, Michael C. 2014. *The Recursive Mind: The Origins of Human Language, Thought, and Civilization*. Updated edition with a new foreword. Princeton, NJ: Princeton University Press.

Currie, Mark. (2007) 2010. *About Time: Narrative, Fiction and the Philosophy of Time*. The Frontiers of Theory. Edinburgh: Edinburgh University Press.

Dannenberg, Hilary. 2008. *Coincidence and Counterfactuality: Plotting Time and Space in Narrative Fiction*. Lincoln, NE: University of Nebraska Press.

Edge of Tomorrow. 2014. Screenplay by Christopher McQuarrie, Jez Butterworth, and John-Henry Butterworth. Dir. Doug Liman. Perf. Tom Cruise, Emily Blunt, Bill Paxton, Brendan Gleeson. Warner Brothers Pictures.

Faraci, Devin. 2011. "How Many Days Was Phil Connors Stuck in GROUNDHOG DAY Loop." *Birth. Movies. Death.* http://birthmoviesdeath.com/2011/02/02/how-many-days-was-phil-connors-stuck-in-a-groundhog-day-loop.

Fraser, J. T. 2007. "That Awesome Gift: Human Freedom." In *Time and Time Again: Reports from a Boundary of the Universe*, 235–53. Leiden: Brill.

Grimwood, Ken. (1986) 2002. *Replay*. New York: Harper.

Groundhog Day. 1993. Screenplay by Danny Rubin and Harold Ramis. Dir. Harold Ramis. Perf. Bill Murray, Andie McDowell, Chris Elliot, Stephen Tobolowsky. Columbia Pictures.

Hermann, Martin. 2011. "Hollywood Goes Computer Game: Narrative Remediation in the Time-Loop Quests *Groundhog Day* and *12:01*." In *Unnatural Narratives – Unnatural Narratology*, edited by Jan Alber and Rüdiger Heinze, 145–61. Freiburg: De Gruyter.

Kakalios, Jim. 2011. "Guest Post: Jim Kakalios on the Quantum Mechanics of Source Code." http://www.preposterousuniverse.com/blog/2011/04/12/guest-post-jim-kakalios-on-the-quantum-mechanics-of-source-code/.

Kermode, Frank. 1967. *The Sense of an Ending: Studies in the Theory of Fiction.* New York: Oxford.

Lupoff, Richard A. (1973) 2013. "12:01 P.M." In *The Time Travel Megapack*, edited by John Betancourt, et al., 144–60. Wildside Press, LLC. (www.wildsidepress.com); orig. published in *The Magazine of Fantasy and Science Fiction*, December 1973.

Navarro-Remesal, Victor and Shaila García-Catalán. 2015. "Try Again: The Time Loop as a Problem-Solving Process in *Save the Date* and *Source Code*." In *Time Travel in Popular Media: Essays on Film, Television, Literature and Video Games*, edited by Matthew Jones and Joan Ormrod, 206–18. Jefferson, NC: McFarland and Company.

Nietzsche, Friedrich. (1888, 2001) 2008. *The Gay Science: With a Prelude in German Rhymes and an Appendix of Songs.* Edited by Bernard Williams. Translated by Josefine Nauckhoff. Poems translated by Adrian Del Caro. Cambridge Texts in the History of Philosophy. Cambridge: Cambridge University Press.

Richardson, Brian. 2012. "Antimimetic, Unnatural, and Postmodern Narrative Theory." In *Narrative Theory: Core Concepts and Critical Debates*, edited by David Herman, James Phelan, Peter J. Rabinowitz, Brian Richardson, and Robin Warhol, 20–28. Theory and Interpretation of Narrative. Columbus: Ohio State University Press.

Rubin, Danny. 2008. "The Magic of Friendship." *Blogus Groundhogus.* http://www.dannyrubin.com/blogusgroundhogus/2008/01/29/the-magic-of-friendship/.

Sakurazaka, Hiroshi. (2004) 2014. *All You Need Is Kill.* Translated by Joseph Reeder with Alexander O. Smith. San Francisco: Haikasoru.

Schachter, Daniel L. and Donna Rose Addis 2009. "Remembering the Past to Imagine the Future: A Cognitive Neuroscience Perspective." *Military Psychology*, 21: (Suppl. 1) S108–S112.

Source Code. 2011. Screenplay by Ben Ripley. Dir. Duncan Jones. Perf. Jake Gyllenhaal, Michelle Monaghan, Vera Farmiga, Jeffrey Wright. Summit Entertainment.

12:01 P.M. 1990. Screenplay by Stephen Tolkin and Jonathan Heap. Dir. Jonathan Heap. Perf. Kurtwood Smith. Chanticleer Films.

Weissert, Thomas. 2013. "Crossing the Junctureless Backloop." Paper presented at the 15th Triennial Conference of the International Society for the Study of Time: Time and Trace. Orthodox Academy, Crete.

Wittenberg, David. 2013. *Time Travel: The Popular Philosophy of Narrative.* New York: Fordham University Press.

CHAPTER 10

Petrotemporality at Siccar Point: James Hutton's Deep Time Narrative

Barry Wood

People think in five generations – two ahead, two behind – with heavy concentration on the one in the middle. Possibly that is tragic, and possibly there is no choice. The human mind may not have evolved enough to be able to comprehend deep time. It may only be able to measure it.

JOHN MCPHEE (1998, 90)

∴

On a nearly perfect summer afternoon in June 1788, Sir James Hall, James Hutton, John Playfair, and several farmhands launched from Hall's ocean front estate near Edinburgh on the coast of Scotland and sailed southeast, scanning the rocky cliffs facing the North Sea. After passing Douglas Burn Beach and Reed Point, they reached the spectacular scene of Pease Bay: a two-hundred-yard long beach backed by a rocky promontory called Siccar Point rising seventy feet above the sand. The bottom fifty feet consisted of layered, gray-colored schist – the strata indicating formation by horizontal sedimentation except that the layers were standing vertically like a set of encyclopedias on a shelf. This was topped by a chaotic layer of fragmented schist; above this lay a nearly horizontal bed of layered red sandstone (see Figure 10.1).

Hutton was ecstatic. In the next few minutes he explained what they were seeing, drawing on his own encyclopedic knowledge of geology and a "theory of the Earth" that he had been working out for years and presented before the Royal Society of Edinburgh three years earlier. What he described was a vast geological history necessary to account for what they were seeing at Siccar Point. This included an immense chronology of separate events resident in various rock strata – the first recognition that the configuration of rocks recorded ancient events that comprised an overarching historical narrative. Of the three, the mathematician John Playfair may have been the most skeptical of Hutton's ideas; as a former Presbyterian minister, he was imbued with

FIGURE 10.1   Siccar Point Unconformity, east coast of Scotland, discovered
by James Hall, James Hutton, and John Playfair, June 1788.
Sedimentary layers of gray schist, standing vertically, topped by
nearly horizontal sedimentary layers of red sandstone.

the prevailing biblical view of a world created roughly 5,800 years earlier. But
Hutton's narrative astonished him; what Playfair went on to describe was a
transforming experience of seeing for the first time the vast depths of time:
"We felt ourselves being carried back ... An epoch still more remote presented
itself ... Revolutions still more remote appeared in the distance of this extraordinary perspective. The mind seemed to grow giddy by looking so far into the
abyss of time; and while we listened with earnestness and admiration to the
philosopher who was now unfolding to us the order and series of these wonderful events, we became sensible how much farther reason may sometimes
go than imagination can venture to follow" (1805, 72–73). On that singular afternoon in 1788, they understood the greater vision of a man who described
his life's work as "reading in the face of rocks the annals of a former world"
(McPhee 1998, 77).

The description of Siccar Point by James Hutton (1726–1797) signaled the
first glimpse of what is now called deep time, a concept made popular by John
McPhee in *Basin and Range* (1982) and now systematically explored by special-
ists in other fields, notably the social sciences (Shyrock and Smail 2011). The
idea was far beyond what anyone can conceive; as McPhee put it, "Numbers do
not seem to work well with regard to deep time. Any number above a couple
of thousand years – fifty thousand, fifty million – will with nearly equal effect awe the imagination to the point of paralysis" (McPhee 1998, 29). Hutton

recognized that no data were available for dating of any kind; thus he avoided any use of paralytic numbers as he organized his observations while laying the groundwork for thinking about the Earth over the next century. It would come to maturity in Charles Lyell's *Principles of Geology* (1830–1833), and Charles Darwin's *Origin of Species* (1859), and would remain fundamental for any subsequent geological history, as indicated by the numerous references to Hutton's insights in, for instance, John Mcphee's Pulitzer Prize winning *Annals of the Former World* (1998). Hutton recognized that the rather spectacular geology at Siccar Point provided visible punctuation points in the story of a place with planet-wide implications. His unique contribution was his hard-won ability to "see" that rocks preserve a record of their past, thus providing snapshots of a history of change. Specific geological events opaque to others were visible to him; moreover, he appears to have been the first who was able to imagine unbelievably long expanses of time necessary to account for the visible configuration of the rocks. Today we know that events he described began at least 440 million years ago and lasted over 100 million years. The measure of that history was made possible by the freezing of time in stratified rocks and their measurable loss of radioactivity which we have defined as *petrotemporality* (Wood 2015b, 166–69).

1   Eighteenth-Century Flood Geology

Through the eighteenth century, objective geological thinking was almost non-existent. Current understanding of the Earth was shaped by the prevailing biblical narrative which eliminated any idea that the Earth had a history. The untrained imagination could not imagine time except in Earthbound units of centuries or millennia, and any view of time stretching beyond the three- or four-thousand-year span of recorded history was simply inconceivable. In the biblical narrative, the world had been created in a one-time event a few thousand years ago and had since existed in roughly the form we find it today. The only subsequent event of relevance was the biblical flood from which the present world had emerged as flood waters subsided. Continents, rivers and oceans, hills and mountains and valleys had existed unchanged from creation until the biblical flood; today's unexplainable anomalies were the aftermath of deluge disruption. Traditional dates derived from the prevailing chronology of the world established by Bishop James Ussher in *Annals of the Old Testament* (*Annales Veteris Testamenti*) published in 1650: his date for creation was 4004 BCE. Counting back from their discovery at Siccar Point, Hutton and his friends were living 3492 years since Noah's ark ran aground on Mount Ararat in 2300

BCE. But Ussher's biblical chronology, as Martin Gorst (2001) has documented, was only one of more than a hundred worked out since the third century, all with slightly different dates for creation. Isaac Newton (1642–1727), now regarded as a paragon of mathematical science, evidently accepted Bishop Ussher's dates; at the end of his life he was working on a "chronology of ancient kingdoms," attempting to re-date classical and pagan dates that we might regard as more reliable so they would match the prevailing chronology of the Bible (Repcheck, 67–68). His acceptance of Ussher's chronology undoubtedly provided it with additional authority because of Newton's prominence in physics and mathematics and the Chair he held at Cambridge University.

Believers in biblical chronology were bereft of any view of time other than the daily, seasonal, and annual cycles determined by the orbit of the Earth on its tilted axis. Nothing in the passage of days, years, and centuries implied any more profound concept of time. The limited idea of a world without a history could be maintained because geological changes could not be observed in a single lifetime or even in the four thousand years of recorded history. Lands, rivers, and mountains named in ancient times occupied the same map locations; ideas of lands shifting, oceans widening, or mountains rising were simply beyond imagination. This did not eliminate speculation, but theories remained shackled to the biblical flood. Thomas Burnet's four-volume *Sacred Theory of the Earth* (1681–1689) postulated for the world the same decline found in the story of the fall of man; the world once had an ideal, paradisal climate that was disrupted by the Flood, thus accounting for the vagaries of storms and climate extremes of present times. John Woodward, a biologist interested in fossils, provided an explanation for their presence on land: his *Essay Toward a Natural History of Earth* (1695) argued that the Flood had destroyed the surface of the Earth, then distributed fossils in the sedimentary layers observable above sea level before the waters receded. In his *New Theory of the Earth* (1696), William Whiston edged toward a scientific explanation for the Flood, suggesting it had been caused by a comet hitting the Earth, an explanation influenced by a new awareness of comets from extended presences of Kirch's Comet (1680) and what later became Halley's comet (1682); but the Flood remained central to the story. The posthumous book by Leibnitz, (1646–1716), *Protegga* (1749), postulated a universal ocean from which lands emerged as it seeped away. In Hutton's time, the charismatic teacher who swayed hundreds of students, Abraham Gottlob Werner (1749–1817), continued to present a personal scenario that began with an initial world-wide ocean from which the various rocks of the world – biblically classified by Werner as pre-Flood, Flood, and post Flood – had precipitated. As Jack Repcheck summarized the situation, "all early studies of the Earth attempted to be scientifically rigorous while still deferring to the time scale dictated by the Bible, and stressed the central role of Noah's Flood and the waters of the newly created Earth" (2003, 95).

One exceptional scientist who still believed in the biblical chronology introduced geological ideas independent of his beliefs. Nicholas Steno (1638–1686), a Danish anatomist who relocated to Italy, was interested in layered rock strata that he examined by descending into mine shafts in Italy. He recognized that these were layered horizontally and were continuous between widely separated mines. In addition to his discovery of strata continuity, he reasoned that lower strata must be older than upper layers. This observation seems so simple and obvious today, but in the seventeenth century it signaled a crucial conceptual change – a move away from a one-time creation event toward a temporal sequence, the foundation for a geohistory.

In addition, one of Steno's discoveries led to an insight that solved a problem centuries old. As Martin Rudwick (1985, 1–48) summarizes, Greek and Roman naturalists puzzled over mysterious stones-within-rock ("fossil objects") embedded in sedimentary strata and put forth improbable scenarios for their formation. A full understanding was still illusive in 1565 when Conrad Gessner published *De Rerum fossilium* (*On Fossil Objects*), by which he meant primarily stones and gems. For the Romans and Renaissance naturalists, *fossilis* meant something "dug up" with no organic overtones. However, Steno recognized some things dug up in ancient times called *glossopetrae* ("tongue stones") by Pliny the Elder as identical to sharks' teeth. Steno's position was that stones that looked like animal parts ought to be considered as animal parts. From this he concluded that teeth, bones, or shells falling into muddy sea bottoms that might turn into rock could gradually mineralize while retaining their distinctive shapes. In 1669, he published his *Dissertation Concerning a Solid Body Enclosed by the Process of Nature Within a Solid,* thus focusing attention on a process basic to geology. Today we identify the process, which remained a mystery for Steno, as fossilization of animal life and petrification of ancient trees. His recognition was unique for the seventeenth century because it provided an explanation independent of the biblical flood. However, progress in geological thinking was delayed for a century until scientists could work free of "flood geology." Meanwhile, curious naturalists and laymen took up fossil collecting in earnest.

The continuance of the flood in eighteenth century studies of the Earth illustrates the hold the biblical narrative retained in Hutton's day. With hindsight, we recognized that most of these "theories" or "studies" were stories invented to support and protect the primary flood story still regarded unquestionably as historical fact. In this imaginative climate, objective geological history was virtually impossible, to the credit of Hutton who successfully severed himself from the prevailing world view to develop the first geologically sound theory of Earth history. Reasons for this imaginative breakthrough are three: first, a remarkably eclectic education, including advanced study on the Continent; second, fourteen years as a farmer in hilly, rough terrain of Berkshire in Scotland; and third, participation in a highly educated circle of thinkers in Edinburgh associated with what is now known as the Scottish Enlightenment.

2   Hutton's Eclectic Education

The rigor of a classical humanistic education in Latin, Greek, logic, and metaphysics that Hutton experienced at University of Edinburgh can hardly be faulted. Hutton's education beyond the humanities included chemistry, possibly through his metaphysics course where the instructor included chemistry demonstrations that motivated Hutton to pursue this field on his own; he is reported to have tinkered with chemistry throughout his life. More pervasive was a rigorous introduction to the science of Isaac Newton by Colin Maclaurin, a memorable influence since Maclaurin had worked with Newton – such discipleship always impressive for students. Newton (1642–1726) was the most famous scientist of the day: author of *Principia Mathematica* (1687), the inventor of calculus, discoverer of the universal laws of gravitation, developer of the first efficient reflecting telescope, and an early student of the science of optics. Maclaurin evidently devoted himself to inculcating his students with Newton's ideal of sound scientific method, beginning with plausible theories developed from careful observation, then testing them with multiple experiments. Mclaurin himself was a formidable intellect. At age fourteen, he defended *The Power of Gravity*, an MA graduation thesis at the University of Glasgow; at nineteen he was elected professor of mathematics at University of Aberdeen, holding the record as the youngest professor in history for nearly three centuries. While Hutton was a student, Maclaurin published his two-volume *Treatise on Fluxions* (1742), a systematic presentation of Newtonian scientific method, a perfect example of the right teacher for the right student. The influence of Newton on Hutton is pervasive in his observations on the changes in the land visible on his inherited farmlands and during his numerous geological excursions through England and Scotland.

Following graduation from University of Edinburgh (1743), Hutton undertook medical studies at Edinburgh, the University of Paris, and Leyden University where he earned the Doctor of Medicine degree (1749) with a thesis titled *The Blood and the Circulation in the Microcosm* (*De sanguine et circulation in microcosm*). As David Oldroyd (2006, 42) observes, Hutton's title indicates that he incorporated a medieval and Renaissance notion in which "the human body ("microcosm") was imagined to have analogies with the Earth or the whole cosmos ("macrocosm")," suggesting that he had already developed a perspective broader than simply biological. Jack Repcheck (2005, 90–91) adds that this thesis of blood circulation was significant for its use of Newton's explanation of cycles; his later appropriation of the broader concept of cycles – both hydrological and petrological – in his theory of the Earth led to the first macrocosmic picture of planetary history.

In 1754, at the age of twenty-eight, Hutton took up residence on *Slighhouses*, a farm acquired by his uncle in 1713 and passed on to his father in 1718. This 140-acre property located less than ten miles from the English border was one of two; the other was a 590-acre hill farm eight miles away called *Nether Monynut* best suited for grazing cattle and sheep. *Nether Monynut* was hilly, rocky land with elevations up to 1,000 feet above sea level; *Slighhouses* was arable land sloping from the north at an elevation of 450 feet to the south at 300 feet. This was well drained by Lintlaw Burn, Fosterland Burn, and five other minor creeks. With an elevation drop of 150 feet, visible erosion occurred with regularity during heavy rains and spring runoff. Undoubtedly, Hutton saw evidence of similar erosion not only on the larger hill farm but also throughout the highlands of Scotland and England on his various excursions. He paid great attention to rocks and geology, but his interest was taken by the visible erosion caused by flowing water – the one geological change rapid enough to be seen in a human lifetime – in fact, in a single season. Erosion was visible throughout the Scottish highlands in ice-shattered boulders, slopes of broken scree at the bases of steep mountains, and rocks along mountain streams washed down in spring floods. It was dramatically visible in *Gutted Haddie*, a hollow near the top of Arthur's Seat, a mound near Edinburgh, where a heavy storm brought down a landslide in 1744 while Hutton was a student in Edinburgh.

3   The Emergence of the Theory

Hutton worked his farms for thirteen years (1754–1767), during which time he applied the best agricultural methods available. He dealt with erosion of the sloping land with perimeter stone walls, drainage ditches, judicious plowing designed to reduce runoff, and rotation of crops of wheat, turnips, and barley, with cattle grazing when the fields were fallow. Meanwhile, he gave much thought to the implications of erosion on a larger scale. Hutton's diverse and advanced education had liberated him from a six-thousand year biblical chronology; he considered that the Earth had existed much longer. But the ongoing process of erosion highlighted a question: If the Earth had a history extending back thousands or even millions of years, how could the rugged landscape of the Scottish highlands, the European Alps, or even the low hills of the English Cotswolds remain? Surprisingly, no one before Hutton had asked a central question: with erosion by wind and water a constant force on hills and mountains, should not all the hills and mountains of the world have long ago washed down to the sea?

The answer that came to Hutton derived from his study of rocks. The layering of sedimentary rock, particularly layers that contained fossils, indicated that most rocks were formed under water from accumulated layers of silt, sand, clay, shells, and animal remains. These were often found high above sea level. The immense thickness of some sedimentary outcrops – hundreds of feet – necessitated inordinate amounts of material and unimaginable expanses of time for their accumulation. From his observation of erosion on his farms and along the creeks and rivers of Scotland, he reasoned that the primary materials in sedimentary rocks were silt and debris washed down from the highlands. Layers of clay on the ocean floor hardened into slate; layers of sand hardened into sandstone; accumulated shells hardened into limestone. During spring floods, rushing water brown with eroded topsoil tumbled rocks downstream where this erosional mixture found its way to the sea bottom as contributions to finer sediment. Stones, gravel, or fragments of granite thus became embedded in composite rocks. Given enough time, even the passage of human feet over rocky footpaths wore away the hardest of rocks. The whole process appeared to move in one direction – from land to sea – until Hutton recognized that vast layerings of sedimentary rock above sea level must have been raised there by enormous underground forces. Hutton seems to have had little interest in volcanoes, but their molten lava provided a clue: heat from deep inside the Earth must eventually cause sea-bottom sediment to rise above the water, thus renewing the land. This narrative provided the essential clue: the Earth was subject to a cycle whereby erosion provided material for sea-bottom sediment, and underground heat provided the energy to lift undersea rocks, sometimes to unbelievable heights. His explanation for sea-bottom uplift was faulty, we now understand, but with his recognition of this cycle, Hutton's theory of the Earth was complete.

3.1   The Theory of the Earth

The late eighteenth century in Scotland was the scene of some of the most powerful intellectuals ever assembled. Now known as the Scottish Enlightenment, the group included several academics and thinkers who have achieved lasting fame: Joseph Black (1728–1799) who discovered carbon dioxide and the mixture of gases that make up the atmosphere; the philosopher David Hume (1711–1776) who argued for the primacy of emotion in human nature; Adam Smith (1723–1790), famous for *The Wealth of Nations* (1776); James Watt (1736–1819), inventor of the steam engine. This circle was not confined to scientists and social scientists: the poet Robert Burns (1759–1796) and novelist Sir Walter Scott (1771–1832) were part of this sterling collection, along with a dozen more associated with the Universities of Edinburgh and Aberdeen. A belated celebration of this circle of intellects by the artist Charles Martin Hardie led to a painting called *The Meeting of Burns and Scott* (1893) that depicted a unique and memorable event at Sciennes, the home of Adam Ferguson. It included fifteen members of the Scottish Enlightenment including recognizable images not only of Robert Burns and Sir Walter Scott but also of Joseph Black, Adam Smith, and James Hutton (see Figure 10.2).

James Hutton became an associate of this intellectual circle in 1757, when he leased out *Slighhouses* and moved back to Edinburgh. He was independently wealthy, enjoying regular income from rents and, as Archibald Clow (1947) has documented, from a chemical manufacturing process he and a partner had designed years earlier to manufacture sal ammoniac from local soot. The academic atmosphere of the times had been established decades earlier with the formation of the Society for the Improvement of Medical Knowledge in 1731 and the Philosophical Society in 1737 by Hutton's former teacher, Colin Maclaurin. In 1754, the Select Club began a tradition of having a member make a formal presentation on their current research. In 1777, Adam Smith, Joseph Black, and Hutton formed the Oyster Club, a weekly dining club that attracted half a dozen others of philosophical bent. Benjamin Franklin, an occasional visitor and participant in various academic groups in Edinburgh, described this circle as "a set of as truly great men ... as have ever appeared in any Age or Country" (McPhee 1998, 72). In 1782, a petition was sent to London to form a branch of the Royal Society; in 1783, a charter was granted for the Royal Society of Edinburgh. Shortly thereafter, Hutton was invited to make the first presentation. His theory of the Earth was delivered in two parts, on March 7 and April 4, 1785.

It is worth pointing out the contrasting predecessors of Hutton's presentation. The temper of the times illustrated in the works of Hooke, Descartes, Newton, and Liebniz, as Paolo Rossi (2000, 165) has pointed out, was to connect observations from nature with "the Biblical narrative and theology ... the Creation and the Apocalypse." The best known example was Thomas Burnet's *Sacred Theory of the Earth* (1680). While "sacred" referred to Burnet's emphasis on placing geological observations within a biblical framework, "theory of the Earth" opened up a philosophical theme that would dominate through

FIGURE 10.2   *The Meeting of Burns and Scott*. Painter: Charles Hardie. To the left
is Adam Ferguson stoking the fire, then Robert Burns standing
opposite Walter Scott, aged 15; Dugald Stewart seated; at the table,
Joseph Black MD, Adam Smith, John Home (author of *Douglas*),
and James Hutton. The painting was inspired by Scott's later
description of the meeting.

the next century. A few years later, the English student of Isaac Newton,
William Whiston – known today for his still-in-print translation of Josephus –
published *A New Theory of the Earth* (1696), a work intended to improve on
Burnet, though still within a biblical framework. These works captivated the
imaginations of philosophers and naturalists; as Rudwick (2014, 61) remarks,
"'Theory of the Earth' was now well launched as a scientific genre," but with
differences that emerged later in the eighteenth century. "First, it absorbed ...
the vastly enlarged sense of the Earth's likely timescale that emerged during
that century.... Second, in the cultural climate of the intellectual movement of
the Enlightenment, 'Theory of the Earth' was generally narrowed down into a
project concerned exclusively with the laws of nature in the physical world."
The attempted linkage of science with the Bible "was generally abandoned
or at least marginalized," a tendency made possible within the widespread
divine-hands-off deism of Enlightenment thinking. Hutton's new "theory of
the Earth" was thus in concordance with the spirit of the times.

Hutton's central narrative was cyclical: erosion by wind and water moved the
land, grain by grain, to the ocean where it settled to become stratified sedimentary rock; subsequently, underground forces raised sea-bottom rocks above sea
level to create new land where the process would begin again. Sergei Tomkeieff

(1946; 1962) provided the term "geostrophic cycle" for Hutton's theory of the Earth, particularly as it was evidenced in the unconformity at Siccar Point. The process Hutton described occurred in the past, continued in the present, and would continue in the future; every particle of rock and every grain of sand underfoot had gone through this cycle numerous times. As Hutton put it at the end of his presentation, "the result, therefore, of our present inquiry is that we find no vestige of a beginning, – no prospect of an end" (1788, 104). In his later expanded version, he clarified his final remark: "this world has neither a beginning nor an end." Despite this obvious repudiation of the biblical creation narrative, his presentation brought forth no significant reaction; one could assume that most members of the audience were enlightened thinkers already liberated from non-scientific traditions.

4   Petrotemporality at Siccar Point

Over the next three years, Hutton made several geological excursions – to Glen Tilt, a river in the Eastern Highlands; Galloway in southwestern Scotland; and the Island of Arran in the Firth of Clyde on the west coast. In the first two, he discovered evidence for some part of his theory. In the third, above North Newton shore on the island of Arran, he discovered a revealing outcrop that turned isolated parts of his theory into a geotemporal narrative. Above the beach at an angle of forty-five degrees rose blocks of what is now known as Dalradian – altered sedimentary rocks from Precambrian times that had been distorted by pressures of folding and metamorphosis; we can now date these to 500 million years ago. Above these lay nearly horizontal strata of sedimentary rock from the Devonian/Carboniferous Era, approximately 360 million years ago. The line of contact is now known as an "unconformity," and this Arran island version holds the record as the earliest discovery of this geological feature. It was Hutton's first sighting of a feature that could be reconstructed as a narrative of three distinct chapters: the Dalradian strata had been laid down horizontally; they had been severely distorted and tilted to a steep angle; the overlying rock of a different kind had been laid down later.

    Through careful observation and remarkable imagination – nurtured by an exceptional education and continual exploration of the Scottish landscape – Hutton seems to have developed an intuitive grasp of temporal events in rock, rock strata, and igneous intrusions into metamorphic or sedimentary rock. Beginning more than forty years ago, the physicist J. T. Fraser (1982) developed a five-part taxonomy of time, each temporality germane to one of several stable levels of reality which he regarded as coexisting in a nested hierarchy.

His most suggestive temporality for historical reconstruction of the universe is *atemporality*, which refers to the absolute freezing of time in massless photons, allowing for the delivery to the present of unaltered past time in light and other forms of radiation. We are thus able to view the universe at various times in the past, all the way back to leftover radiation from the big bang. Obviously, the reconstruction of pre-planetary history would be impossible without atemporality delivering billions of years of cosmic history for present viewing. To Fraser's taxonomy we have suggested the addition of *petrotemporality* (rock-time), for various kinds of freezing or semi-freezing of time are evident in fossils, rock strata, and radioactive isotopes that decay on a measureable schedule (Wood 2015b). It is, in fact, the freezing of past time in rocks – the preservation of ancient temporalities – that makes a geotemporal history of the planet possible. The original horizontality of the upper strata and the lower distorted and tilted outcrop Hutton discovered on the Isle of Arran represent two distinct petrotemporalities separated by a considerable expanse of time – with the intervening period adding a third. The overlaid horizontal strata would thus represent a fourth petrotemporality. Hutton recognized the larger story emerging from this conjunction of geological features from different times; as he wrote, "It may now be observed, that the present history with regard to the island of Arran, is ... an example of Cosmogony, where it may be proper to see the connection of various things, or the several parts which enter the constitution of the globe" (Baxter, 156).

Hutton had now acquired solid evidence for his theory of the Earth. What remained was an appropriate setting for his dramatic narrative. That setting was Siccar Point, a coastal outcrop that Hutton may have anticipated from observations on his farms not far away. *Nether Monynut*, his upland farm, was located on what he called "schistus"; the fertile soil of *Slighhouses*, his lower tract, was derived from red sandstone. The contact between these two rock formations was partly visible along the eastern boundary of *Nether Monynut*, running north toward the coast. Hutton's 1785 boating expedition with Playfair and Hall was a search for this schistus-sandstone contact where it might be exposed. As we have seen, their search was successful in more ways than one. They discovered the contact, but more importantly, Hutton discovered the proof for the theory of the Earth he had described three years earlier – proof virtually on the doorstep of Edinburgh that the Earth's past extended in an incomprehensibly long "deep time."

Hutton is not renowned for his writing; neither *The Theory of the Earth* (1788), the text of his 1785 talk before the Royal Society of Edinburgh, nor his two-volume *Theory of the Earth with Proofs and Illustrations* (1795) earned great praise. But his speech was colorful, sometimes laced with invective, and persuasive enough to convert James Hall, a mathematician, and John Playfair, a chemist, to a new understanding of the Earth. Indeed, Playfair's eloquent Illustrations in *The Huttonian Theory of the Earth* (1802) reveals the extent of his influence. For a decade following their discovery at Siccar Point, he followed Hutton's lead and became as thoroughly educated in geology as Hutton himself.

Hutton's excited narrative at Siccar Point unfolded in seven chapters (see Figure 10.3), each revealing a distinct petrotemporality, a piece of frozen time clearly evident in the rocky headland. The first chapter was the most ancient: over millennia, erosion carried silt downstream to the ocean where it sank to the bottom; over unimaginable reaches of time it built up, layer by layer to form strata of grayish rock known as greywacke. Hutton understood the process; direct observation of sediment washing into the ocean along with layered strata was evidence enough that the process had occurred. His second chapter focused on a process of folding caused by unknown lateral forces, much like the folding of a carpet when the ends are pushed together. The result of this was an alteration of horizontal layers until they were compressed into vertical waves. Again, this was a case of simple observation: Hutton had seen numerous examples of folded sedimentary layers throughout Scotland and England, though he was unable to explain what these forces were.

His third chapter entailed a change of sea level. With knowledge of plate tectonics 150 years away, Hutton could only imagine some deep underground force, possibly heat, raising the sea bottom, though our knowledge of sea-level changes – three to four hundred feet during the last ice age – suggests periods of glaciation as an alternate explanation. Hutton's fourth chapter was an erosional process that wore off the upper folds of the greywacke, leaving a topping layer of chaotic fragments. In his fifth chapter, a reversal occurred; the eroded greywacke settled below sea level, the result of changes Hutton could only surmise. In the sixth chapter, new layers of erosional sediment settled atop the rubble, gradually building up layers of strata, but this time it was red sandstone. A seventh chapter occurred after the entire structure was forced again above sea level with very slight tilting from the horizontal. Over a period of time, some sandstone erosion occurred, but not enough to remove all of the overlay. What finally developed above this was a layer of topsoil supporting typical groundcover – grass, moss, and lichen.

Comparing Hutton's narrative with the prevailing flood geology, the difference is profound: earlier writers committed to preserving the biblical narrative necessarily postulated a world that achieved its present form from a catastrophe: this form of explanation was known as Catastrophism or, if the role of the sea as precipitating landforms was emphasized, they were called

1 Sedimentary layers of greywacke
   deposites on seafloor of Iapetus
   Ocean 440–420 million years ago.

2 Lateral pressure from closing of
   Iapetus Ocean causes tectonic
   folding of sedimentary layers.

3 Sea-level drop or sea-bottom
   upthrust raises vertical folds of
   greywake above sea level.

4 Erosion removes upper folds to
   sea level, leaving broken rubble
   above vertical folds.

5 Sea-level rise or sea-bottom
   subsidence lowers greywacke
   and rubble below sea level.

6 Red sandstone sedimentary
   layers collect above rubble
   360–340 million years ago.

7 After upthrust and partial
   erosion, topsoil and plants
   collect above red sandstone.

FIGURE 10.3

Neptunists after the Greek sea-god Neptune. Hutton's narrative suggests a
vast expanse of time during which present observable processes operated
with changes occurring uniformly and gradually: his view became known as
Uniformitarianism or Gradualism. The crucial difference was temporality; as
noted earlier, the biblical narrative extended no more than 6,000 years into the
past, and this diminished chronology left no room for any meaningful planetary history. Hutton's narrative laid out the past like chapters of a longer story,
each one entailing an immense expanse of time, setting out a plausible history

based on careful observation of the visible geology of the Earth. Each feature of Siccar Point displayed a unique petrotemporality – two layers of strata, one much older; the stony rubble separating them; the dramatic unconformity, combined with the knowledge that sedimentary rock formation is an undersea process and erosion occurs on the land, provided the basic punctuation points of the larger story.

5   Dating Hutton's Siccar Point Narrative

With the exception of John Playfair, no one took up Hutton's theory of the Earth until Charles Lyell (1797–1875) undertook his ground-breaking exploration of geology published in his *Principles of Geology* (1830–1833), which went through eleven editions. Hutton's fundamental assumption that the world had an immensely long history was adopted by Lyell, whose explorations uncovered further evidence. Hutton's explorations had been limited to Scotland and England; Lyell's took in not only the British Islands but also the Auvergne volcanic district of southern France, Mount Vesuvius in Italy, Mount Etna in Sicily, the United States, and the Canadian Joggins Fossil Cliffs in the Bay of Fundy. Charles Darwin (1809–1882) read Lyell's *Principles of Geology* during his voyage on the *Beagle* and he, too, adopted the extended chronology established by Hutton, a necessity for the biological process of speciation outlined in *The Origin of Species* (1859). Neither Lyell nor Darwin ventured any authoritative age for the Earth, and anyone in the nineteenth century who did lacked sound evidence; in this they followed Hutton who, as Playfair had noted, "did not attempt to explain the first origins of things. He was too well skilled in the rules of sound philosophy for such an attempt" (1805, 52). An accurate dating of Hutton's narrative would have to wait well over a century.

The breakthrough came with the work of Marie and Pierre Curie, beginning with their discovery and exploration of "radioactivity" – the word coined by Marie – then the discovery of various elements containing radioactive isotopes. Through the early years of the twentieth century, chronologies for radioactive decay were established, resulting in the familiar idea of the half-life and thus a way of establishing the age of rocks. By 1911 measurement of isotopic decay had advanced to the point where Ernest Rutherford could suggest that radiometric dating of the world's most ancient rocks would provide a way to measure the age of the Earth. When radioactive dating was extended to meteorites, which were formed at the same time as the planets, the age of the Earth was set at 4.6 billion years (Dalrymple, 404). This date was settled in the 1950s; nothing since has altered it.

Applying radiometric dating to rocks at Siccar Point suggests that the formation of the underlying greywacke occurred during the Silurian Period approximately 435 million years ago, though the process of strata formation probably occurred over as much as thirty million years. The overlying red sandstone strata are younger, formed approximately 370 million years ago during the Devonian Period. These formations represent widely separated events in Hutton's narrative. The intervening chapters – the folding and bending of the greywacke strata, its elevation above sea level, and the erosion of its upper folds – were not well understood until the 1950s and 1960s with "a revolution in the Earth sciences" (Hallam 1973). Hints of the theory began with the idea of continental drift suggested by Arthur Wegener (1912) later affirmed with the discovery of crustal movements now known as plate tectonics, supported by studies in paleomagnetism. Tectonic theory matured with the work of S. K. Runcorn (1962), E. Irving (1964), and Xavier Le Pichon (1968).

A reconstruction of tectonic plate movement makes sense of what happened at Siccar Point and, in fact, throughout Scotland. The role of plate tectonics in the formation of the North Atlantic Ocean and the geology of the British Isles developed with the work of Tuzo Wilson (1966), John F. Dewey, and John M. Bird (1971). Before the assembly of the supercontinent of Pangea five hundred million years ago, the Iapetus Ocean separated the continent of Laurentia, corresponding to North America and Greenland, from Baltica, which is now northern Europe, and Avalonia, a drifting microcontinent to the south. Coastal fossil records from this era – trilobites and brachiopods, for instance – reveal a faunal realm in Laurentia characterized biologically as "strikingly uniform and clearly distinct" from those in Baltica and Avalonia (Sullivan, 259). As L. R. M. Cocks and R. A. Fortey (1982) have shown, these faunal variations are primary evidence for the era of the Iapetus Ocean. At that time, northwestern Scotland was part of Laurentia while England and Wales were part of Avalonia. The Siccar Point graywacke strata was formed as sea-bottom at the northern margin of the Iapetus Ocean. Between 490 and 390 million years ago, plate movement gradually closed and eliminated this body of water – one of the planet-wide changes that joined disparate terranes to form the supercontinent of Pangea. Avalonia collided with Baltica, then Laurentia: in the British Isles, Scotland and England were brought together at what is known as the Iapetus Suture which runs through Scotland from southwest to northeast. This suture, in fact, crosses the eastern coast of Scotland south of Siccar Point and close enough to provide some insight into the jumbled geology discovered by Hutton and his friends. The result of this collision was extensive orogeny and the formation of the Caledonides, a mountain range paralleling the Suture and continuing to the coastal mountains of Scandinavia. To the north of the Iapetus Suture, as Phil Stone (2012) has shown, the collisions of these terranes is recorded in a complex band of ridges, much distorted from collisional folds. Siccar Point reveals this geological turmoil at the coast.

Meanwhile subduction of Iapetus seabed resulted in extensive inland volcanic activity, responsible for prominent landforms across Scotland. When volcanic eruptions waned millennia later, a series of volcanic plugs remained that have resisted erosion by wind, water, and ice while much of the surrounding mountain material eroded away. Some are bare granite plugs; others have the appearance of high hills; all are prominent landmarks through the highland region – Arthur's Seat (823 feet) and Castle Rock (433 feet) in Edinburgh; Dechmont Hills (712 feet); Berkwick Law (613 feet); Largo Law (950 feet); Dunbarton Rock (240 feet); and the offshore island, Bass Rock (351 feet).

Initially, the Caledones may have been mountains of Alpine or even Himalayan dimensions, but 400 million years of erosion have reduced them to rugged highlands of 3,000 to 4,000 feet elevation. Along the eastern coast of Scotland, massive tectonic pressure crushed and folded the undersea greywacke layers, shifting horizontal strata to the vertical, and raised them above sea level, and subsequently submerging them. These rapid and violent changes took place in a geologically brief and highly chaotic period of 65 million years – no more than 1.5% of the history of the Earth. By 350 million years ago, the British Isles had formed and the supercontinent Pangea had assembled. The final phase at Siccar Point – the laying down of red sandstone strata and its elevation to its present position seventy feet above sea level – occurred as Pangea achieved its maximum dimensions.

The entire formation at Siccar Point, including the most famous unconformity in the world, is thus an imprint of a collision that brought an end to an ocean and created an ancient continent. Historical reconstruction from paleomagnetic analysis reveals that the entire process of collision took place far from its present location – in the southern hemisphere; and the accumulation of the upper red sandstone layers occurred in a warm region near the Earth's Equator as the entire complex of terranes was drifting north, thus providing an additional perspective on the wide-ranging travels of drifting continents in times past.

When the breakup of Pangea occurred around 225 million years ago, plate separations did not correspond precisely to the Caledonian Suture. To the east, a slight separation created the relatively narrow North Sea and a disjunction between the Caledonian mountains of Scotland and their continuation along the Scandinavian coast. A much greater separation occurred to the west, giving rise to the Atlantic Ocean (Hallam 1971). The northwestern portions of

Scotland and Ireland remained with the European plate, fused together as the Atlantic Ocean opened up while parts of Avalonia were carried off as part of New England and Newfoundland. Today, the Atlantic, now thousands of miles across, continues to widen at approximately a centimeter per year.

What Hutton provided was the first extended narrative of geohistory. Its span of one hundred million years is beyond human imagining, though it adds up to little more than two percent of Earth's 4.5-billion-year history. Its importance, however, is profound, for it provides a narrative emblematic of the whole of Earth history, the most important component outside of life itself for the grand narrative of cosmic history. In the form Hutton narrated, it speaks to the non-scientist who can appreciate Earth history if cast as geo-narratives, biology if presented as bio-narratives, and cosmology if formatted as cosmic narratives. Hutton's presentation was the earliest example of narrative as bridging what C. P. Snow called "the two cultures," a pedagogical method still rare in the teaching of science though easily adaptable for every level of education from grade school to grad school (Wood 2013, 2015a).

6   The Father of Modern Geology

Hutton's Siccar Point narrative, later verified by radiometric dating and plate tectonics, reveals the power of his imagination, particularly his ability to connect the separate dots of observation. Unfortunately, he was appreciated by few during his own lifetime and, were it not for Playfair's presentation of his theory in well-crafted prose, we might never have known him beyond an influence on later, more articulate geologists. Flood geology continued well into the nineteenth century. Most of Werner's students dismissed Hutton's Gradualism. It is perhaps emblematic of failure to appreciate what he had accomplished that no one knows where Hutton was buried in Greyfriars Kirkyard in Edinburgh, where a belated plaque was placed in 1947 on the 150th anniversary of his death. Also emblematic is the attrition of what must have been a spectacular collection of fossils and rocks as they were passed from hand to hand; today, no single item in any of several museum collections can be attributed to Hutton's relentless collecting. The same is true of his manuscripts, notebooks, letters, and drafts, many of which were lost or destroyed; what survives today is meagre for a man now celebrated as the father of modern geology. The whole disappointing story is told by Stephen Baxter in his Epilogue to *Ages in Chaos* (2003).

It was well into the twentieth century before a full appreciation of the long-term effects of erosion were realized and the immense depths of ocean-bottom sedimentary rock were discovered through the drilling of deep ocean cores. When the full history of drifting continents was charted and colliding plates were understood as the power that raised the Alps and Andes and Himalayas, Hutton's theory of the restoration of land from ocean-bottom rocks suddenly took on a long unrecognized elegance. With clues gleaned from limited evidence from surface observation, he was able to reconstruct a remarkable history of Earth's cycles of destruction and renewal.

The delay in his recognition means that Hutton has earned the credit due him only in the late twentieth century, and monuments to his achievement are surprisingly recent. Belatedly, he has earned a place in Scottish intellectual history, though contemporary records of his life are scant. A 1787 portrait by Sir Henry Raeburn hangs in the National Portrait Gallery in Edinburgh. A professional frieze in the Scottish National Portrait Gallery includes a dozen participants in the Scottish Enlightenment, including Raeburn, the novelist Scott, the poet Robert Burns, the inventor James Watt, and Hutton, as well as the lesser known surgeon John Hunter, Judge Lord Jeffrey, and civil engineer Thomas Telford. John Playfair (1805) provided a biographical sketch in *Transactions of the Royal Society of Edinburgh*, reprinted in Playfair's collected works (1814), but it has not yet been reprinted.

Appreciation for Hutton's place in the history of geology was long delayed. His stature as one of the greatest intellects of the Scottish Enlightenment arrived with the James Hutton Memorial Garden, constructed in 2001, which marks the location of his Edinburgh home on St. John's Hill. The garden includes a display of five boulders: two from Glen Tilt showing veins of intrusive granite penetrating sedimentary rock, and three conglomerates showing pre-existing pebbles and rocks from earlier phases of erosion embedded in later sedimentary strata. These are not the boulders Hutton transported back to Edinburgh but they tell the same story. The James Hutton Borders Trail, which opened in 2006, begins at the display at the Riever Country Farm Food Shop, passes Hutton's home at *Slighhouses* and the pit where he obtained calcareous marl for fertilizer, then proceeds to Siccar Point, and returns to his upland property, *Nether Monynut*. Siccar Point is the highlight of the trail, symbolizing the triumphant proof of his theory of the Earth and is now considered one of the most important geological sites in the world. But recognition of Hutton as "a man ahead of his time" and interpretative pamphlets summarizing what he saw and explained at Siccar Point are surprisingly recent (Porteus and Browne 2014; Miller 2015), leaving earlier visitors and students bereft of accessible information on this complex geological formation.

On the bicentenary of Hutton's death, March 26, 1997, Donald McIntyre made the following remarks:

FIGURE 10.4
James Hutton (1726–1797). Portrait
by Sir Henry Raeburn, National
Portrait Gallery, Edinburgh

Fifty years ago, Arthur Holmes, the most distinguished geologist of his
time, said: 'To the geologists a rock is a page in the Earth's autobiography,
with a story to unfold.' Hutton showed how to read it; doing so he
disclosed the marvel of deep time.... Today we have come to know that
living creatures evolve, that continents drift, that the stars and galaxies
are born, mature, grow old and die. We salute the memory of James
Hutton who opened our minds to these wondrous possibilities.

BUTCHER, 5

The fame of Siccar Point draws local schoolchildren on field trips and geologists from around the world to view the story in the rocks. The refinements of
petrotemporality allow us to fill in the story of this remarkable formation and,
by extension, several dramatic chapters of the Earth's grand narrative.

References

Baxter, Stephen. 2003. *Ages in Chaos: James Hutton and the Discovery of Deep Time*. New
York: Tom Doherty Associates.
Burnet, Sir Thomas. 2016. *The Sacred Theory of the Earth: Containing an Account of the
Original of the Earth and All the General Changes Which it Hath Already Undergone,
or Is to Undergo, Till the Consummation of All Things*. Palala Press.

Butcher, Norman. 1997. "James Hutton, Charles Lyell, and the Edinburgh Geological Society." *The Edinburgh Geologist*, No. 30: 2–6. [Delivered at the Bicentennial Conference, Edinburgh.]

Clow, Archibald. 1947. "Dr. James Hutton and the Manufacture of Sal Ammoniac." *Nature* 159: 425–27.

Cocks, L. R. M. and R. A. Fortey. 1982. "Faunal Evidence for Oceanic Separations in the Palaeozoic of Britain." *Journal of the Geological Society, London* 139: 465–78.

Dalrymple, C. Brent. 1991. *The Age of the Earth*. Stanford: Stanford University Press.

Darwin, Charles. 1859. *Origin of Species*. London: John Murray.

Dewey, John F., and John M. Bird. 1970. "Mountain Belts and the New Global Tectonics." *Journal of Geophysical Research* 75, May 10: 2625–47.

Fraser, J. T. 1982. *The Genesis and Evolution of Time: A Critique of Interpretations in Physics*. Amherst: University of Massachusetts Press.

Gorst, Martin. 2001. *Measuring Eternity: The Search for the Beginning of Time*. New York: Broadway Books.

Hallam, A. 1971. "Mesozoic Geology and the Opening of the North Atlantic." *Journal of Geology* 79: 129–57.

Hallam, A. 1973. *A Revolution in the Earth Sciences: From Continental Drift to Plate Tectonics*. London: Oxford University Press.

Hutton, James. 1788. "Theory of the Earth: or an Investigation of the Laws Observable in the Composition, Dissolution, and Restoration of Land upon the Globe." *Transactions of the Royal Society of Edinburgh*, Vol. 1: 209–305.

Irving, E. 1964. *Paleomagnetism and Its Application to Geological and Geophysical Problems*. New York: Wiley.

Le Pichon, X. 1968. "Sea Floor Spreading and Continental Drift." *Journal of Geophysical Research* 73: 3661–97.

Lyell, Charles. 1930–1933. *Principles of Geology*. 3 vols. London: John Murray.

Miller, Angus. 2015. *Siccar Point*. Edinburgh: Lothian and Borders GeoConservation/ Edinburgh Geological Society.

McPhee, John. 1982. *Basin and Range*. In McPhee 1998: 19–143.

McPhee, John. 1998. *Annals of the Former World*. New York: Farrar, Strauss and Giroux.

Oldroyd, David. 2006. *Earth Cycles*. Westport, Connecticut: Greenwood Press.

Playfair, John. 1802. *Illustrations of the Huttonian Theory of the Earth*. Edinburgh: William Creech.

Playfair, John. 1805. "Biographical Account of the Late Dr. James Hutton." *Transactions of the Royal Society of Edinburgh*, Vol. 5: 39–99. Edinburgh: Royal Society of Edinburgh.

Porteus, Cliff and Mike Browne. n.d. *James Hutton: A Man Ahead of His Time*. Edinburgh: Lothian and Borders RIGS Group, British Geological Survey.

Repcheck, Jack. 2003. *The Man Who Found Time*. Cambridge, Massachusetts: Perseus Books.

Rossi, Paolo. 2000. *The Birth of Modern Science*. Trans. Cynthia De Nardi Ipsen. Oxford: Blackwell Publishing.

Rudwick, Martin J. S. 1976. *The Meaning of Fossils: Episodes in the History of Palaeontology*. Chicago: University of Chicago Press.

Rudwick, Martin J. S. 2014. *Earth's Deep History: How It Was Discovered and Why It Matters*. Chicago: University of Chicago Press.

Runcorn, S. K. 1962. "Paleomagnetic Evidence for Continental Drift and Its Geophysical Cause." In *Continental Drift*. Ed. by S. K. Runcorn. New York: Academic Press.

Shyrock, Andrew and Daniel Lord Smail, eds. 2011. *Deep History: the Architecture of Past and Present*. Berkeley: University of California Press.

Steno, Nicolaus. 1968. *The Prodromus of Nicolaus Steno's Dissertation Concerning a Solid Body Enclosed by Process of Nature within a Solid*. Trans. John Garret Winter. New York / London: Hafner Publishing.

Stone, Phil. 2012. "The Demise of the Iapetus Ocean as Recorded in the Rocks of Southern Scotland." *Open University Geological Society Journal* 33: 29–36.

Sullivan, Walter. 1974. *Continents in Motion: The New Earth Debates*. New York: McGraw-Hill.

Tomkeieff, Sergei. 1946. "James Hutton's 'Theory of the Earth,' 1795." *Proceedings of the Geologists' Association* 57: 322–28.

Tomkeieff, Sergei. 1962. "Unconformity – An Historical Study." *Proceedings of the Geologists' Association* 73: 383–417.

Wegener, Arthur. 1912. *The Origin of Continents and Oceans*. Trans. [1966] by J. Biram. London: Methuen.

Whiston, William. 2015. *A New Theory of the Earth*. Arkose Press.

Wilson, J. Tuzo. 1966. "Did the Atlantic Close and Then Reopen?" *Nature* 211. August 13: 676–81.

Wood, Barry. 2013. "Bridging the 'two cultures': the Humanities, the Sciences, and the Grand Narrative." *The International Journal of Humanities Education* 10: 53–63.

Wood, Barry. 2015a. "Big Story Narratives: Reframing K-12 Science Education." *Proceedings, 13th Annual Hawaii International Conference on Education* (HICE): 1966–91.

Wood, Barry. 2015b. "Underlying Temporalities of Big History." *KronoScope* 15: 157–78.

CHAPTER 11

Time's Urgency Ritualized: The Centrality and Authority of Mayan Calendars

Margaret K. Devinney

Abstract

The K'iche' Maya, a Mesoamerican people still living in their ancestral lands, have for centuries regulated their sacred, social, business, and agricultural activities by adhering to their ingenious astronomical calendars. Mayan hieroglyphic writing records the dates of historical events and of the reigns of royalty as well as the worship of deities; this dating derives from mathematical processes applied to astronomical observations that focus on the mythological and agricultural deities associated with the Sun, the Moon, and the planet Venus, primarily as the Morning Star. The complete calendar consists of more than three distinctive "counts" and the combination and interpretation of these by especially revered diviners in the community allow for recording the past, advising on current activities, and predicting the future. A historical overview of the culture and examples from the *Popol vuh*, the K'iche' Mayan sacred creation myth, will help illustrate the depth of the calendar's significance in the community's recovered holistic culture.

Keywords

timekeeping – calendars – K'iche' Maya – Mesoamerican cultures – ethnography – archeoastronomy – agricultural practices – anthropology – *Popol vuh*

1   Introduction

It is not unusual for agrarian cultures to set planting times according to natural occurrences such as floods or celestial events. Nor is it unusual for people to formulate calendars that divide longer temporal periods into discrete units to mark rituals or celebrations. But for the K'iche' Maya, who live in the highlands of Guatemala, interpretation of the calendars by shaman-keepers is the most enduring and significant aspect of their ancestral heritage, and an integral facet of current culture. Daykeeping as ritually proscribed informs their quotidian activities, their ethnic customs and sacred beliefs, as well as their recording the past and prognosticating the future.

Extensive ethnographic and archeological work has uncovered thousands of stone and clay artifacts and several important documents referring to their ancestral cultural and spiritual beliefs and practices; and a revival of public practice of these rites since the 1970s reifies the centrality and authority of the calendar in most aspects of modern village life. A brief overview of K'iche' history illustrates the tenacity of many to protect sacred beliefs and artifacts, which are interpreted currently by successors of the ancestral keepers.

Central to this discussion is the time-keeping process, which embodies not only the dating mechanisms but also the time-space connection at the core of Maya spirituality. Fortunately, modern scholars have access to detailed records of celestial observations that K'iche' astronomers had preserved. Interpreting these produced an integrated cosmology that provided the basis for both practical and spiritual dimensions of life.

Almost five centuries of colonial rule, extended civil war and struggle for human rights saw the K'iche' and their culture repressed, attacked, and significantly diminished in political power and authority. Despite this, many hidden documents and oral traditions involving the calendars and their keepers survived, providing a strong impetus for the Maya recovery in the later twentieth century. All of this comes together in their ancient creation myth, the *Popol vuh*. Such a broad application of meaningful integration is noteworthy among world communities.

A brief history of the highland Maya, followed by a description of their calendars and the significance of observations derived from archeologists, ethnographers, anthropologists, and astronomers, will lay the groundwork for a discussion of how the K'iche' epic *Popol vuh* embodies the world view of the engendering culture.

2   Brief History of K'iche' Maya Culture

Mayan people living in the highlands and lowland rain forests of Guatemala and Mexico's Yucatán continue a culture that archeologists have traced to at least 3500 BC, with large cities rising as early as 500 BC. And following an eventual collapse of the civilization around 900 AD due to warfare, overpopulation, and agricultural disasters, a distinctive Mayan culture rebounded: they

TABLE 11.1   Major periods of Maya history

2000 BC–300 AD	Preclassic
300–600	Early Classic
600–900	Late Classic
900–1200	Early Postclassic
1200 to the European Invasion (1524)	Late Postclassic
1524–1821	Colonial/National Period

established urban centers, a glyph-based phonetic writing system, forms of creative artistic expression, and a well-developed social structure. Over the centuries, the culture diversified, and the kingdom of K'iche' rose and flourished from about 1200 AD until the arrival of the Spanish Conquistadors in Mesoamerica in 1524.[1]

From the time of conquest, with their mission to convert indigenous people, as well as claim their fabled gold for Spain, Catholic missionaries – for the most part – tried to eliminate or replace native beliefs and customs.[2] Generally, the imposition of Spanish rule and oppressive control by its colonial representatives caused locals to hide their ancestral treasure, documents, and ritual practice from public view. With the passage of time, however, changes occurred: syncretism with Catholic (and for the past 150 years also mainline Protestant and evangelical) beliefs; political turmoil that caused widespread uprooting and migrations in which people had to leave precious artifacts hidden or buried; and additionally, some (expected) loss of orally transmitted information.[3]

---

1   The remoteness of the Guatemalan highlands protected the K'iche' for a few years, allowing time to copy documents and hide treasures before the Spanish conquered them in 1541.
2   A notable exception to the missionaries' destroying written works as satanic: Between 1701 and 1703, parish priest Francisco Ximénez made a copy of the *Popol vuh* and added his Spanish translation. Today, this is the sole surviving Mayan-language version.
3   When Allen Christenson lived among the K'iche', he learned that various families kept sacred bundles of their ancestors' belongings. He describes a ritual of opening the bundles on specific dates, praying, apologizing for disturbing them, and then restoring them to their designated sacred places (17).

TABLE 11.2   Some major events in more recent K'iche' history

1821	Independence
Mid-19th century	Rise of coffee plantations/loss of land/subjugation of local population
1944–54	Unsuccessful Guatemalan democracy movement
1952	Agrarian reforms that negatively affect K'iche'
1954	US-backed coup of Guatemalan government
1960–96	Civil conflict
Late 1960s	Oil, nickel deposits found on Maya lands
1966	UN-sponsored peace accords
5.29.1978	Massacre at Panzós over ursurped lands
12.29.1996	"Accord for a Firm and Lasting Peace" signed

3   Contemporary K'iche' Social Issues

Today, like most indigenous Mesoamerican peoples, and despite their notewor-thy heritage, the highland K'iche' community is poor and underserved socially and economically. However, their adherence to the history and values of their ancestors is significant. Traditional communities regularly observe ancestral feasts and rituals, often mixed with Christian/Catholic symbolism. Although most K'iche' people inherited Roman Catholicism from the Spanish, and many practice a form of evangelical Protestantism, the ancient beliefs remain deeply rooted, and rituals proscribed by the calendars meld elements of multiple faith traditions. Such syncretism, however, might be expected, since the religious practices share many common elements: priests, ritual use of incense, proces-sions, pilgrimages to sacred places, reverence of images, and belief in a god who died and was resurrected.

The centrality of ancestral religious and social roots is also apparent in po-litical events, especially since the mid-1980s, after decades of armed rebellion and civil war caused massive emigration to Mexico, the United States, and Canada. Although K'iche' constitute at least 30% of all Maya in the highlands, and the Maya in Guatemala constitute 60–80% of the total population, politi-cal, economic, and civil rights participation long remained the privilege of the ladino population.[4] Quite recently, however, highland Maya, like many ethnic groups, have actively promoted their culture, participating more in the broader community, and strategizing on ways to assert authority concerning their own cultural interests.

In her comprehensive study of contemporary highland Maya spirituality, ethnologist Jean Molesky-Poz describes her participation in the highland Maya community in Momostenango at the start of the pan-Maya movement. Her observations illustrate the centrality of their ancestral heritage as they revalued their ethnic roots: "Individuals and communities are articulating their differences and constructing new political and social space" (15) while maintaining connections to nature and the ancients. With more men and women becoming daykeepers (*aj q'jab*) and more daykeepers holding public office, she found a significant reclamation and daily use of the *Chol Q'ij* (the 260-day calendar).

However, at that time (2006), the community was far from unified in its resolve, and deeply conflicted. In her interview with an *aj q'ij* who advocated the return to ancestral spiritual ways, she reports his description of the negative impact of competing Christian practitioners with minimum or no concern for indigenous practices. He lamented also that many local K'iche' had converted in order to participate effectively in business and politics; and he revealed that those who continued the ancestral path did so hidden from public observation.

In this context, it is productive to consider how current-day Maya are grappling with an obvious intrusion into their culture and the concurrent economic gain it provides: tourism.[5] In recent decades, increasingly greater numbers of individual visitors as well as organized groups have sought to view and/or participate in Mesoamerican indigenous cultures, engendering profits as well as problems for the highland Maya. As so often happens, tourist companies generate revenue by appropriating symbols of traditional cultures while failing to ground the tourists' experience in any meaningful understanding. Where ancestral Mayan symbols and traditional attire were once mocked, and at

---

4   Previous legislation, over centuries, sometimes actually intended to improve the lot of the Maya, however had unintentionally caused the loss of ancestral land including sacred sites, weakening of local political institutions, and widening class divisions.

5   The source article consulted for this data, "Pirates of Our Spirituality," focuses on the Kaqchikel, an associated highland group second in size to the K'iche'; their geographical and societal proximity would indicate that the issue described holds for both groups.

times forbidden, Mayas of Guatemala currently participate in national tourism programs via a 1992 agreement among Central American governments.

According to Elizabeth Bell, however, this is a reasoned decision by the locals, providing opportunities for them to assert authority based on their value to the broader community. The practical advantage for the participating Maya consists mainly in revenue from the tourist companies and minimal profit from sales of craft items. What many consider the most important benefit is the opportunity for locals to demonstrate their authority in matters of belief, ritual, and ethnic values. And here also, as daykeepers demonstrate and explain ceremonies to visitors, they pass on inherited indigenous knowledge and spiritual practices, thus asserting authority over their contemporary cultures as well as applications of the rituals that visitors adopt/adapt in their own communities.

Linked to the goal to prove their value to the broader business and political communities, the Maya daykeepers stress that they must be permitted to present only authentic rituals, rather than versions designed to enhance the spectacle aspects of tourism. However, as Bell explains, this stress on authenticity is rooted in the expected effect of assigning value, which ultimately values the practitioners and consequently helps protect them and their beliefs from extinction. Although this kind of symbiotic relationship seems fraught with risks of identity-loss and continued political inequality for the Maya, Bell sees necessity and even optimism:

> [A]s long as the Mayas themselves are able to construct the message presented, having more participants in Maya spirituality is both economically and ideologically advantageous, helping to validate Maya practices in a nation-State that does not recognize their value. (105)

4   Contemporary K'iche' Culture vis-à-vis Daykeepers

Anthropologists, archeologists, and ethnographers often describe the significant, seamlessly integrated interaction between current culture and traditions related to the sacred calendars. This is due primarily to the authority of the shaman-priest daykeeper (*Aq'jab*), who maintains his/her traditional role as spiritual – and often also practical – leader, serving as adviser and mediator between the old beliefs and the contemporary community. Most are farmers or small business owners with the usual economic and human-rights concerns of Western societies. Although the official language of government and business is Spanish, the K'iche' actively maintain their ancient language and traditions, most of which are rooted in the time-keeping tradition. Adherence to these key beliefs in the midst of strong external influences engenders a vibrant syncretism in religious and cultural observances.[6]

Observations by scholars who have lived among the K'iche' also attest to the authority of today's daykeepers' activities. For example, Allen Christenson reports that his research compiling a K'iche' dictionary and translating ancestral documents occasioned his "work[ing] with a number of Quiche [K'iche'] *aj'q'ijab'*, who continue to conduct calendric and divinatory rites in a manner little different from that practiced at the time the *Popol vuh* was compiled"(25). Testifying to the widespread participation in daykeepers' rituals, Michael Coe reports a K'iche' event celebrated at dawn every 260 days in Momostenango involving "tens of thousands of Indians" and "over 200 shamans act[ing] as intermediaries between individual practitioners and the supreme deity Dios Mundo."[7] Participants offer prayers of atonement and petition, as well as adoration and thanksgiving – much like rituals described elsewhere, while adding a potshard to the huge pile on the site to mark their presence.

And Barbara Tedlock, anthropologist and trained shaman-daykeeper, reifies the integral connection between current ritual practices and the 260-day calendar. She describes illnesses brought to the shaman-priests, including physical ailments as well as inebriation and money loss, noting the role of the daykeeper: "... one who burns incense and offers prayers at shrines on designated days of the *tzolk'in*, the 260-day sacred calendar" (123).

In ancient tradition as well as in contemporary K'iche' Mayan culture, daykeeper is the premier position of respect in the community. Allen Christenson and others report on the contemporary prestige of daykeepers, who, as in the ancient belief system, today still include matchmakers, midwives, and others who are charged with maintaining divine order in the community. There is no formal educational requirement; candidates are recognized by the community for their expertise and appropriateness. And daykeepers also take on other leadership roles besides the shamanic divination, such as prayer leaders, herbalists, and bonesetters. Each new year, celebrations include ritual initiation of daykeepers, who may actually serve indefinitely.

---

6   For example, anthropologists have observed that often, before a daykeeper performs a reckoning, he/she precedes the K'iche' ritual with a Catholic "Hail Mary" or "Our Father."

7   This is an example of syncretism, in the Maya culture often between Christian/Catholic and ancestral concepts: "Dios Mundo" is the Spanish epithet "God of the World." (254).

An integral part of the role of daykeeper requires interpreting the calendar by divination; and among contemporary practitioners the ritual involves casting an arbitrary handful of maize kernels or of *tz'ite* (red seeds of the coral tree) and then counting off the number of kernels corresponding to days on the calendar. The day at the conclusion of the count is then interpreted as either favorable or inauspicious for the situation at hand. This might involve personal issues concerning marriage, predictions for babies' fortunes depending on birthdays, settling business deals, or planting and harvesting crops. The daykeeper also reckons specific days throughout the year that are set aside for various feasts and observances, as well as providing information on ancient historical events.

This ritual counting and interpretation serves social, economic, and spiritual fundamentals of the community: it allows for recording the past, advising on current activities, and predicting the future; while the daykeeper acts as intermediary between the people and the gods, transmitting offerings, petitions, and gratitude; and also bringing to the people advice and demands from the gods.

5   Mayan Astronomy and Calendar Calculations

Archeological finds and evidence obtained from codes of the glyphs reveal the superiority of the Maya to other Mesoamerican cultures in mathematics, astronomy, and writing. Early written evidence of a calendar by ancient Maya astronomers dates from the sixth century BC, and by the Classical period, their descendants had devised ways of establishing significations and linkages among days, events, and ruling lineages (Thompson, 5–12). In addition, the achievements of the Maya in these areas reveal their actual history as well as their belief in divine aid in personal and ritual activities. For example, Mayan hieroglyphic writing on stelae, in almanacs, and in surviving natural-fiber fold-out books records the dates of historical events and of the reigns of royalty as well as names and functions of deities; this dating derives from mathematical processes applied to astronomical observations that focused on the deities associated with the Sun, the Moon, and the planet Venus, primarily as the Morning Star rising in the east.

Although their early ancestors, the Olmec people, had devised calendars, and other tribes of the region had derived dating systems from those, the Maya calendar is known to be the most sophisticated. It consists of several cycles or counts of different lengths. The 260-day count is known to scholars as the *Tzolk'in*, which was combined with a 365-day "vague" solar year known as the *Haab'* to form a synchronized cycle lasting for 52 *Haab'*, known as the Calendar Round. All of these are still in use by many groups in the Guatemalan highlands.

5.1 Tzolk'in / Chol Q'ij

The *Tzolk'in* (meaning "count of days") is the name commonly used for the Maya Sacred Round divinatory calendar; however, the specific K'iche' word is *Chol Q'ij*. It combines twenty day-names with thirteen day-numbers to produce 260 unique days used to determine the time of religious and ceremonial events and for divination. The 20 day-names and 13 numbers may be modeled as discrete interlocking discs, which allow for progress through sequential dates as the day disc rotates through all of the 13 number designations. [See Figure 11.1.] From the start of the new year, then, the first day would be 1 Imix,[8] followed by 2 Imix, and completing the cycle with 13 Imix; the process then moves to the second day name, Ik': 1 Ik', 2 Ik', etc. Thus the 20-day cycle is processed 13 times, for a full count of 260 days.

This 260-day lunar calendar can be traced to at least 500 BC. Among many applications are forecasting an individual's life experience based on the day of birth and predicting appropriate days for the planting cycle. It survives today among the K'iche', and current daykeepers still rely on it, as explained above. 260 days corresponds generally to the human gestation period and also approximates the Mesoamerican agricultural cycle.

5.2 Haab'

The *Haab'* count, also called the "vague" year, comprises eighteen months of twenty days each, plus a period of five days at the end of the year that synchronizes (although imperfectly) with the 365-day solar year. These five days, called *Wayeb*, are thought to be a dangerous time, with no boundaries between the earthly world and the Underworld. Nothing prevents malevolent deities from causing disasters such as disease, death, political upheaval, and war. An event in the *Popol vuh* illustrates the danger of this period mythologically, reifying the interpretations of the daykeepers: After Hunahpu and Xbalanque the Hero Twins, have been killed and thrown into the water by the Xibalbans, lords of the underworld, they reappear on the fifth day and begin their mission leading to the ultimate creation of humans and the current world order. The period of

---

8 Day names have been modified or changed by various groups over time. Day and month names in this article are designations identified by Lynn Foster in *Handbook to Life in the Ancient Maya World* as "16th-century Yucatec Mayan," (276).

trials they have endured at the hands of the Xibalbans, culminating in their "death," corresponds to the five "vague" days at the end of the solar year.

The *Haab'* calendar includes the *Tzolk'in* and thus shares 20 day-names; discrete days are identified by a day number in the month, followed by the name of the month and its sequence number. For example, Figure 11.1 shows the date 4 Imix 8 Kumk'u, the fourth iteration of the day called Imix, the eighth iteration in Kumk'u, the eighteenth month on the *Haab'* calendar. The names of the day and month glyphs, like their progression in the interlocking system, follow a set sequential pattern, and are recognized by their distinctive individual glyphs.

5.3   *Calendar Round*

Like other early Mesoamericans, the Maya have a 52-year calendar called the Calendar Round, formed by an interlocking cycle of the 260-day ritual *Chol Q'ij* calendar and a 365-day solar (also called "vague") year. A Calendar Round date meshes *Tzolk'in* and *Haab'* days, and because of its structure, the individual days will not repeat until the completion of 52 *Haab'* years consisting of 365 days each – or 18,980 days: thus the name "Calendar Round." In accordance with the preeminent Maya belief that existence, and therefore time, are cyclical, the conclusion of the 52-year *Haab'*count signals not conclusion, but rather a continuance into the next phase of the Long Count, which itself is cyclical.

5.4   *Long Count*

Yet another type of calendar tracks periods of time longer than 52 years, and was used for refining calendar dates (i.e., identifying when one event occurred in relation to others). Called the Long Count, this marks days since a mythological starting-point, the successful creation of humans, as explained in the *Popol vuh*. Many experts maintain that this success, after previous, less auspicious versions, started on 4 Ahau[9] 8 Kumk'u. According to the correlation between the Long Count and Western calendars accepted by the great majority of Maya researchers, this initial date, marking the successful creation of humans, is equivalent to August 11, 3114 BC when correlated with the proleptic Gregorian calendar.[10] By its linear nature, the Long Count could be extended to refer to any date far into the past or future. Such dates are numerous on ancient stelae recording historical events, disasters, and kings' achievements; and although

---

9   Ahau (also Ahaw or ajaw) is the twentieth day name; and appropriately for its connection
       with creation, it means "Lord."
10   Foster, 225.

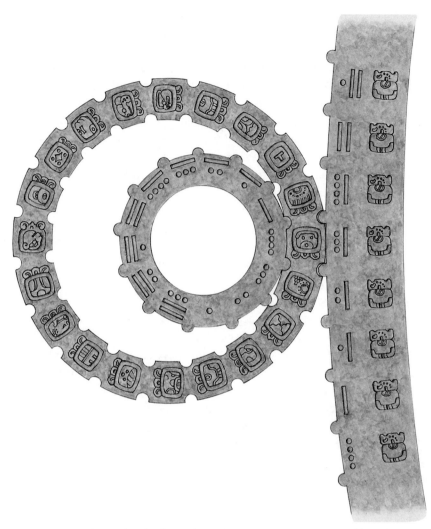

FIGURE 11.1   The *Haab'* Calendar model. In this interlocking cycle of three components of
Mayan dating, the smaller two represent the 260-day *Tzolk'in*, or *Chol Q'ij*; the
larger, the 365-solar year. As the *Tzolk'in* components rotate clockwise, they
progress through the 13 numbers assigned sequentially to the 20 day-names;
and as the combined *Tzolk'in* interlocks sequentially with the solar year
cycle (moving counter-clockwise), it produces the *Haab'* date. http://www.
dkfindout.com/us/history/mayans/mayan-calendar

its use seems to have been discontinued by the early tenth century, the Long Count dates provide reliable information for Mayanists.

One Long Count event in particular, December 20, 2012, prompted wide attention in the international press and especially among New Age adherents and believers in astrological significations. Many of these practitioners predicted the end of the world on that date, basing it on misinterpreted readings of the Mayan calendar. Scholars insist that there are no predictions nor prophecies related to that date in Mayan writings. What they do explain is that the Mayan date corresponding to 12.20.2012 is the end of a Long Count period, 13 *bak'tuns*[11] after the beginning date that corresponds to 8.11.3114 BC. Like the seamless continuation of dates at the completed cycle of the Calendar Round, the Long Count continued uninterrupted to begin a new cycle.

6   The Dresden Codex: Key to Deciphering the Calendars

Although use of the calendars has been consistently handed down, and continuously practiced, even if often secretly, its structural rationale is not immediately obvious. However, the most useful recordings of astronomical data that inform the calendars is available in three almanacs from the late sixteenth century: the Dresden, Madrid, and Berlin Codices, so named for their current locations. The Dresden Codex is central to this discussion because it is the most comprehensive, including extensive records of observations of Venus as well as mythical and ceremonial references to the *Popol vuh*.

While the astronomical records correspond with many made by ancient astronomers and astrologers, these relate specifically to fundamental Maya beliefs, especially concepts of resurrection and the cyclical nature of everything. Observers of Venus calculated a cycle divided into periods of visibility in various positions in the sky and periods of disappearance when its orbit takes it behind the Sun and when the Sun's brilliance obliterates it as it passes in front. Venus's alternating appearance as Morning Star and Evening Star depends on its advancing position in its orbit in relationship to observers on Earth. Because the planets' orbital speeds differ, their relative positions vary with respect to each other. For example, the 584-day cycle of Venus includes an 8-day disappearance as it passes in front of the Sun, a brief appearance

---

11   Bak'tun: in the Mayan calendar, a period of 20 K'atuns, each equivalent to 7,200, or 144,000 days.

in the morning sky prior to sunrise (Morning Star), increased brilliance before finally fading, and obliteration by solar light as it moves closer to the Sun
(seven weeks).

The Dresden Codex also features glyphs depicting deities and their activities, and also lists of numerical records of astronomical observations. These
led the observers to anticipate lunar and solar eclipses, phases of the Moon,
and movements of the planets. Archeoastronomer Anthony Aveni explains the
critical importance of three particular pages that record the phases of Venus
along with a list of *tzolk'in* dates. The associated glyphs, because they refer to
omens and augury, link to Venus, especially in reference to the Maya belief that
the planet/god was associated with bad luck at certain times. Along with observation tables that have been certified as correct, these features of the codex
attest to the astronomical and spiritual foundations of Mayan timekeeping. All
of these data led to the formulation of the particularized time-keeping evident
in the calendars described above. Shamans could predict favorable times for
planting and expected harvest dates, as well as for determining appropriate
dates for specific rituals.

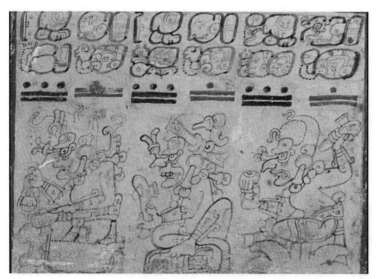

FIGURE 11.2   A typical page from the Dresden Codex, considered the oldest
                book produced in the Americas. It contains glyphs, astronomical
                information, and actual drawings representing events in the
                Popol vuh. Additional pages record astronomical observational
                data and describe rituals associated with the planet Venus.
                Notice the integration of dates (glyphs and numbers above
                them, consisting of dots and bars) and stylized figure drawings,
                which were painted in vivid natural-dye colors.

7   *Popol vuh* vis-à-vis Astronomy and the Calendars

Because the K'iche' had very few natural resources like the gold and silver desired by the Conquistadors, they were among the later Mesoamerican popula-
tions to come under Spanish control (1541). This allowed time for members of
noble families to transcribe their old glyph texts, using a modified Latin script,
for preservation and safe-keeping. Among such documents that survived
destruction by the Spanish are the Western-script *Popol vuh*, the traditional
Mayan creation epic, likely completed between 1554 and 1558.

Although it is difficult to summarize the *Popol vuh*, it suffices here to note
that it contains the story of the creation of deities and humans (forefathers of
the K'iche' clan patriarchs), as well as their struggle to escape the evils of the
underworld and find a permanent homeland. And because the stated purpose
in the opening lines,

> This is the beginning of the Ancient Word, here in this place called Quiché
> [K'iche']. Here we shall inscribe, we shall implant the Ancient Word, the
> potential and source for everything done in the citadel of Quiché, in the
> nation of Quiché people,
>
> TEDLOCK, 63

is to preserve their ethnic values, the writers often link events in the story to
customs and religious practices of their time. More than five centuries later,
their descendants continue these beliefs and practices, confirmed for them by
the same texts.

A significant aspect of the *Popol vuh* for both ancient and current K'iche' is
the transmission of crucial explanations concerning the community's place
in the universe. And to a significant extent, this is accomplished via stories
whose characters and actions relate directly to astronomical and teleological
interpretations. Furthermore, in this context, it is helpful to note the direct interplay of the stories with the calendars and consequently, with rituals, in addition to the fact that many character names are also day names of the calendar,
and most day names are related to various places and events.

When daykeepers prescribe days for ritual events, the actual day name may
be associated with the ritualized remembrance. For example, when the Hero
Twins have defeated the Xibalbans in the Underworld, they try to revive their
murdered uncle, Seven Hunahpu. Although they fail, they promise him that
human beings will always remember him by coming to pray on this day (the
day named for Hunahpu) at this place where his remains are believed to rest.

Currently, Hunahpu days are dedicated to ritual remembrance of ancestors by visiting cemeteries and leaving gifts such as incense, food, drink, or flowers; and perhaps even more interesting in terms of the centrality of the *Popol vuh* to the culture, graveyards are called by the same word (*jom*) as the sacrificial ball courts in the epic.

8 Celestial/Astronomical Correspondencese: *Popol vuh*

At this point, a few examples from the myth will illustrate some of the relationships between our remarks about the calendar's reliance on astronomy and agriculture, as well as providing a rationale for various rituals that are still practiced.

Dennis Tedlock presents a compelling analysis of the following story and its consequences: Blood Moon, daughter of a Xibalban (Underworld) lord, is the mother of the Hero Twins, Hunahpu and Xbalanque, and the daughter-in-law of the creator goddess Xmucane. Xmucane's sons (One Hunahpu and Seven Hunahpu) have lost at the ball court in Xibalba, and Seven Hunahpu's head (a calabash, by trickery) has been placed in a tree. Blood Moon happens by and is directed by the head to hold out her hand while the head spits into it, causing her to conceive Seven Hunahpu's twin sons, Hunahpu and Xbalanque.

The story works on several levels to fulfill the K'iche' aims, as Tedlock explains: at the astronomical level, Blood Moon corresponds to the Moon, appearing in the west at night, corresponding to her arrival at Seven Hunahpu's skull at the Place of the Ball Game Sacrifice, when she became pregnant. On her release from the Underworld, she visits Xmucane, claiming to be her daughter-in-law. Xmucane sets a test involving gathering corn, which Blood Moon completes by using a magical net, proving that she is telling the truth and carrying Xmucane's grandchildren. Xmucane, adept in the interpretation of the Mayan calendar, knows that one of the twenty day-names in the calendar is "Net." Thus, as a daykeeper herself, she interprets the events within the context of the calendar – a typical example of the integration of events and characters in the *Popol vuh* with contemporary belief and ritual (33–37).

This analysis of the situation clarifies the essential relationship of the epic and astronomical reckoning: From a calendrical point of view, the story so far is that Venus rose as the Morning Star on a day named Hunahpu,[12] corresponding

---

12 Note that the day Hunahpu is eponymous for the *Popol vuh* character, which is the basis for naming the day.

to the ball playing of Xmucane's sons, One Hunahpu and Seven Hunahpu; then, after being out of sight in Xibalba [the Underworld], Venus (Blood Moon figure) reappeared as the Evening Star on the day named Death,[13] corresponding to the defeat of her sons One Hunahpu and Seven Hunahpu and the placement of Seven Hunahpu's head in a tree in the west. The next sky-related event in the story is the rebirth of Venus as the Morning Star, which should fall, as Xmucane already knows, on a day named Net. When she sees the imprint of the net in the cornfield, she takes it as a sign that this event is near and that the sons born to Blood Moon will make it possible.

Further episodes correspond to an appearance of Venus in the east, the direction of morning and life, thus prefiguring the cycles of Venus and preparing for the first rising of the Sun, as part of the creation story. For example, as the Hero Twins are fighting for their lives in the Death Bat's house in Xibalba, "the horizon of the sky began to redden, for it was about to dawn"[14] and the possum deity, who is in charge of dawn and addressed as "old man," is directed to delay it, to make time for Hunahpu's head to be restored. Eventually, the old-man possum allows the Sun to rise for the first time. Thereafter, the "old man" of the myth will initiate each new solar year, as contemporary K'iché daykeepers reckon the event.

To enhance the discussion, above, of the cyclical nature of timekeeping among the Maya peoples, we might investigate the relationship of the phenomena to K'iche' traditions related to astronomical and calendrical calculations. For example, on their way to answer the invitation of the Xibalbans to come to the Underworld, Hunahpu and Xbalanque (the Hero Twins) visit their grandmother, Xmucane. They ritually sow corn at the center of her house and tell her that when the corn dies, it will be a sign that they have died; but when it sprouts again, she will know that they are again alive. This is central to the K'iche' ideas about agricultural – and cosmological – regeneration cycles, which are also reflected in the calendrical systems. Planting corn in the house prefigures a harvest ritual that persists among the K'iche': In fact, contemporary K'iche' adhere to the tradition of keeping consecrated corn in the house, as part of the stored harvest, and they plant according to agricultural observations and daykeepers' reckonings.

This particular situation provides a further link to the calendar and astronomical observations: A play on the words for "corn" (*aj*) and "house" (*ja*) produces the day name *Aj*. And this play is familiar to contemporary K'iche' daykeepers, who use it when they explain to clients that the day Aj is

---

13   Death is actually a day name in the *Tzolk'in* calendar.

14   Christenson, 173.

portentous in matters affecting households. As Tedlock explains: If the twins planted their corn ears in the house on this day, then their expected arrival in Xibalba (the Underworld) seven days later, would fall on the day named Hunahpu. This fits the Mayan Venus calendar perfectly: whenever Venus rises as the Morning Star on a day named Net, corresponding to the appearance of Hunahpu and Xbalanue on earth, its next descent into the Underworld will always fall on a day named Hunahpu (39). The cyclical operation of the calendar system causes such accurate predictions and reliability of proper, timely celebration of events.

From the point of view of astronomy, the visit of Hunahpu and Xbalanque to Seven Hunahpu's grave reflects additional interconnection of the myth on the cycles of Venus. First, the Morning Star first appears on a day named Hunahpu. Then, after the twins have promised the dead Seven Hunahpu, that he will never be forgotten, they rise as the Sun and Moon. To the contemporary K'iche', this is the origin of equally important daylight and nocturnal illumination; they regard the full Moon as an equivalent of the Sun, because, like the Sun, it is a full disk, is bright enough to see by, and transverses the sky in the same time it takes the Sun to do so.[15]

9   Celestial/Astronomical Correspondences: Agricultural Phenomena

While the *Popol vuh* essentially incorporates the spiritual beliefs of the K'iche', it also demonstrates the crucial interlocking of agricultural concepts with a significantly deeper teleology. That an agriculture-based society should describe its astronomical observations in terms of farming practices is not unusual, but the K'iche' seem to have expanded the biological processes of agriculture into a complex metaphor that extends to astronomical and spiritual aspects of the passage of time by relating observable, recurring phenomena to calendrical data. The pervasive significance of the cyclic operation of nature for them is revealed especially in the interconnections of the concepts "sowing," "sprouting," and "dawning." While sowing obviously involves planting – preparation for seeds sprouting – this "sowing" prepares directly for a kind of "dawning" – the emergence as something new but closely related. Descriptions of creation in the *Popol vuh* might help us understand the fundamental concepts, which apply almost universally; that is, far beyond the realm of agriculture. For

---

15   Tedlock considers that, most likely, the twin who became the Moon is to be understood specifically as the full Moon, whereas Blood Moon, the mother of the twins, would account for the other phases of the Moon" (43).

example, the narrative arc of the *Popol vuh* posits a natural cycle that begins in the sky as well as in agriculture, and moves inexorably to death and subsequent rebirth – sprouting (the death of the seed) and dawning.

Careful and frequent observations of the sky led early Mayans to consider that celestial bodies, many of which represented deities, disappeared into an underworld only to rise again, to "grow" as they traversed the sky, before traveling downward (from the observer's perspective) once again.[16] They noted that after harvest, corn died; but when dried kernels were planted in the earth, below the surface, they eventually sprouted and grew to be harvested. These conclusions lead directly to the belief in human regeneration after death and, crucially, to the emphasis on the cyclical nature. of time and the centrality of timekeeping to the community ethos. Even if more sophisticated scientific and arcane knowledge brings more mature thinking, ritual and belief may remain, as folklorist and ethnologist Alan Dundes maintains: myths are, for their progenitors, sacred narratives.

Human creation in the *Popol vuh* is a metaphorical example of planting, rather than copulation. For example, the spittle from Seven Death's head hanging in the calabash tree impregnates Blood Moon with the Hero Twins when she holds out her hand as he requests: "Right away something was generated in her belly, from saliva alone, and this was the generation of Hunahpu and Xbalanque [the Hero Twins]". Then, fulfilling the required opposite of dawning, in Part Four the K'iche' gods who become Venus, Sun, and Moon must first descend into the Underworld of the Xibalba before rising – "dawning."

10   Conclusion

Measuring and recording time are at the core of K'iche' culture. For them, records of celestial observations and mathematical systems tracking temporal units continues to promise fulfillment of practical and spiritual community imperatives. The extraordinary continued significance of shaman daykeepers and ancient texts that linked their deities directly to timekeeping attest to this. And while it is not unusual for a community to record historical events, to schedule ritual celebrations and memorials, the continuing comprehensive centrality and authority of the K'iche' timekeeping system may be regarded as exceptional.

---

16   The cycle of Venus as observed by Mayans is a good example of this concept of death and resurrection as described above.

References

Aveni, Anthony F. "Venus and the Maya: Interdisciplinary Studies of Maya Myth, Building Orientations, and Written Records Indicate That Astronomers of the Pre-Columbian World Developed a Sophisticated, If Distinctive, Cosmology." *American Scientist* 67, no. 3 (1979): 274–85. http://www.jstor.org/stable/27849219.

Aveni, Anthony. *Skywatchers*. Austin: University of Texas Press, 2001.

Bell, Elizabeth R. "Pirates of Our Spirituality: The 2012 Apocalypse and the Value of Heritage in Guatemala." *Latin American Perspectives* 39, no. 6 (2012): 96–108. http://www.jstor.org/stable/41702297.

Christenson, Allen J. *Popol vuh. Translated from the original Mayan Text*. Norman: University of Oklahoma Press, 2007.

Coe, Michael D. *The Maya*. New York: Thames & Hudson, 2011.

Foster, Lynn V. *Handbook to Life in the Ancient Maya World*. New York: Oxford University Press, 2002.

Gates, William. *The Dresden Codex: Reproduced Tracings of the Original*. Baltimore: The Maya Society at Johns Hopkins University, 1932.

Milbrath, Susan. *Star Gods of the Maya: Astronomy in Art, Folklore, and Calendars*. Austin: University of Texas Press, 1999.

Molesky-Poz, Jean. *Contemporary Maya Spirituality: The Ancient Ways Are Not Lost*. Austin: University of Texas Press, 2006.

Schele, Linda and David A. Friedel. *A Forest of Kings: The Untold Story of the Ancient Maya*. New York: William Morrow & Company Inc., 1990.

Tedlock, Barbara. *Time and the Highland Maya*. Albuquerque: University of New Mexico Press, 1982.

Tedlock, Dennis. *Popol vuh. Definitive Edition of the Mayan Book of the Dawn of Life and the Glories of Gods and Kings*. Simon & Schuster: New York, 1996.

Thompson, Eric S. *The Rise and Fall of Maya Civilization*. Norman: University of Oklahoma Press, 1954.

CHAPTER 12

Telling Time: Literary Rituals and Trauma

Daniela Tan

1 Introduction

In March 2019, eight years had passed since the tsunami and the nuclear catastrophe in North-Eastern Japan on March 11, 2011. How has literature been responding to and reflecting the immediate impacts and the more long term consequences of this disaster so far? By analyzing the phases of this process, this article discusses literature about Fukushima, which is not only about natural disaster but also about man made nuclear threat. This specific event will then put in relation to the more general subject of catastrophic literature in Japan in order to contextualise it within a broader frame and carve out literary patterns in the process of dealing with catastrophic impacts and trauma. The focus of this discussion will be on literary texts from 1945 up to the present.

A-bomb literature constitutes a genre of its own in Japanese postwar literature under the label *genbaku bungaku*. Literature on this topic differs several phases, from the urgent need to testify as an immediate reaction until the discussion of the responsability of writers at a later point (cf. Kawanishi 2001 or Treat 1995). I will argue that these phases can be observed in literature dealing with other catastrophic events as well, and therefore can be contextualized within a broader framework of a literary pattern of texts concerned with catastrophic events.

Further, the time passed between the experience and the act of telling is in a clear relation to how traumatic events are eventually told. Therefore it will be argued that in analyzing this process, literary patterns show a similarity to the clearly defined steps of a ritual. This approach will help to understand the shift that occurs through the confrontation with events that exceed the human capacity of understanding towards the role of literature as a tool in the process of coming to terms with what happened.

The following chapter gives an overview on the recent discussion of literature dealing with the trauma of nuclear threat and provides a framework for the discussion of the primary text sources.

2 The Nuclear Age: From Hiroshima and Nagasaki to Fukushima

In many ways the nuclear accident in the Fukushima power plant reawakened the discussion about atomic energy and nuclear threat in Japan. The fact that both, literature about Fukushima and about the atomic bombs had to deal with radioactivity led to the term *kaku-bungaku* (nuclear literature). This put the existing genre of A-bomb literature in a broader context and served as a connecting line. It has been argued that literary texts about both Hiroshima and Nagasaki as well as Fukushima can be discussed within the broader framework of literature dealing with nuclear power (Kawamura 2011). The post-Fukushima literature is presently discussed within the broader context of contemporary literature of the era defined as nuclear age (*kaku-jidai*) (Kuroko 2018), which draws a bow from atomic bomb literature until the actual day and also brings along political implications regarding the positioning of writers within the discourse about atomic energy.

Another parallel is the phase of latency – on an individual, personal level as well as on an administrative level of censorship. Most literary works dealing with the A-bombs could only be published with a huge delay of at least nine years – what made a public discussion about what really happened in the war impossible (Komori 2014: 66). Similar efforts have been undertaken also after 3/11, although witnessed facts, rumours and distribution of various and sometims even contradicting information was processed unequally faster through the advanced communication technologies of the 21st century. However, in the same way the impacts of radioactive contamination and its effects on health and environment was revealed and understood in pieces and much after the bombs, the long-term effect of Fukushima is yet to be revealed in the future.

Although it might seem problematic to draw a comparison between a-bomb literature which started to be written more than 70 years ago and post-Fukushima literature about 10 years ago, this article tries to give a model for further research in analyzing the patterns in the process of telling traumatic events and catastrophes.

3 Fukushima March 2011 to August 2018: From Immediate Reactions to Present Discussion

On March 11, 2011, an earthquake in Northeast Japan with magnitude 9.0 and the following tsunami triggered the worst nuclear accident since Chernobyl. The Fukushima Daiichi power station located on the Pacific coast was damaged, and the destruction of power supplies led to a leakage of radiactive materials from the power plant.

While auxiliary forces tried their best to regain control over the situation, an incredible amount of misinformation and hyperbole circulated in the internet and through social networks. The government tried to keep the panic down, irritating the people and delaying desperately needed help.

"'A loss of words' or 'a lack of words' – these phrases most accurately illustrate the initial response of writers, poets and literary critics to the disaster. Many of those who work in literary arenas said that they did not know what to say or how to say what they felt" (Yuki 2015: 215).

In the midst of this, writers such as Murakami Ryū (*1952) chose to write on the situation from their personal point of view, and give an account of what happened.

> The earthquake hit just as I entered my room. Thinking I might end up trapped beneath rubble, I grabbed a container of water, a carton of cookies and a bottle of brandy and dived beneath the sturdily built writing desk. Now that I think about it, I don't suppose there would have been time to savour a last taste of brandy if the 30-story hotel had fallen down around me. But taking even this much of a countermeasure kept sheer panic at bay. [...]
>
> At present, though, our greatest concern is the crisis at the nuclear reactors in Fukushima. There is a mass of confused and conflicting information. Some say the situation is worse than Three Mile Island, but not as bad as Chernobyl; others say that winds carrying radioactive iodine are headed for Tokyo, [...]
>
> MURAKAMI 2011

Poet Wagō Ryōichi (*1968), a resident of Fukushima, chose to a more radical approach and used the short format of twitter to publish local news and impressions in a direct yet poetic language. The long tradition of condensing thoughts, feelings and landscape in the short and strictly defined shape of poetry such as Haiku is well suited to the 140 character restriction of a tweet. His eywitness reports about the situation in and around Fukushima quickly attracted attention all over Japan and even abroad. The mixture of powerlessness and urgency implied in the lines below serves as an example for the density of Wagō's imagery.

放射能が	*Hōshanō ga*	Radioactivity
降っています	*futte imasu*	is falling.
静かな夜です	*shizukana yoru desu*	A quiet evening.
(Wagō 2011)		

Another example of this way of dealing with the immediate shock and the questionable status of much of the available information can be seen in the novelist Yū Miri's (*1968) approach, who entered the exclusion zone and wrote down what she saw in a kind of undercover report titled *Keikai kuiki*, Caution zone. She started it in 2012, when she began working as a host on a radio talk show, which was part of a city-sponsored disaster-preparedness effort. In 2015, she even moved to Minami Soma, a city located only 25 kilometers north of the crippled Fukushima nuclear power plant (cf. Honda 2015), where she started a book store in 2018. In describing the paradox of the first days of spring with the beauty of blossoming cherry trees in the evacuated landscape with its empty schools, Yū tries to grasp the lives of the people who have lived here until recently. "The houses in this zone are not damaged at all, you can see blooming trees and laundry that was still hanging to dry. The destruction was of invisible nature, the people had to leave because of the radioactivity," she says in an interview (cf. Iwata-Weickgenannt 2012).

> The trees strike roots at a certain place, reach out their branches, let grow their trunks, and carve their own time.
>
> YŪ 2012: 260

In adding her memories of a seemingly unharmed past to the scenery, Yū creates a double layer of times, contrasting the ghostly emptiness to the vivid impression she had only a year ago:

> Last year about this time, under the cherry trees in full bloom, families and lovers walk by hand in hand and enjoy their beauty, walk under the falling cherry blossoms, broken-hearted about the short moment with the beloved and bereaved – I imagine how they walk under the cherry blossoms together.
>
> YŪ 2012: 261

The literary technique of combining two layers of time and creating a split-screen effect of simultaneousness is encountered often in literary texts dealing with traumatic events in an early stage. It serves as a means to emphasize the contrast between and – often nostalgically idealized – past and the injured presence. The motive of ruins is typically used to depict the vast scenery of an aftermath. It also is a popular metaphor for the destructive power of war or catastrophes that goes far beyond material loss It oscillates between the association with both emptiness and the imperative to rebuild the world, and hosts the invisible and untold fear of those who have survived, but will have to continue their lives with the memory of what had happened. In the case of a nuclear catastrophe, the invisibility of radioactive contamination is expressed by this means of contrasting evocation of a time that does not exist anymore.

In an unpublished interview in September 2016, Yū described her commitment in the stroken aera as a deeply compassionate task, serving as a mere container (*utsuwa*) for the stories of loss and pain people intrust to her. Beside her documentary work she is completing the volume *Keikai kuiki*, extracts of which were published serialized in the literary magazine *g2*.

It is the "surreal experience of unending images unfolding accross innumerable screens; feeling part of scenes that should not be" (Slaymaker 2015) that can be set alongside with the vital images described as being essential in constituting trauma:

> 'Vital images' are the enduring products of overpowering experiences that vividly and accurately 'freeze,' embody and visualize the conflicted terms of traumatic experience so that they can subsequentedly be accessed, 'thawed,' worked over and through, (re)constituted, known and acted upon ...' The image is thus that cognition of the truth of what every other discourse rushes to deny.'
>
> DAVIS 2001: 13–14

The transformation of the looping traumatic memory into the process of assimilation can be realized through the means of narration (cf. Stahl 2016: 172). Whereas this process may require a long period including latency on a personal level, shortly after the first shock, an increasing number of books and magazines began to be published in rcsponse to the catastrophe:

> A large number of writers, poets, and literary critics started to write about the 3/11 disaster, either in a direct manner as represented in reportages such as *Tsunami to genpatsu* (Tsunami and Nuclear Power Plant) written by Sano Shin'ichi (*1947) or in a more subtle, indirect manner as exemplified by the novel *Suuīto hiaafutā* (Sweet Hereafter) by Yoshimoto Banana (*1964).
>
> YUKI 2015: 215

When the immediate shock and the difficult situation which information source one could rely on was confronted with the urge to testify and use literature as a means to report what one had witnessed, slowly an other phase started. Based on the assumption that shock is followed by realization (Tachibana 1998) is is restrained by becoming aware of oneself as a survivor – and the confrontation with loss and pain. This can be substantiated by texts dealing with loss and parting from the deceased. In his 2013 novel *Sōzō rajio*, Imagination

Radio, Itō Seikō (*1961) unfolds the narrative of those washed away by the tsunami. Narrator and protagonist DJ Arc presents his late night show in Imagination Radio, chatting and playing music. The strange thing about his radio station is, that it can be received by anyone. During the process, DJ Arc starts realizing that he himself is one of those swept away in the flood and that he is dead, yet still unable to detach completely from the world of the living.

> In the small town that spreads out in front of my eyes upside down, there is not one single human being! So to speak, you could say that I must have started to talk incessantly to escape from this terrifying feeling and this imaginary radio is a desperate measure against my loneliness. What the heck has happened in this world I am now? If I start to think like that it makes me crazy.
>
> ITŌ 2013: 25

In her 2014 novel *Zazen gāru*, Zazen girl, Taguchi Randy (*1959) approaches the topic of loss and displacement of memory from the point of view of Yōko, a writer in her forties, who is confronted with the situation of loss on various levels. She adopts Rinko, a young woman who shows strong signs of dissociation. She has problems adapting to her social environment and is obviously suffering from partial memory loss. In the course of the story, it becomes clear that Rinko has lost her boyfriend in the tsunami and is about to give up on herself due to this grief. At first, Yōko feels the urge to take care of Rinko, but at the same time she is repelled by her behaviour. Struggling with this internal conflict, she integrates Rinko in her family and social environment, and finally inspires her to start practicing Zen meditation. The loss of beloved ones relates to the complete social network one had until the catastrophe and the consequences such as evacuation destroyed local communities and exposed many individuals to a degree of isolation they had not experienced before.

Unlike other writers responding to the disaster, Taguchi Randy has been dealing with nuclear issues including those of Chernobyl and Fukushima since the late nineties, when a serious criticality accident occurred in JCO, an uranium reprocessing facility in Tōkaimura, which is 120 kilometers north of Tokyo (cf. Taguchi 2011: 28). She has also been visiting Hiroshima many times henceforward, resulting in the volume entitled *Hibaku no Maria*, Maria, nuked (2009) with four stories that are set in Hiroshima. Taguchi avoids getting trapped within any catastrophe romanticism in order to feel oneself real and alive through the authentic contact with a disturbing reality. Rather she builds up a critical distance and defies control and instrumentalization by ideology, enabling herself to broach the issue of the "sway between a logic of reality and a logic of simulacra and simulation" (Yuki 2015: 225).

In the present, literature as a means of coping with the past is still in an on-going process. But in the meantime a discourse has evolved among the literary critics about the role of the author. Literary criticism has become dominated by the imperative of the current, and although literature is not expected to de-liver answers to present problems, it is expected to give interpretations of our time and even anticipate forthcoming tendencies.

The question of whether, and if so, in which way, literature has to engage with contemporary issues, is discussed intensely within Japanese literary criticism. Although the boundaries between novel, report and essay are being blurred, some critics tend to the opinion that literature has to be more than a means to transport timely information.

This evolution of discourse is repeating itself in many ways if we compare it to the literature and history of literary criticism in the post war era, as I will show in the next section.

4   Hiroshima August 6, 1945: Shock and the Telling of Trauma

When on August 6 of 1945 shortly after eight, the morning of a steaming hot summer day in western Japan like the day before and the day before that, a uranium bomb exploded 600 meters over the center of Hiroshima city, the world knew nothing about nuclear violence and the destructive power of the atomic bomb. *Pikadon*, the onomatopoetic term for the dazzling flash of light and the thunder that followed, was one of the first words for what had just happened. About 80,000 people lost their lives in that split-second, and other tens of thousands died later from the after-effects of radiation. The poet Hara Tamiki (1905–1951), who was only a mile from the epicenter, wrote that it was "as if the skin of the world around me was peeled off in an instant."[1]

A bit more than three weeks later, a young writer's record of her personal experience was published in the *Asahi shimbun* newspaper. In the first liter-ary treatment of the atomic bomb, Ōta Yōko (1906–1963) describes the drop-ping of the atomic bomb and the immediate consequences for the people of Hiroshima.[2] The young woman asked her relatives in the countryside, where she had fled lightly injured, for a pencil stub and scraps of paper and wrote down her impressions. She was awakened by the atomic flash, and she saw the

---

1   Hara, Tamiki quoted in Itō Narihiko, Sigfried Schaarschmidt and Wolfgang Schamoni (eds.): Seit jenem Tag. Hiroshima und Nagasaki in der japanischen Literatur. Frankfurt am Main: Fischer, 1984, p. 201.

2   Ōta Yōko 大田洋子. *Kaitei no yō na hikari. Genshi bakudan no kūshū ni atte* 海底のよう な光。原子爆弾の空襲に遭って, 1945. (In: *Ōta Yōko zenshū* 2, Nihon tosho sentā, 2001, p. 275–280).

images of her dream mix with the greenish blue light, a "light as on the bottom of the sea." Her brutal awakening is followed by the nightmare, the chaos of the streams of burned and confused people, corpses, the fragmentary information, snippets of conversations about rumors of the unknown weapon, and medical information. All this is put together like a collage. The word *genbaku* – atomic bomb – does not appear yet. The brief text saw the light of day for one reason only: in the short time between the Japanese surrender on August 15 and the enactment of the press law of the American occupiers on September 19, there was no censorship. All later literary texts dealing with the atomic bomb could see print only long after the fact.

The genre of atomic bomb literature (*genbaku bungaku*) within Japanese literature of the twentieth century is unique, yet one can discern clear parallels to the literature of catastrophe in a wider sense: Shock is followed by the eye-witnessing and collage-like gathering of the available and fragmentary information. After an interval of time, the role of literature is discussed and critics begin to disparaghe the testimonial. In a later period, when the passing of time reveals long term consequences, an ideological discourse assumes shape. At this point, writers and critics that have not been directly affected start to take part in the discourse, as for example writer Ōe Kenzaburō (*1935) did with his Hiroshima notes in 1965, where he reported on the situation of locals injured by the bomb, legal and medical issues and the peace movement. Another example would be the novel Black rain by Ibuse Masuji. Atomic bomb literature understands the existence of nuclear weapons as a central problem of society and human civilization.

It can be divided roughly into four groups (cf. Thornber 2001, Treat 1995, Tan 2017).[3] The first is the works of authors who were directly affected, who witnessed the dropping of the bomb themselves and wrote about it in detail, like Hara Tamiki, Ōta Yōko or Agawa Hiroyuki (*1920). In the second group belong the works of authors who witnessed the bombs as children but whose autobiographical narratives often have the character of a requiem for victims of radiation sickness – family, neighbors, such as Nagasaki born Hayashi Kyōko (1930–2017) and manga artist Nakazawa Keiji (1939–2012), who documented what he witnessed in *Hadashi no Gen*. Ōba Minako's (1930–2007) novel *Urashimasō*[4] (1977) lets the reader participate in this search to back-track to

---

3 Kawanishi 2001 classifies theree groups of hibakusha writers: those who witnessed the bomb as adults and gave direct literary testimony; those who did not eye-witness the bonbs but wrote about it, and last those who witnessed the bombings in their childhood or youth and only started to write about it after a 20–30 year long period of latency.

4 English translation by Yu Oba: *Urashimaso*. Saitama-ken: Josai University. Center for Inter-Cultural Studies and Education, 1995.

the witnessed horror. Like most of these authors, it took many years of latency before she started to write about the traumatic events.

The third group includes texts whose authors did not witness the the bombs themselves but who work with the aid of data and interview material from victims of the atomic bomb. These texts were often published much later, such as Ibuse Masuji's (1898–1993) *Kuroi ame* (Black Rain),[5] in 1966. The narrative is structured through the fictional figure of Shigematsu Shizuma, who copies passages from the diary of his niece Yasuko to send them to a marriage broker. As rumors circulate about Yasuko's infection with radiation sickness, it proves to be almost impossible to arrange a marriage for her. Shigematsu is keen to give a most authentic account of the everyday life of those affected by the atomic bomb. For this reason, he inserts passages from his wife's diary as well, who mainly records her sadness and her attempts to maintain a minimal quality of life by the daily search for food. In basing his narrative on various accounts and diary entries, Ibuse Masuji creates a polyphonic piece of work where the shock and uncertainty of the atomic bomb affect life in the weeks afterward. Within the frame of the story, these events are reflected and modified, and the result is a thematic accentuation and compression. In Japan at various times *Black Rain* and with it all atomic bomb literature have been heavily criticized: that these accounts lack literary qualities and serve as a cheap means for the authors to distinguish themselves, that they are descriptions mainly of political events.

Even Ōe Kenzaburō, winner of the Nobel Prize for literature in 1994, was confronted with a similar critique for his *Hiroshima nōto*, Hiroshima notes,[6] published in 1965.

The fourth group includes authors such as Tsuji Hitonari (*1959) or Murakami Ryū, who employ the historical setting of nuclear disaster as a background to their novels or even create dystopic scenarios, such as in *Gofungo no sekai*. The world five minutes later (1994), where Japan's present is split into humanity living beyond earth to escape the radioactive contaminated surface, and communities which returned to an archaic way of living in the woods. In a similar way, the historic past of the atomic bomb shines through in the novel *TIMELESS* (2018) by Asabuki Mariko (*1984), where one protagonists' grandmother has survived the atomic bomb but has to suffer the medical consequences. These writers were born in the post war era and deal with the subject of the A-Bombs from a stance beyond historical or political intentions.

---

5   An english translation by John Bester appeared in 2012 at Kodansha America under the title *Black Rain*.
6   English translation by David L. Swain and Toshi Yonezawa appeared 1995 at Grove Press under the title *Hiroshima notes*.

If priority is given to witnessing, the consequence is the first person perspective of eye-witnesses. For example, Hara Tamiki wrote *Natsu no hana*, Summer flowers (1945)[7] while suffering radiation sickness. The short novel should have been published in January 1946 under the title "The Atomic Bomb," but censorship blocked it. It deals with the time immediately after the bomb on August 6 in Hiroshima and the first person narrator's flight from the approaching flames. His way down to the river leads him among the injured: "At first sight, rather than pity, I felt my hair stand on end." Encounters with bleeding and injured friends and relatives give him only brief hope for their survival; his anxiety increases with every line. The I-narrator reflects, "Here everything human had been obliterated."[8] In his memoir *Oboegaki* (1945), Tōge Sankichi (1917–1953) left a kaleidoscope of horrors. He depicted impressions from many perspectives that occurred in the brief period between August 6 and Japan's surrender. He recounts in a fragmentary manner visits of the first person narrator to a camp for the injured, visits to a dying acquaintance, and impressions of third parties. It is difficult at first for readers to orient themselves, for too many unstructured impressions overwhelm them. Certain images recur in the writings of other authors, for example, the description of a box of onions upset on the river bank, or a draft horse. The writers do not give priority to ordering the events; the only thing that counts is the attempt to hold the horror at arm's length and fix it within what the writers thought up to that point was reality. The act of witnessing is followed by the expression of solidarity with the victims and with the movement against war. In the intervening years, knowledge about the long term effects of radiation sickness has grown. In turn this has produced an increased number of victims who have to fight for acknowledgement.

5   Comparison

In order to draw a comparison between the ways literature responded to the events discussed above in different times, it has to be stated that this is a process in various phases, of which those of immediate urgency do show correspondence.

---

7   An English translation by Richard Minear of the whole text is available in the volume *Hiroshima. Three witnesses*, edited by Richard Minear. Princeton University Press, 1990.

8   Translation in Minear, Hiroshima, pp. 52, 57. *"Natsu no hana"* is the title both of the first section of the book, which appeared Hara completed in 1945 (it appeared in 1947), and of the book as a whole, which appeared in Mita bungaku, June 1947.

The choice of a radical subjective point of view is an other common element. When only conflicting information is available, the only solid stance can be found in radical subjectivity. This act of telling has been designated as counter-memory (cf. Tachibana 1998: 197). The narration of autobiographical aspects is deeply rooted in the literary tradition of Japan, but at present, a new intention can be seen. The narrative from a subjective and personal point of view blurs the border between fiction and personal account, and serves as a means to create immediacy – and urgency.

But in contrast to immediate postwar literature, it also serves to create authenticity, which is highly valued by present readers. In creating literary immediacy, the subjectivity suggests to the reader to be part of something real (*riaru*) and therefore fulfills two demands of present society: authenticity and real-time.

The problem of such an imaginary participation in catastrophic events is obvious: The hunger for reality is an highly emotional issue, and readers may turn to the next shocking news in order to feed it, and easily get manipulated for nationalistic reasons.[9] The desire to be part of something bigger that exceeds one's own existence could be seen as the search to escape the isolation of the individual by linking it to (the illusion of?) a collective memory.

Then, producing an overall picture with the fragmentary information available (scenery, health, immediate measures, information spread by government, rumours) is achieved by assembling the pieces of information available into a collage.

Finally, another element in common is collecting the voices of people. The author can either function as a medium, channeling the voices of the dead, creating a polyphonic narrative. It is characteristic of these texts that they are not structured by an omniscient narrator, and perspective keeps shifting between the protagonists, between internal and external.

6   Conclusion

Writing about the unspeakable means to speak out and by doing so to fall short – a paradox that authors of atomic bomb literature and dealing with disaster do confront. Memory is not linear. Occasionally, the separation into the

---

9   The *kizuna*-movement shows this ambivalency accurately. While contributing to a wave of solidarity and mobilizing material and active support for the damaged areas, it did also bear a nationalistic spirit in affirming the cohesion of all the Japanese in order to overcome the impacts of the disaster. In the chapter "A Brief History of the Antinuclear Power Movement in Japan" Oguma Eiji describes closely the various shapes of this community bound by ideological agenda (cf. Oguma 2016).

narrating and the experiencing I, both aspects of the same self but located at different points along the spatiotemporal continuum, challenges the orientation. This phenomenon prevails especially among the authors who describe events that took place in their childhood. There is no path that leads to this space of memory, the place where trauma is located deeply in one's mind. Much is inchoate, and approach through language seems impossible. With the aid of narrative techniques such as overlap and consolidation, witnesses can shape the experience into a verbal form that is broad enough and does not implode or freeze. Overlapping impressions result in a semantic accumulation, a condensation that creates the ambivalence significant for the narration of trauma. On a lexical level expressions of inchoateness and amorphousness serve the depiction of things that are hard to verbalize. The disposition for a full comparison to the Post-Fukushima literature is not yet given, as the intervals of time are not equal. Nevertheless, parallels can be seen in the immediate and mid-term literary reaction, such as immediate descriptions of the caution zone and the discussion about the role and responsability of literature. As the long term consequences may still be in a state of latency, the topic will remain a research topic of interest in the future. Recent research and critique in Japan has contextualised Fukushima in the broader discussion of a literature of the nuclear age. Catastrophic literature of the present therefore has to deal not only with the unpredictability of natural desasters, but also the invisibility of radioactive threat and its long-term effects. The process of realization and integration of traumatic events does show a certain pattern. However, it has to be object of future research to abserve the literary discourse yet to come in the next decades.

The search for reality beyond real-time has been taking shape not only in dismay, but also in nostalgia for a past that can be imagined as easier to understand as the present. But to tell the untellable does also mean to recognize silence. Silence in terms of respect for the dead and as a source of power for the alive. The ghostly, haunted experience of reading catastrophe literature comes from an associative approach, wherein the reader opens the door to memory and enters a place that does not exist.

References

Bohn, Thomas M., Feldhoff, Thomas, Gebhardt, Lisette, Graf, Arndt (eds.). *The Impact of Disaster: Social and Cultural Approaches to Fukushima and Chernobyl*. Berlin: EB Verlag, 2015.

Davis, Walter. *Deracination: Historicity, Hiroshima, and the Tragic Imperative*. Albany: State University of New York Press, 2001.

Geilhorn, Barbara and Iwata-Weickgenannt, Kristina (eds.). *Fukushima and the Arts. Negotiating nuclear disaster*. London/New York: Routledge, 2017.

Hirano, Keiichiro. "The Bees that Disappeared." *Granta Magazine* 127 (2014). Accessed June 15, 2016. http://granta.com/the-bees-that-disappeared/.

Honda, Masakazu. "Away from home but where the heart is – novelist moves to Fukushima." *Le blog de Fukushima-is-still-news*, June 5, 2015. Accessed June 15, 2016. http://www.fukushima-is-still-news.com/2015/06/author-yu-miri-moves-to-mina mi-soma.html.

Hutchinson, Rachael, and Morton, Leith (eds.). *Routledge Handbook of Modern Japanese Literature*. London/New York: Routledge, 2016.

Itō, Seikō いとうせいこう. *Sōzō rajio* 想像ラジオ. Tokyo: Kawade shobō, 2013.

Iwata-Weickgenannt, Kristina. "'Vieles wird man nur begreifen, wenn man es langfristig verfolgt.' Interview mit der Autorin Yū Miri zur Atomkatastrophe von Fukushima." *Minikomi* 81 (2012): 34–41.

Karashima, David, and Elmer, Luke (eds.). *March was Made of Yarn. Writers Respond to Japan's Earthquake and Tsunami and Nuclear Meltdown*. London: Harvill Secker, 2012.

Kawamura, Minato. *Genpatsu to genbaku. "Kaku" no sengo seishin shi*. Kawade books, 2011.

Kawanishi, Masaaki. *Shōwa bungaku-shi. Gekan*. Tokyo: Kodansha. 2001.

Komori, Yōichi. *Shisha no koe, seija no kotoba – bungaku de tou genpatsu no nihon*. Tokyo: Shinnihon shuppansha. 2014.

Kuroko, Kazuo. *Genbaku bungaku-shi/ron. Zetsubōtekina "kaku (genpatsu)" jōkyō ni kōshite*. Shakai hyōron sha, 2018.

Minear, Richard (ed.). *Hiroshima. Three witnesses*. Princeton: Princeton University Press, 1990.

Murakami, Ryu. "Amid Shortages, a Surplus of Hope." Translated by Ralph F. McCarthy. *The New York Times*, March 16, 2011. Accessed June 15, 2016. http://www.nytimes.com/2011/03/17/opinion/17Murakami.html?_r=4.

Ōba Minako. *Urashimaso*. Translated by Yu Oba. Tokyo: Center for Inter-Cultural Studies and Education Josai University, 1995.

Ōe, Kenzaburō (ed.). *Atomic Aftermath: Short Stories about Hiroshima and Nagasaki*. Tokyo: Shueisha, 1984.

Oguma, Eiji. "A New Wave Against the Rock: New social movements in Japan since the Fukushima nuclear meltdown." *The Asia-Pacific Journal* 14/13:2. Accessed July 29, 2016. http://apjjf.org/2016/13/Oguma.html.

Sklar, Marty (ed.). *Nuke Rebuke: Writers and Artists against Nuclear Energy and Weapons*. Iowa City: The Spirit That Moves Us Press, 1984.

Slaymaker, Doug. "Horses, Horses, In the Innocence of Light (Execrpts). Furukawa Hideo. Translated and with an introduction by Doug Slaymaker." In: *Japan Focus*

13/10:3 (2015). Accessed March 31, 2016. URL: http://apjjf.org/2015/13/10/Doug-Slaymaker/4298.html.

Stevens, Carolyn. "Images of Suffering, Resilience and Compassion in Post 3/11 Japan." In: *Japan Focus* 13/7:9 (2015). Accessed October 27, 2016. URL: http://apjjf.org/2015/13/6/Carolyn-Stevens/4285.html.

Stahl, David. "Critical Postwar War Literature. Trauma, narrative memory and responsible history." In: Hutchinson and Morton (eds.): 169–83.

Tachibana, Reiko. *Narrative as Counter-Memory: A Half-Century of Postwar Writing in Germany and Japan.* Albany: State University of New York Press, 1998.

Taguchi, Randy 田口ランディ. *Hiroshima, Nagasaki, Fukushima: Genshiryoku o ukeireta Nihon* ヒロシマ、ナガサキ、フクシマ 原子力を受け入れた日本 (Japan Went Nuclear: Hiroshima, Nagasaki, Fukushima). Tokyo: Chikuma purimā shinsho, 2011.

Taguchi, Randy 田口ランディ. *Zazen gāru* 座禅ガール. Tokyo: Shōden sha, 2014.

Tan, Daniela. "Literature and The Trauma of Hiroshima and Nagasaki." *Japan Focus* 12/40:3 (2014). Accessed September 29, 2018. URL: http://apjjf.org/2014/12/40/Daniela-Tan/4197.html.

Tan, Daniela. *ZwischenWelten. Ōba Minako im Kontext der Introvertierten Generation.* Berlin: EB Verlag, 2017.

Thornber, Karen L. "Atomic Bomb Writers." In: Modern Japanese Writers, ed. Jay Rubin. New York/London: Charles Scribner's Sons. 2001: 49–70.

Treat, John Whittier. *Writing Ground Zero: Japanese Literature and the Atomic Bomb.* Chicago: The University of Chicago Press. 1995.

Wagō Ryōichi 和合亮一. *Shi no tsubute* 詩の礫. Tokyo: Tokuma shoten, 2011.

Wagō Ryōichi 和合亮一. "Shi no mokurei" 詩ノ黙礼. *Youtube* (August 25, 2013). Accessed October 27, 2016. URL: http://www.youtube.com/watch?v=JbsFIA9tvOg.

Yū, Miri 柳美里. "Keikai kuiki" 警戒区域. *G2* 9 (2012): 258–71. Accessed June 15, 2016. http://g2.kodansha.co.jp/10955/11421/11422/11423.html.

Yuki, Masami. "Language and Imagination Before and After Fukushima: A Concept of Zone as a New Theoretical Framework in Taguchi Randy's Works." In: Bohn, Feldhoff et al. (eds.): 215–26.

CHAPTER 13

Sequence and Duration in Graphic Novels

Ileana da Silva and Marc Wolterbeek

1 Introduction: A Traditional Contemporary Narrative

Sequential art is a term used to identify a broad range of visual narratives, from ancient cave paintings to contemporary graphic novels, and it is distinctive in its combination of temporal and spatial arts – that is, literature and painting.[1] The term itself indicates the dual nature of this form: sequence implies narrative, which is diachronic, taking place over time, while pictorial art is synchronic, presenting a single frozen image. The "brilliant antithesis" made in Antiquity by Simonides of Ceos – that "painting is mute poetry and poetry a speaking painting"[2] – should not be oversimplified, however; Lessing himself was aware that literature is often illustrative and painting can be narrative.[3]

Because of its nature, sequential art often concerns itself with temporal and spatial dimensions and pays particular attention to two fundamental concepts about time: sequence and duration. These characteristics of time were evident to the earliest philosophers, and "in scholastic authors familiar to Descartes, such as Aquinas and Suarez, duration is simply persistence in being while tempus is the measure or numbering of beings that are 'successive' or composed of parts existing one after the other."[4] In other words, time "lasts" or "endures" – a task may take an hour to complete – and time can also be divided or numbered into seconds, minutes, and hours (a sequence).[5]

---

1 The term "sequential art" was coined by Will Eisner in *Comics and Sequential Art: Principles and Practices from the Legendary Cartoonist* (New York: Norton, 2008) and it is thoroughly defined by Scott McCloud in *Understanding Comics: The Invisible Art* (New York: HarperCollins, 1993), 5.

2 Gotthold Ephraim Lessing, *Laocoön: An Essay on the Limits of Painting and Poetry* (Trans. Edward McCormick. Baltimore: Johns Hopkins University Press, 1962), 4.

3 See Lessing, *Laocoön*, 91: "succession in time is the province of the poet just as space is that of the painter."

4 Geoffrey Gorham, "Descartes on Time and Duration," *Early Sciences and Modern Medicine* 12 (2007): 43.

5 For classic concepts of time, see S. Alexander, S., "Spinoza and Time," *Studies in Spinoza*, ed. S. Paul Kashap (Berkeley: University of California, 1974); Christopher Drohan, "A Timely Encounter: Dr. Manhattan and Henry Bergson," in Mark White, ed. *Watchmen and Philosophy: A Rorschach Test*. Blackwell Philosophy and Pop Culture Series (Hoboken, New Jersey: John

In analyzing the temporal nature of sequential art, it is important to distinguish different types of time, in particular reading and narrative time. Reading time (including "reading" artwork) tends to be immediate and is determined by microstructural elements, such as the panels and page layouts. Narrative time, including plot, is more complex and consists of several elements, and in many respects resembles the way time functions in conventional literature.[6]

On a microstructural level sequence is established in this form of art by panels or frames that set off one illustration from another, moving both reading and narrative time in a linear, progressive fashion. Theoretically each panel represents a "snapshot in time," though even visual elements, such as motion lines, indicate duration. Further, most panels include words, usually set off in text boxes or speech balloons, and hence introduce temporality into the reading process.

Other microstructural aspects of sequential art reveal its unique ability to present time and space. As Scott McCloud explains, large panels generally slow down reading time while narrow panels speed it up.[7] Other formal dimensions, like the gutters separating panels or page lay-outs, affect reading time. The sequencing of panels also influences narrative time: panels representing one moment followed by another (called "moment-to-moment panels" by McCloud) not only move the reader's eyes quickly (like panes of film) but also capture the narrative present.

On the macro level narrative time belongs to the main plot of the story, and it may be represented diegetically and mimetically. For example, a story told through the eyes of a first person narrator is often recounted diegetically, as the character describes his or her past to the reader retrospectively, using the past tense. On a microstructural level this narration is often contained within text boxes in sequential art. But a story is also told mimetically, as happening in the here and now, and this rendition occurs on a micro level both visually and verbally, with speech balloons capturing actual present speech. As in conventional literature, the plot of a graphic novel may be analyzed in Aristotelian terms (exposition, rising action, climax, falling action, and dénouement) that

Wiley and Sons, 2009); and Gideon Yaffe, "Locke on Consciousness, Personal Identity and the Idea of Duration," *Noûs* 45, no. 3 (2011).

6 See Gérard Genette, *Narrative Discourse: An Essay in Method*, trans. Jane Lewin (New York: Cornell University Press, 1980), 27, for the distinction of "narrative" from "story" and "narrating": the term "story" designates "the signified or narrative content"; "narrative" indicates "the signifier, statement, discourse, or narrative text itself"; while "the word narrating" pertains to "the producing narrative action and, by extension, the whole of the real or fictional situation in which the action takes place."

7 See Scott, *Understanding Comics*, 101–2.

are both sequential and durational, as each segment of the structure lasts a certain amount of time. Poststructuralist concepts of structure and narrative, such as those enunciated by Todorov, Barthes, Genette, and others, may also be applied to this art form.[8]

Perhaps on the largest level, time pertains to character and theme. Characters may be static (flat) or dynamic (round) as they perform actions over time; they may possess a fixed identity or they may change. Whereas plot (*fabula*) emphasizes sequence, what Russian formalists call the *syuzhet* (theme or subject matter) of a story is, in a sense, timeless, possessing a kind of eternal duration. As we shall see, time may become theme.[9]

Two critically acclaimed graphic novels – Neil Gaiman's *Sandman* and Alan Moore and Dave Gibbons' *Watchmen* – explore the sequential and durational aspects of time both in their formal structures and in their literary content – that is, character and theme. While the theme of time may not be as explicit in *Sandman* as in *Watchmen*, one of its chapters, entitled "24 Hours," reveals an obsession with time, yet the temporal exigencies of narrative derail the author's original intention to establish a rigid sequence. In contrast, in *Watchmen* time is an explicit concern, and characters envision time in radically different ways. For Jon Osterman (or Dr. Manhattan), who is omnipresent, time is asynchronous – as in eternity, it ceases to exist – yet the duration of memory establishes his sense of self; for Laurie Juspeczyk (the second Silk Spectre), time is, above all, the present, yet she finds her sense of identity through memory in the form of flashbacks; and for Adrian Veidt (Ozymandias), time is historical, moving from Antique models to a megalomaniacal present and ending in near apocalypse.

2   "Twenty-Four Hours": Sequence vs. Duration

"Twenty-Four Hours," the sixth chapter of *The Sandman: Preludes and Nocturnes*, is a twenty-four page story about the murderous rampage of Dr. D (or Dr. Destiny) in a diner. Originally Gaiman planned to allocate a single

---

8   See Roland Barthes, "The Structuralist Activity," in *Critical Essays*, trans. Richard Howard (New York: Farrar, Straus, and Giroux, 1972); Genette, *Narrative Discourse*; Tzvetan Todorov, "The Structuralist Analysis of Literature: The Tales of Henry James," in *Structuralism: An Introduction*, ed. by David Robey (Oxford: Clarendon Press, 1973), pp. 73–103.

9   One may also distinguish "the 'surface structure' of narratives – the sequence of actions, or syntagmatic dimension," from "deep structures," which are abstract and atemporal; see Wallace Martin, *Recent Theories of Narrative* (Ithaca, NY: Cornell University Press, 1986), 99–100.

page to each hour, but he states that he gave up once he realized "that the first few pages had to be devoted to introducing the characters properly."[10] In other words, Gaiman's plan to develop a structural tour de force was thwarted by narrative necessities, especially the need for exposition. As a result, the first hour is the "longest" of the twenty-four, "lasting" five pages.[11]

While time functions, or malfunctions, most obviously in relation to the narrative structure of this chapter, time is also important at the textual level, the page lay-outs that affect the reading process, as well as at thematic and symbolic levels. The first page of this chapter demonstrates how the graphic novelist may manipulate the reading process through paneling (see Figure 13.1).

The first panel, consisting of a large rectangle that arrests reading time, establishes setting: the exterior of the "All Nite Diner," "24 Hours." This panel depicts space without motion. The second panel, also a large rectangle, shows the interior of the restaurant: Bette, the waitress, is wiping the counter (motion lines indicate movement and time), and speech balloons capture a request by Judy, a customer, for a coffee refill and a tuna on rye. Even though the reader's eye is slowed down by the size of the panel, there is motion and language – both temporal components. Motion speeds up in the last moment-to-moment panels at the bottom of the page that show Bette preparing the sandwich, picking up a coffee pot, and carrying both to Judy's booth; the effect of this layout is cinematic: the panels function like the panes of film, moving the reader's eye, and again motion lines capture Bette's movements.

The next four pages of the expository first hour introduce the primary characters, and typically new pages introduce new characters, indicating a relation between space and time within this hour. Bette, the waitress, is a closet writer who gathers her raw material from her customers, creating mental fictions about their lives; thus a parallel world with its own time dimensions co-exists with the primary fictional world. An unnamed young man, who still has "an hour to kill" before an interview, asks Bette for a coffee refill; a married couple, Gerry and Kate, enter the diner, followed by Marsh, a trucker; and the villainous Dr. D sits in the back of the diner observing the others. Hence the exposition is primarily devoted to establishing character, and all six move through time. Large "slow" panels alternate with small "quick" ones, and two iconic images – a clock and a picture of sheep – appear unobtrusively.

---

10   Hy Bender, *The Sandman Companion: A Dreamer's Guide to the Award-winning Comic Series* (New York: Vertigo/DC Comics, 1999), 37.

11   Gaiman states that he was inspired by the British film *Drowning by Numbers*, in which three women drown their husbands along with a coroner; during the course of the film, the numbers one to one hundred appear randomly with no apparent relationship to the plot. See Bender, *Sandman Companion*, 35–36.

FIGURE 13.1   Gaiman, *Sandman*, p. 159

After the exposition, the rising action constitutes the major portion of the chapter, spanning hours two to fifteen, and consists of several conflicts controlled by Dr. D, whose ruby enables him to control the characters' minds. Individual hours occupy full pages or half pages until the nineteenth hour, which consists of a single panel in which Dr. D is recounting the story of Snow White to the women, and the twentieth, in which the three women sing a song for Dr. D. These hours consisting of single panels demonstrate time's urgency as Gaiman attempts to telescope hours into narrow spaces, and they raise issues about narrative time and reading time. If visual art is truly spatial, simply representing a single moment, time cannot pass within a single illustration. Yet Lessing himself recognized that "painting too can imitate actions, but only by suggestion through bodies."[12] Moreover, the language contained in these panels signifies a passage of time: it takes time to recount the story of Snow White; it takes time to sing a song. The reader, however, experiences a disjuncture between narrative time and reading time, and this discrepancy occurs throughout the chapter.

The paradox of a single panel representing an hour of time is more acute in the turn toward the denouement. Hour twenty-two consists of a single, full-page panel depicting Dr. D standing over the dead diners. Language is restricted to the text box (simply stating "Hour 22," with no title for the chapter) and a small sign stating, "please and thank you are magic words." The entire page is rendered in deep blues, except for the bluish-green figure of Kate, giving a feeling of stillness, and ironically a table clock indicates the same time (12:10) in this hour as in hour fourteen (page 174, panels 2 and 4), suggesting that time is frozen. It takes time to study this page, just as this analysis takes time.

Not only is the formal structure of the chapter compromised by hours of varying length, but one of the central symbols, the clock, presents temporal difficulties. In fact there are multiple clocks in this chapter, some of them within the paneling (and therefore part of the narrative), others appearing in splash pages outside the paneling (and therefore external to the narrative and purely iconic). The biggest problem is with the first appearance of a clock, indicating that it is 12:00 at the first hour (160, 1), presumably noon, but two hours later (hour three) this clock indicates that only one hour has passed (showing 1:00; 165, 8). At hour twelve, a wall clock shows 10:05 (172, 1), and at hour thirteen it is 11:00 (173, 5), both times revealing that the first clock of hour one should have indicated 10 a.m., not noon. To complicate the matter further, an antique table clock indicates the same time (12:10) at hours fourteen (174, 2 and 4) and twenty-two (180, 1). This third clock may be broken, but its final time of 12:10

---

12   Lessing, *Laocoön*, 78.

would ironically coincide with the 12:00 temporal sign of the first hour. These clocks lack mimetic consistency and serve iconic rather than functional ends. This is also true about clocks appearing outside the panels on splash pages.

Gaiman's overall structural plan to devote a page to each hour fails due to narrative exigencies, and even the horometric iconography lacks consistency. The formal function of temporality and spatiality in the paneling is characteristic of all sequential art, but its relation with the hourly segments raises issues about narrative duration and the reading experience. Gaiman's attempted tour de force does not work because he cannot fit the dimensions of storytelling into a rigid temporal sequence.

3   *Watchmen* and the Dynamic Dr. Manhattan

Whereas Gaiman's use of sequential time in "24 Hours" is primarily structural and separate from narrative or durational time, Moore's concept of time in *Watchmen* is much more varied and complex. In this graphic novel, time informs structure in a pervasive, cohesive manner; it is a central concern of the characters, especially Dr. Manhattan, who obsesses over it; and it constitutes a major theme, time's urgency, as humanity heads towards a nuclear apocalypse. For three of the six central characters – Dr. Manhattan, Laurie Juspeczyk, and Adrian Veidt – personal identity and selfhood are intimately connected with radically different notions of time.

The character most affected by time is Jon Osterman, a nuclear physicist who is physically destroyed in an "intrinsic field" chamber and reincarnated as Dr. Manhattan, capable of traveling through time and space. Chapter Four of the graphic novel, entitled "Watchmaker," recounts this event from the perspective of the reincarnated Dr. Manhattan, whose diegetic narration enables him to lead the reader to experience his omniscient viewpoint. The "macro" narrative of this chapter – that is, the segments constituting the plot – is chronological, telling the story of Jon and his girlfriends Janey Slater and Laurie Juspeczyk, also known as the Silk Spectre II, but on the "micro" level – the individual panels on each page – scenes leap achronistically among past, present, and future times, reflecting Jon's omnipresent mind.

The first two pages of the chapter constitute the exposition, depicting Dr. Manhattan on Mars remembering past events. Two temporal processes are at work: negative sequencing (as he counts backwards in seconds toward a future event) and omnipresence (as he remembers past events atemporally). In the first panel he holds a photograph of Janey Slater, his first girlfriend, and himself at a New Jersey amusement park in 1959, and in the second panel he states in a diegetic textbox that he will drop the photo twelve seconds in the future (see Figure 13.2).

His use of present tense verbs to describe the photo ("they are at an amusement park, in 1959") and his actions twelve seconds into the future ("I drop") reveals his presentism, and his statement that the photo is "already lying there, twelve seconds into the future," indicates his proleptic ability to envision the future as "already" occurring.[13] A second textbox in panel two begins the countdown, "ten seconds now," showing that two seconds have elapsed within a single panel, aligning reading time, which is both temporal and visual, with diegetic narrative time.

Jon then remembers his recent visit to the Bestiary: the picture "is still there, twenty-seven hours into the past … I'm still looking at it," he thinks in panel four, which depicts him visiting the bar – he has traveled into the past (which he describes as the present) and teleported himself from Mars to earth in the gutter between panels three and four. The countdown continues, and in panel six he exists in the narrative "present" – October 1985 on Mars – and in the past – July 1959 at the Palisades Amusement Park, even though this latter scene is not depicted, suggesting a disjunction between the temporal (1985 and 1959) and spatial (Mars only). He then drops the photograph onto the pink Martian sand. The action on the page lasts twelve seconds, about the time it takes to read it.

Not only does time flow in two directions for Dr. Manhattan, but it also stands still, frozen in the photograph of Janey and him at a New Jersey amusement park in 1959. When her watchband breaks, a "fat man" steps on the watch and breaks it, freezing time at 8:17 (116, 9), the exact moment of the Hiroshima explosion, illustrated by another watch later in the chapter (134, 8). The photograph, as David Barnes explains, "becomes the visual home base to which the narration constantly returns and refers," signifying Manhattan's "inability to change or respond" to a "pre-determined structure of events" (55). The watch belongs to a much larger symbolic nexus of the novel, the central iconic image being the Doomsday Clock that ends every chapter and progresses, minute by minute, toward midnight, reiterating the theme of time's urgency that

---

13   Genette, *Narrative Discourse*, 40, defines "prolepsis" as "any narrative maneuver that consists of narrating or evoking in advance an event that will take place later," while "analepsis" is "any evocation after the fact of an event that took place earlier than the point in the story where we are at any given moment." "Anachrony" designates "all forms of discordance between the two temporal orders of story and narrative." On presentism, see Sean Power, "A Philosophical Introduction to the Experience of Time," *NeuroQuantology* 7, no. 1 (March 2009), and on time travel, see Bradley Monton, "Time Travel without Causal Loops," *The Philosophical Quarterly* 59 (January 2009): 234.

FIGURE 13.2   Moore and Gibbons, *Watchmen*, p. 111

culminates in the novel's apocalyptic climax. Time becomes static in these visual images; it moves only in the narration, and in the case of Dr. Manhattan, it is both proleptic (moving toward the future) and analeptic (moving toward the past).

The central, life-changing event in Dr. Manhattan's existence occurs in the next scene at Gila Flats, when Jon gets trapped in an "intrinsic field chamber." Light from "particle cannons" (118, 1) vaporizes Jon's body and his colleagues assume he is dead, but three months later a floating "circulatory system" (119, 5) starts haunting the facility and soon his body reassembles itself like a watch. This event is explained in academic terms by Professor Milton Glass, who states in an interstitial article separating chapters four and five that "a form of electromagnetic pattern resembling consciousness survived, and was able, in time, to rebuild an approximation of the body it has lost" (140). Jon is reincarnated as Dr. Manhattan, who is not only gifted with omnipresence, but also possesses the ability to change in size and produce multiple versions of himself.

The second major plot development of the chapter depicts Manhattan's relationship with Janey and his work as a crime fighter, and within this section a single panel raises important issues about identity and time. When Manhattan assumes a new identity by engraving a hydrogen atom on his forehead, he remembers when his father handed him a newspaper article announcing the Hiroshima bombing in a panel (122, 8) that is visually the same as the earlier one depicting this action (113, 3) but verbally different (the second panel contains a diegetic textbox; the first one presents mimetic speech balloons). Manhattan's ability to remember and even relive past events before his disaster suggests that his identity has remained intact despite his bodily disintegration. Such a view is consistent with Locke's belief that "our memories tie together our identities" and Descartes' concept of the "disembodied mind."[14] From a philosophical perspective, if identity is predicated on memory, then Manhattan possesses a "diachronic identity," which is a contradiction in terms, since Jon Osterman and Dr. Manhattan do not exist simultaneously, at least in physical form.[15] And if one accepts Parfit's "physical criterion" – that "some significant part" of a person's "material body (specifically the brain)" must continue to exist,[16] then Manhattan is not Jon, for he has changed physically: his skin is now blue, he varies in size, and he splits himself into three. Manhattan has also changed

---

14   James DiGiovanna, "Dr. Manhattan, I Presume?" In Mark White, ed., *Watchmen and Philosophy: A Rorschach Test*. Blackwell Philosophy and Pop Culture Series (Hoboken, New Jersey: John Wiley and Sons, 2009), 105–6.

15   See Erwin Tegtmeier, "Three Flawed Distinctions in the Philosophy of Time," *International Journal for Ontology and Metaphysics* 8 (2007): 56.

16   DiGiovanna, "Dr. Manhattan," 108.

temperamentally – he is now detached from others, emotionally dead – and he later changes his mind about the significance of human life when he realizes that it is a "thermodynamic miracle" (306, 4; 307, 1). Further, when he suffers a second disintegration and reincarnation at the hands of Ozymandias in the novel's last chapter, he differentiates his present self from his former one, saying that restructuring himself after "the subtraction of my intrinsic field was the first trick I learned. It didn't kill Osterman ... do you think it would kill me?" (400, 2). It is possible that Manhattan is not remembering the past at all, but reliving it by travelling through time and that panels depicting past events are actual (that is, mimetic), not mental (or diegetic). This solution is supported by the fact that Manhattan uses present tense verbs when describing past (and future) events.

Even though Manhattan can travel through time, he cannot change events; therefore, his time traveling abilities do not violate the "grandfather paradox" – the premise that if one travels backward in time and kills his grandfather, then he could never have been born. The issue of using time travel to prevent crimes does not, however, pertain to travel to the past; instead, the issue arises in regards to future events. Janey criticizes Manhattan for not doing anything to prevent President Kennedy's assassination since he knew it was going to happen (126, 1), and Manhattan replies, "I can't prevent the future. To me, it's already happening" (126, 2). The Comedian also criticizes Manhattan for not changing a future event when the Comedian kills a Vietnamese woman who is pregnant with his child: "you coulda changed the gun into steam or the bullets into mercury," the Comedian says to Jon, "but you didn't lift a finger" (57, 4). And when Manhattan realizes that Ozymandias has launched a destructive monster on New York City, he thinks that "tachyons" – theoretical sub-atomic particles that can supposedly travel backwards in time faster than the speed of light – have interfered with his ability to see into the future (389, 3). Jon's inability to change future events is a reversal of the grandfather paradox, for the question is not about changing past events, but future ones. He lacks free will because "everything is preordained," even his responses (285, 3). "We're all puppets," he says to Laurie, "I'm just a puppet who can see the strings" (285, 4).

The main narrative ends with two pages (134–35) that enable the reader to experience Jon's omnipresence. The diegetic narrative present – that is, the narration by Dr. Manhattan on Mars – describes the deterioration of Manhattan and Laurie's relationship, presented in three mimetic panels on each page, and these analeptic scenes are antedated by pluperfect ones, Janey and Jon at the amusement park, Janey handing Jon a beer at the Bestiary, the Comedian's funeral, Manhattan entering the deserted bar before his departure. Present, past, and pluperfect scenes are juxtaposed as non sequiturs, but the reader, now familiar with each of them, is capable of making the non-chronological leaps from panel to panel.[17]

For Dr. Manhattan "there is no future. There is no past … time is simultaneous, an intricately structured jewel that humans insist on viewing one edge at a time" (286, 5–6). In other words, there is no time at all, only an intricate object occupying space. This is the perspective of eternity, in which time has ceased to exist. Moore and Gibbons convey this profound insight most effectively by using the medium of sequential art, enabling them to allow the reader to experience Jon's perspective while at the same time maintaining a foot in human reality. They do this through clever manipulation of paneling, repeating images until the reader is familiar with the fragments of Manhattan's life, and through traditional narrative, which moves in a general chronological direction.

4   Laurie Juspeczyk: The Present Everywoman

In the eyes of Dr. Manhattan, the division between past, present, and future is irrelevant, for he exists in all three temporal states – simultaneously. The synchronic manner in which Manhattan experiences time dazzles the imagination and raises intriguing philosophical questions, but his experience is intangible to most readers. Moore and Gibbons, for the sake of narrative fidelity, must craft a character to whom the common reader may relate. That character is Laurie Juspeczyk, Silk Spectre II. Much the way Chapter Four is structured and designed to be Manhattan's "minute," Laurie's "minute" is necessary in terms of exposition and character development, transforming her from a flat character into a round one. As previously noted, the splash pages introducing *Watchmen*'s chapters correspond to minutes passing on the recurring image of the Doomsday clock, and Laurie's chapter is no exception. However, whereas Manhattan's "minute" is esoteric and conveys the perspectives of academics like the fictional Milton Glass, Laurie's "minute" functions as a mouthpiece for the reader, voicing his or her experiential observations of time and space, sequence and duration, and the semiotic distinctions most people draw between past, present, and future.

Moore and Gibbons achieve these purposes in Chapter Nine, "The Darkness of Mere Being," by making deliberate formal and content-level decisions to reinforce the theme of time as well as the reader's temporal experience with

---

17   McCloud, *Understanding Comics*, 72, defines non sequiturs as having "no logical relationships between panels whatsoever."

the text. During this chapter, Laurie reflects on her childhood. She recalls her parents' volatile marriage and laments her inability to shape her own destiny, being forced by her mother, the original Silk Spectre, to inherit the role as Silk Spectre II. Most importantly, she discovers the identity of her biological father by revisiting a series of repressed memories. Past events are critical in this chapter because as Laurie seeks to define her identity, she must navigate prior, formative events. The consequence of this process compels Laurie and Manhattan to act "against" the Doomsday Clock, reinforcing the relationships among time's urgency, the meaning of existence, and the will to power.[18]

Chapter Nine begins with a full-page splash of an open perfume bottle, emblazoned with "N" for Nostalgia, the brand of unisex fragrance that Adrian Veidt, or Ozymandias, and his corporate empire markets and sells (280–81; see Figure 13.3). "Nostalgia," the fragrance, functions as an objective correlative throughout the text of *Watchmen*, appearing as a perfume bottle (247), magazine advertisement (345), TV spot (226), and billboard (83, 271). It works to connect the spatial and temporal, linking the events of the "present" to the past in both visual and semiotic terms. While "Nostalgia" makes multiple appearances in *Watchmen*, it is most prominent in Chapter Nine, Laurie's backstory. The bottle of perfume spinning in the air, with its contents spilling out until it ultimately shatters like the snow globe from her childhood, is depicted as a scene-to-scene transition (288, 3–4) that transports readers across significant distances of time and space. Additionally, the scene-to-scene transition resembles a graphic match cut, a cinematic technique that provides continuity by juxtaposing two shots featuring figures of similar shape and form.[19]

The match cut provides visual continuity between scenes since the shape of the snow globe and the perfume bottle are both circular, but more importantly, it highlights metaphorical relationships between these objects in both shots (288, 3–4; 304, 1 and 9). The match cut of the scene-to-scene transition allows both Laurie and the reader to connect the events of the past (the snow globe) to the present (the bottle of Nostalgia), yet both objects persist as analeptic vehicles that transport Laurie from the present to the past, and from one version of the past to another, as she remembers, reimagines, and reassembles the pieces of her past and her fragmented present self through memory and dialogue. Her "present" identity and self-image are formulated by conceptions

---

18   The driving force of humanity or the need to realize one's highest aspirations. See
       Nietzsche, *The Will to Power*.

19   "The shape, color, or texture of the two figures matches across the edit, providing continu-
       ity" Barsam and Monahan, *Looking at Movies*, 498.

FIGURE 13.3   Moore and Gibbons, *Watchmen*, p. 288, demonstrating visual and metaphorical
continuity between the snow globe and bottle of Nostalgia perfume

that have developed diachronically; that which has already transpired, up
through the present, forms an individual's current identity.

In *Watchmen*, Moore and Gibbons suggest to readers that the meaning of existence is linked to an acute awareness of time and the urgency to "strike while
the iron is hot," a perspective that is informed by a long tradition of Western
meditations on the nature of "time" and its undeniable influence on what it
means "to be." Moore and Gibbons work to show readers that the two ideas go
hand-in-hand, much the way Heidegger's aptly titled *Being and Time* suggests
a connection. According to Heidegger, since Aristotle's *Physics*, discussions of
time have prioritized the present. The future is the "now-not-yet," and the past
is the "now-no-longer," leaving people with only one relevant alternative: the

present, for it is the only moment in time over which an individual may exert his or her will, or control.[20] This belief is echoed when Laurie chooses to focus on free will and the present as she refutes Dr. Manhattan's rationale for non-intervention despite the imminent threat of the world's destruction. The confrontation presents an extreme example, yet one that epitomizes the urgency and necessity for Laurie (and in a sense, for all people) to act before she runs "out of time."

To Laurie, like most people, incentive to act is based on the ticking clock. "Clock time" reminds individuals that time flows "forward," requiring that tasks be accomplished by a certain hour, day, week, month, year, and so on. Some tasks are based on routine, such as waking up in the morning for school or work or meeting a deadline for a report. Other tasks are elevated to the status of dreams, goals, and aspirations. Achieving these higher-order tasks creates meaning in one's existence. Therefore, the idea of "wasting time" and the fear of failing to take action in the "present" to achieve a desired effect in the "future" impel people to make the best "use" of it *now*. If an individual is unaware of time or has an endless supply of it, then, according to conventional wisdom, a person has no reason to act, for nothing will be wasted or lost. When people do not act "in time," we often call this a "missed opportunity," and when people try their best to complete a task, regardless of its relative importance, we often say, "you tried your best to get it done [on time]." Thus, time's urgency, whether it is real or imagined, is constantly reinforced and reiterated in even simple speech and social exchanges, not to mention in a text like *Watchmen* in which varying forms and perspectives on time are showcased. Owing to this deeply ingrained cultural emphasis on "clock time" and its influence on human desire to act, Laurie's awareness of the Doomsday Clock (250) and her knowledge that the world will be destroyed in a matter of moments motivates her to prevent the apocalypse.

In stark contrast, Dr. Manhattan focuses on synchronicity and predestination. He states, "Everything is preordained. Even my responses ... we're all puppets, Laurie" (285, 4). Laurie issues a challenge in response: "what happens if I just stay down here and screw all your *predictions*, huh? What happens then?" (285, 5). Manhattan refuses to help save Earth from the impending apocalypse, which Laurie cannot understand because she views time as a linear sequence represented as past, present, and future. She tells Manhattan that "humanity is about to become *extinct*" and wants to know, "... doesn't it *bother* you? All those people *dead* ..." (290, 5). The conversation evolves into one that goes beyond time but considers the meaning (or meaninglessness) of existence. Manhattan

---

20   Heidegger, *Being and Time*, 386.

channels an almost Sisyphean sentiment: "All those generations of *struggle*, what *purpose* did they ever achieve? All that effort, and what did it ever lead to?" (290, 5–6). Manhattan insults Laurie's view of time as "narrow; life insisting on life's viewpoint, when alternatives exist" (293, 7); Dr. Manhattan is the voice of the theorist who has become too detached from humanity to understand its preoccupation with the present, someone who "won't see existence in human terms" (303, 1). Although Manhattan's meditations certainly provide food for thought, his endless drifting in existential ennui is disconnected from the reality of the human experience – he is quite literally on another planet, in outer space.

Far more down-to-earth is Laurie's impassioned defense of humanity, its purpose, and ability to determine its future. Laurie, as the "everywoman," expresses a conventional retort: "In those terms, sure, mankind hasn't *helped* the *environment*, but *against* that, you have to measure the lives of *artists, scientists, poets* ... hell, even *my* life. That has to be worth *something* ... look, about the *environment*: without *life*, there wouldn't even be an environment!" (293, 5–6). Moore and Gibbons, through Laurie's character, present a perspective on existence and time using colloquial language meant to resonate with the "*ordinary people*" who are reading *Watchmen*, a point further highlighted when Laurie demands an answer from Manhattan: "... all the things that *happen* to them ['*ordinary people*']... doesn't *that* move you more than a bunch of *rubble*?" (296, 9). Not only is Laurie critiquing the cold rationalism of academicians like Dr. Manhattan, but she is also expressing a classic existentialist sentiment that inextricably links the meaning of life to the passing and utilization of time itself.

5   Adrian Veidt: Ancient Futurism

Adrian Veidt's perspective on time is similar to Laurie's in that it is linear, yet that is where the similarities end. For Laurie, the present is paramount, but for Adrian, it is the imagined future. Unfortunately, Adrian's perspective on time is inherently flawed not only because the future has not happened yet, but also because his vision of the "future" depends heavily on history, which carries with it numerous temporal paradoxes.[21] One paradox of historicity is that elements of the past may exist in the present. For instance, the ruins of the Acropolis objectively exist today, but they are a product of former events, and thus they also belong to the past. In other words, the Acropolis, in its current state, contains aspects of the past, even now. Another paradox of historicity is

---

21   For more on the paradox of historicity, see Heidegger, *Being and Time*, 347.

FIGURE 13.4   Moore and Gibbons, *Watchmen*, p. 349: "a miraculous bubble of Tropicana"

the idea that anything that has a history is in a stage of becoming; that is, the subject in question is in a developmental phase. However, development may be predicated on a rise or a fall and may not necessarily denote progress, although people have the tendency to equate the future with progress. A third paradox of historicity lies in the way history is recorded and transferred. History is liable to manipulation due to politics, culture, society, and other human institutions.[22] Whatever is handed down is considered "historical," without much consideration given to whether this "history" is the construction of events that have occurred – or omissions of them. Adrian believes in the objective factuality of the past, and as a result he seeks to establish or exceed the standards set by "history."

From the beginning of his "minute" in Chapter Eleven, bitingly titled "Look on My Works, Ye Mighty ..." Adrian begins the process of the apocalypse by saying, "... there's no point in putting things off any longer ... and no time like the present" (352, 4–5) as he presses the "doomsday button" to unleash his weapon on New York City (353, 1–7). Everything that surrounds him is a construct – the glass dome houses "a miraculous bubble of Tropicana set into endless subzero wastes ..." (355, 2; see Figure 13.4). Butterflies, tree frogs, flowing water, palm trees, lily pads, hanging vines, fish, toadstools, rose bushes, clusters of moss – tropical, prehistoric flora and fauna that are out-of-place within the dome, surrounded by the barren, frozen landscape of modern-day Antarctica. It is a seemingly impossible construction, yet one that Adrian manages to accomplish, nothing short of a miracle (355). Adrian's "present" is in itself a paradox of historicity.

Lacking any present-day, real-life sources of inspiration, Adrian has admired ancient individuals, civilizations, and the mythic, orientalist representations of them. Adrian's home base, Karnak, despite its Egyptian name, evokes

---

22   See Heidegger, *Being and Time*, 345–48.

images of Coleridge's Xanadu, with its lush forests inside a "stately pleasure-dome" surrounded by darkness. Adrian's Karnak is an amalgamation of ancient essentialism; the aesthetics are dominated by Alexandrian frescoes and Egyptian obelisks, random ankhs and pyramids, and the Eye of Horus. This all-seeing eye is also symbolic, for Adrian uses technology in the form of security cameras (all of which are connected to a massive surveillance room in Karnak) to allow him to see into the lives of others, contributing to his delusions of grandeur and helping him achieve his aspiration to become truly omniscient – a god (342, 7).

Adrian's monologue later in the chapter reveals that he has idolized Alexander the Great, ruler of "the civilized world" (356), and found the Egyptian pharaohs and "their wisdom truly immortal" (358, 7). He claims that his life is worth celebrating because "it represents the culmination of a dream more than two thousand years old" (355, 3). Not only does this reveal his egomaniacal nature, but it also serves to demonstrate the view that time is progressive and evolving; thus "culmination" is the operative word in this context. He credits Alexander's ingenuity of slicing through the Gordian Knot rather than seeking a way to untie it as a symbol of "lateral thinking ... *centuries ahead of his time* [emphasis added]" (358, 2). Yet Adrian sees himself as even superseding Alexander by concocting a plan to end the current phase of civilization and being able to see his own will implemented (369–70).

Inspired by the visionaries of legend, Adrian takes it upon himself to exceed the standards of the past and construct the future as he sees fit. This is evident in Adrian's fitness guide, *The Veidt Method*, which advises its readers to "create a new you." In it he writes, "You will learn that one can either surrender responsibility for one's actions to the rest of the social organism, to be pulled this way and that way by society's predominating tensions, or that one can take control by flexing the muscles of the *will* [emphasis added] common to us all" (346). In the "present day" of *Watchmen*, Adrian observes that the political climate is unsustainable, predicting that the entire world will crumble into chaos and that world war is inevitable (369, 4–7). To "save the world," Adrian concocts a plan to unleash a biological weapon on New York City, intended to kill "half the city" upon impact (374, 4). Adrian believes his intentions are magnanimous, but to most readers, his plan is the product of insanity and megalomania. While Adrian argues that he is doing what is best for humanity, the reader wonders if this is merely the next step in completing his "transformation" from man to god. In many ways, Adrian is the exact antithesis to Dr. Manhattan: while Manhattan sees the future as one that is predestined, Adrian is the ultimate champion of free will and believes that the future is mutable. Whereas Manhattan wallows in apathy and indecision, Adrian takes it upon himself to construct the identity he wishes to embody and attempts, in a godlike manner, to destroy and rebuild the world much the same way.

The metanarrative postscript of Chapter Eleven takes the form of an interview with Adrian by a journalist named Doug Roth for the fictitious magazine *Nova*. Adrian introduces the concept of futurology in his answer to the question "What sort of world do you see it being in the future?"

> That depends upon us ... each and every one of us. Futurology interests me perhaps more than any other single subject, and as such I devote a great deal of time to its study. I would say without hesitation that a new world is within our grasp, filled with unimaginable experiences and possibilities, if only we want it badly enough ... it all depends on us, on whether we, individually, want Armageddon or a new world of fabulous, limitless potential. I see twentieth century society as a sort of race between enlightenment and extinction. In one lane you have the four horsemen of the apocalypse ... [in the other] the seventh cavalry. (379–80)

Adrian Veidt's ambitions are steeped in an obsessive, almost fanatical belief that regards ancient knowledge and culture as ideals to be attained and then exceeded. His unyielding reverence for antiquity and insistence on a linear model of time views events merely as those that have happened, those of the present, and those of the future, which are imagined but can be realized if an individual wills them to be. Like Manhattan, Adrian views life as an all-or-nothing endeavor; both men place their principles and ideals above the complex, real-life needs of humanity. The only meaningful difference between them is that Adrian acts while Manhattan refuses to act, yet the results are equally disastrous.

Dr. Manhattan and Adrian Veidt are "godlike" characters; in Manhattan's case, such a role is imposed by the environment via a nuclear accident, and in Adrian's case, it is one that he has chosen and crafted for himself. In a sense, both characters are as abstract as the ideas they espouse – they are two sides of the same coin. Manhattan, by choosing not to act or being unable to do so, becomes arrested and stagnant, and Adrian, representing the other extreme, stops at nothing to achieve his goal, even if that ends in total apocalypse. Laurie, whose chapter is strategically situated between the two, is the most humanistic and balanced, the median between the two extremes. She values the present, yet she recognizes and accepts the importance of the past as well as the future.

The intersection of time, space, and being is echoed throughout *Watchmen*, as the tensions between varying models of time, both in terms of diegetic and mimetic expression as well as the interplay between visual and textual, present a wide range of philosophical positions, from the classical to the postmodern. For all its explorations and manipulations of spatiotemporal relationships in both form and content, perhaps the most resonant dichotomy in *Watchmen* can be reduced to the age-old debate surrounding free will and predestination, and "time" – whatever that means to the reader – is a measure of an individual's ability to navigate circumstance.

6   Graphic Novels: Ready-Made Time and Space

Graphic novels are inherently multimodal, for the intersection of art and text coalesce into a narrative whole. Temporality in literature and art is not a new concept, but the emergence of the graphic novel compels readers to reevaluate visual and textual codes, as they are inseparable in this art form. "Static" art, that is, isolated images, and traditional literary texts feature narrative elements, but the most inimitable aspect of sequential art involves its sequencing, which combines the visual and the textual to indicate both the stasis and passage of time, and its expression of duration, in particular the way panels are spaced, sized, arranged, and colored. The graphic novel, being a narrative composite of the literary and visual, requires conscious, deliberate decisions by author and illustrator concerning both microstructural and macrostructural levels of the text. The manner in which these elements fuse may create narratives in which disjunctions in time and space transpire, depending on authorial intentions and the consequent executions of such objectives. Disjunctions and anachronisms may occur as planned by the author, or may transpire due to unforeseen factors, as in Neil Gaiman's *Sandman*, in which the demands of the narrative create disruptions between the fidelity of pure sequence and the subjectivities characteristic of duration.

However, depending on the theme and purpose of a work, time's intricacies may demand synchronicity, as in the case of Alan Moore and Dave Gibbons' *Watchmen*, in which time is not only manipulated in its form, but is also a fundamental theme. Moore and Gibbons attempt to unify multiple forms of time with a spatiotemporal alignment within the text and outside of it, creating an overlapping, concentric vision of "common" time, in which time flows forward from the past, to the present, and the as-yet-to-be future. In *Watchmen,* layers of temporality are formally manifested as narrative time and reading time, and the thematic discussions of phenomenological and cosmological time remind readers that regardless of how one perceives it, time is inextricably linked to existence and the search for meaning in which humans are fated to participate. The motif of the Doomsday Clock, as dramatic as counting down to the end of the world would appear to be, symbolizes the urgency of time and the role this urgency plays in inciting action, not merely for everyday tasks, but in existential decisions. Even in a medium many would consider relatively recent, temporality in graphic novels harkens back to an ancient core of meditations on time, space, existence, and the complexities conjoining them – if one considers it possible for anything to be "ancient" at all.

References

Primary Works

Gaiman, Neil. *The Sandman. Preludes, Nocturnes.* Vol. 1. New York: DC Comics, 1991.

Moore, Alan, and Dave Gibbons. *Watchmen.* New York: DC Comics, 1987.

Secondary Works

Alexander, S. "Spinoza and Time." *Studies in Spinoza.* Edited by S. Paul Kashap. Berkeley: University of California, 1974.

Barnes, David. "Time in the Gutter: Temporal Structures in Watchmen." *KronoScope* 9, nos. 1–2 (2008): 51–60.

Barsam, Richard, and Dave Monahan. *Looking at Movies: An Introduction to Film.* 5th ed. New York: W.W. Norton, 2016.

Barthes, Roland. "The Structuralist Activity." In *Critical Essays.* Translated by Richard Howard. New York: Farrar, Straus, and Giroux, 1972.

Bender, Hy. *The Sandman Companion: A Dreamer's Guide to the Award-winning Comic Series.* New York: Vertigo (DC Comics), 1999.

DiGiovanna, James. "Dr. Manhattan, I Presume?" In White, *Watchmen and Philosophy,* 103–14.

Drohan, Christopher. "A Timely Encounter: Dr. Manhattan and Henry Bergson." In White, *Watchmen and Philosophy,* 115–24.

Eisner, Will. *Comics and Sequential Art: Principles and Practices from the Legendary Cartoonist.* New York: Norton, 2008.

Genette, Gérard. *Narrative Discourse: An Essay in Method.* Translated by Jane Lewin. New York: Cornell University Press, 1980.

Gilbert-Walsh, James. "Revisiting the Concept of Time: Archaic Perplexity in Bergson and Heidegger." *Human Studies* 22 (2010): 173–90.

Gorham, Geoffrey. "Descartes on Time and Duration." *Early Science and Modern Medicine* 12 (2007): 28–54.

Heidegger, Martin. *Basic Writings: From Being and Time (1927) to The Task of Thinking (1964)*. Translated by David Farrell Krell. San Francisco: HarperCollins, 1993.

Heidegger, Martin. *Being and Time: A Translation of Sein Und Zeit*. Translated by Joan Stambaugh. Albany, NY: State University of New York Press, 1996.

Lessing, Gotthold Ephraim. *Laocoön: An Essay on the Limits of Painting and Poetry*. Translated by Edward McCormick. Baltimore: Johns Hopkins University Press 1962.

Martin, Wallace. *Recent Theories of Narrative*. Ithaca, NY: Cornell University Press, 1986.

McCloud, Scott. *Understanding Comics: The Invisible Art*. New York: HarperCollins, 1993.

Monton, Bradley. "Time Travel without Causal Loops." *The Philosophical Quarterly* 59, no. 234 (January 2009): 54–67.

Nietzsche, Friedrich Wilhelm. *The Will to Power*. Translated by Walter Kaufmann and R. J. Hollingdale. New York: Vintage Books, 1968.

Power, Sean. "A Philosophical Introduction to the Experience of Time." *NeuroQuantology* 7, no. 1 (March 2009): 16–29.

Tegtmeier, Erwin. "Three Flawed Distinctions in the Philosophy of Time." *International Journal for Ontology and Metaphysics* 8 (2007): 53–59.

Todorov, Tzvetan. "The Structuralist Analysis of Literature: The Tales of Henry James." In *Structuralism: An Introduction*, edited by David Robey, 73–103. Oxford: Clarendon Press, 1973.

White, Mark, ed. *Watchmen and Philosophy: A Rorschach Test*. Blackwell Philosophy and Pop Culture Series. Hoboken, New Jersey: John Wiley and Sons, 2009.

Yaffe, Gideon. "Locke on Consciousness, Personal Identity and the Idea of Duration." *Noûs* 45, no. 3 (2011): 387–408.

CHAPTER 14

"There's More Than One of Everything": Time Complexity in *Fringe*

Sonia Front

1   Speeding Worlds

In his analysis of contemporary mainstream American cinema, David Bordwell has noted a set of stylistic changes that have evolved into what he calls "intensified continuity": a steady increase of tempo in Hollywood feature films, "traditional continuity amped up, raised to a higher pitch of emphasis."[1] This intensified continuity is enacted by means of four strategies: more rapid editing (from between 300 and 700 shots per film and an average shot length of eight to eleven seconds between the years of 1930 and 1960, to between 3000 and 4000 shots per film with shot length between three and six seconds, and many films below three seconds, at the end of the century), combining various lens lengths within one film, the predominance of closer views, especially of faces, and the use of free-ranging cameras, or what Shane Denson calls "crazy cameras,"[2] which perform unmotivated movements, divorced from any consistent spectating position. It is significant that the enumerated techniques, which used to be utilized in special moments, have become the norm for all types of scenes within the film.[3] For chase scenes, which, according to Siegfried Kracauer, have become a "cinematic subject ... par excellence,"[4] even more frantic editing is used; what Bordwell calls "spasmodic editing,"[5] in which rapid cuts are accompanied by the strategy of constantly shifting visual and spatial positioning in reference to the vectors of evolving action.

---

1   David Bordwell, "Intensified Continuity: Visual Style in Contemporary American Film," *Film Quarterly* 55.3 (2002): 16.

2   See Shane Denson, "Crazy Cameras, Discorrelated Images, and the Post-Perceptual Mediation of Post-Cinematic Affect," in *Post-Cinema: Theorizing 21st-Century Film*, eds. Shane Denson and Julia Leyda (Falmer: REFRAME Books, 2016), http://reframe.sussex.ac.uk/post-cinema/2-5-denson/.

3   Bordwell, "Intensified Continuity," 23–24.

4   Siegfried Kracauer, *Theory of Film: The Redemption of Physical Reality* (Princeton, NJ: Princeton University Press, 1960), 42.

5   David Bordwell, "Unsteadicam Chronicles," *Observations on Film Art Blog*, http://www.david bordwell.net/blog/2007/08/17/unsteadicam-chronicles/.

In Bordwell's estimation, the stylistic changes contribute to an intensification of traditional techniques, rather than their rejection; in other words, film directors still follow the rules of continuity and depict coherent worlds, even if the *sjuzet* is nonchronological. Many film scholars do not agree with his assessment, claiming that continuity, understood as the fundamental orienting structure of narrative cinema as well as the coherent organization of the story and the uniformity of space and time, has been fractured in twenty-first century cinema.[6] Mattias Stork discusses an extreme form of accelerated cinema, which he calls "chaos cinema," and notes its rejection of a coherent course of events in space and time in an imitation of

> the illiteracy of the modern movie trailer. It consists of a barrage of high-voltage scenes. Every single frame runs on adrenaline. Every shot feels like the hysterical climax of a scene which an earlier movie might have spent several minutes building toward. Chaos cinema is a never-ending crescendo of flair and spectacle. It's a shotgun aesthetic, firing a wide swath of sensationalistic technique that tears the old classical filmmaking style to bits.... it barely matters if you know what's happening onscreen. The new action films are fast, florid, volatile audiovisual war zones.[7]

Steven Shaviro has called the phenomenon "post-continuity" and defined it as cinema in which the focus on immediate effects replaces continuity:

> In recent action blockbusters ... there no longer seems to be any concern for delineating the geography of action, by clearly anchoring it in time and space. Instead, gunfights, martial arts battles, and car chases are rendered through sequences involving shaky handheld cameras, extreme or even impossible camera angles, and much composited digital material – all stitched together with rapid cuts, frequently involving deliberately mismatched shots. The sequence becomes a jagged collage of fragments of explosions, crashes, physical lunges, and violently accelerated motions. There is no sense of spatiotemporal continuity; all that matters is delivering a continual series of shocks to the audience.[8]

---

6   Steven Shaviro, "Post-Continuity: full text of my talk," http://www.shaviro.com/Blog/?p=1034.
7   See Mattias Stork, *Chaos Cinema Part 1* https://vimeo.com/28016047 and *Chaos Cinema Part 2* https://vimeo.com/28016704.
8   Shaviro, "Post-Continuity."

One of the reasons for the lack of spatio-temporal coherence is the fact that classical cinema places the narrative in Newtonian immovable space and absolute mathematical time which is believed to pass uniformly without regard to anything external, whereas post-continuity cinema introduces space-time understood in accordance with the theories of quantum physics which defines it as relative, malleable and heterogeneous. In the past, continuity was supposed to correlate with realism – "the illusion of life passing at normal speed,"[9] whereas now the notion of reality is interrogated and continuity is replaced by simultaneously coexistent temporalities and an intricate network of temporal relations.

Television accelerated its pace at the same time as did cinema. Bordwell speculates that the influence might have been reciprocal: early 1960s films with fast cutting rates might have impacted television, and television directors who also made feature films might have transferred to cinema some techniques, such as the usage of video-assisted shots, deficient in detail, full of singles and close views, aimed for the television format, because cinema films eventually ended up on small screen anyway. Another reason for the acceleration in the pace of films Bordwell mentions was multiple-camera filming, which started to be required by the demand for faster work in the climate of general social acceleration. The producers asked for very dynamic trailers and many alternative takes to be available in the postproduction process.[10]

The goal of these stylistic changes is the increased dynamism of film, the aim of which is to maintain the viewer's interest. James Gleick traces the beginning of the erosion of the viewer's attention to remote controls which were invented in the 1950s. As viewers engaged in channel-hopping, programmers started to provide a more accelerated flow of images in an attempt to hold the viewer's attention, because fast editing forces them to "assemble discrete pieces of information, and it sets a commanding pace: look away and you might miss a key point."[11] Gleick enumerates the tactics used to maximize television time: commercials pushed inside the show rather than between the shows so as not to lose the viewer, abandonment of the traditional half-minute of opening titles in favor of opening the film with the story, 'promos,' 'bumpers,' 'opens,' etc. filling in the gaps that were dead air before, and playing the last gag in comedy shows together with the credits.[12] These changes were fuelled by the

9   James Gleick, *Faster. An Acceleration of Just about Everything* (New York: Vintage Books, 2000), 177.

10   Bordwell, "Intensified Continuity," 21–23.

11   Bordwell, "Intensified Continuity," 24.

12   Gleick, *Faster*, 182–85.

conviction that an audience which was trained on channel-hopping, comput-er games and multitasking on the Internet was able, and desired, to consume films and programs cut at a delirious tempo, enabled by new technologies that accelerate the pace of art and life. Gleick claims that the audience has not only become fluent in a visual language consisting of images and move-ments, rather than words like in the past, but also that we have been trained to comprehend "the most convoluted syntax at a speed that would formerly have been blinding."[13]

Yet a problem is that both sped-up cinema and television provide visual fast food that invites mindless consumption of images without processing them. The result is content full of – in Gleick's words – "visual candy and vi-sual popcorn" with "the sinews of plot and character melting away in a boil of visceral gratification."[14] Narratives that impose a heart-thumping velocity and thus promote fast-paced mindless violence, superficially presented charac-ters and vapid spectacle are considered to be devoid of artistic merit, as Laura Thompson points out; therefore their critical reception, particularly that of ac-tion cinema, has been negative.[15]

2   Complex Narratives

However, there is a trend in contemporary cinema described as complex narra-tives or "puzzle films"[16] that utilizes high-speed editing together with elaborate narrative maneuvers and simultaneously meets the markers of aesthetic and cultural value. Rather than pursue the hectic rhythm of mindless velocity for its own sake, speed in these films has an artistic value; it is employed to pic-ture, for example, the frenetic activity of a mind sliding into madness (David Fincher's *Fight Club*, 1999), the experience of drug taking and the ensuing dis-tortion of time (so-called "hip-hop montage" in Darren Aronofsky's *Requiem for a Dream*, 2000), short-term memory loss (Christopher Nolan's *Memento*,

---

13   Gleick, *Faster*, 178.

14   Gleick, *Faster*, 175.

15   Laura Thompson, "In Praise of Speed: The Value of Velocity in Contemporary Cinema," *Dandelion Journal* 2.1 (2011), http://www.dandelionjournal.org/index.php/dandelion/ article/view/35/84.

16   See David Bordwell, *The Way Hollywood Tells It: Story and Style in Modern Movies* (Berkeley, Los Angeles, London: University of California Press, 2006); Warren Buckland, ed., *Puzzle Films: Complex Storytelling in Contemporary Cinema* (Malden and Oxford: Blackwell Publishing, 2009) and Warren Buckland ed., *Hollywood Puzzle Films* (New York, London: Routledge, 2014).

2000), the experience of parallel universes (Eric Bress's and J. Mackye Gruber's *The Butterfly Effect*, 2004), and so on. Although events in such films unfold at an accelerated tempo, they do not merely provide "visual candy and visual popcorn," but allow room for philosophical reflection. They engage the viewer in intellectual and emotional games, scattering intertextual references, twists and temporal manipulations that wow the audience and leave them dazed and confused, inviting multiple viewings and active modes of spectatorship. In case of these fast-paced films, contemplation and reflection are intended to take place *after* the act of watching, which is enabled by digital technologies. Complex narratives are intended to be "like a splinter in your mind" – watched many times, stopped, rerun, manipulated, and broken into segments to extract their meaning(s). The films demand a sophisticated viewer who watches actively with their eyes wide open so as not to miss any critical detail. Speed, therefore, refers here not only to the length of the shot but also to the amount of visual and auditory information one has to absorb: the often machine-gun velocity of the conversation and details placed in the margins of the frame, the sensory overload, many things happening simultaneously (multiple threads within the plot or split screen), and images flashing upon the screen at a pace almost too fast to be perceptible. Speed is also emphasized in contemporary films by plots in which the characters run against the clock to complete a task before a deadline. There is never an abundance of time; the characters are always involved in a struggle to cheat time. In the culture of the nanosecond,[17] it is the split second that matters: before the bomb goes off, before a murderer picks another victim, or, as in Andrew Niccol's *In Time* (2011), before people literally run out of time uploaded onto their wrists as a payment for work.

It is not only complex cinema narratives that offer "brain candy"[18] in the twenty-first century; television narratives are also becoming increasingly complex and experimental. Late twentieth and early twenty-first century cinema paved the way for the television narratives that foreground playing with time. Although they have not replaced traditional storytelling structures, they have come to play a very important role in mainstream television. Jason Mittel has come up with the term "narrative complexity" to denote this type of serial television.[19] Narrative complexity encompasses serialized plots within episodic television with mini-stories concluded in each episode

---

17   Gleick, *Faster*, 6.
18   The phrase comes from the text by Malcolm Gladwell, "Brain Candy. Is Pop Culture Dumbing Us Down or Smartening Us Up?," *The New Yorker*, May 16, 2005. http://www .newyorker.com/magazine/2005/05/16/brain-candy.
19   See Jason Mittell, "Narrative Complexity in Contemporary American Television," *The Velvet Light Trap* 58 (2006): 29–40.

and a story arc carrying across the season(s), along with "an elaborate, interconnected network of characters, actions, locations, props, and plots."[20] The characteristic features of the narratively complex serials are "complexity of the look and televisuality, characters, story arcs, time, genre, intertextuality, self-reference and transmediality," although, obviously, not every show needs to execute all of them.[21] The complexity of time, in turn, involves what Paul Booth calls "temporal displacement,"[22] that is, the emphasis on the aesthetics of manipulating chronometric time by means of flashbacks, flash-forwards, time-travel, parallel timelines and other non-linearities, introduced with few orienting markers. This non-coherent narrative time combined with the visceral fever of speed is instrumental in the depiction of worlds that lack spatio-temporal integrity, and as such belong to the post-continuity landscape of contemporary media.

3   Narrative Complexity in *Fringe*

One of the examples of the embodiment of time's urgency and narrative complexity that underlines temporal displacement is the television series *Fringe* (J. J. Abrams, Alex Kurtzman, Roberto Orci, 2008–2013). At the beginning the show assumes the form of a police procedural in which FBI agent Olivia Dunham (Anna Torv) of Fringe Division, together with civilian consultant Peter Bishop (Joshua Jackson) and his father, a brilliant scientist, Walter Bishop (John Noble), investigate bizarre cases of scientific experimentation that involve nanotechnology, AI, cybernetics, neuroscience, genetic engineering, etc. These phenomena, located on the fringes of science, mix science fact with science fiction, pushing the boundaries of what is possible. Always rationalized within the diegesis of the show, they constitute a twenty-first century update on traditional science fiction tropes, mainly concentrating on the ways in which technology can harm and/or kill us.

The format of a television series allows *Fringe* to pour out a cornucopia of scientific theories whose abuse in criminal cases, "as if someone out there is experimenting, only the whole world is a lab"[23] makes up The Pattern that

---

20   Paul Booth, "Memories, Temporalities, Fictions: Temporal Displacement in Contemporary Television," *Television&New Media* 12.4 (2011): 370, 371.

21   Christine Piepiorka, "You're Supposed to Be Confused. (Dis)Orienting Narrative Mazes in Televisual Complex Narrations," in *(Dis)Orienting Media and Narrative Mazes*, ed. Julia Eckel et al. (Bielefeld: Transcript, 2013), 183.

22   Booth, "Memories, Temporalities, Fictions," 371.

23   *Fringe*, "Pilot," Fox, 2008, written J. J. Abrams, Alex Kurtzman and Roberto Orci.

"rupture[s] the fundamental constants of nature"[24] and reveals the existence of a parallel universe. The two universes occupy the same space-time but exist in different dimensions, held apart by decoherence, a law in physics that ensures their separation. However, Walter's tampering with that law by building a door, The Gate, which would allow passage to another universe twenty five years before, along with The Pattern incidents, have led to the disturbance of the equilibrium between the two realities and the subsequent formation of "soft spots," that is, places where the membrane between the universes is thinner. These spots or rifts can potentially become a vortex that could ultimately endanger the very existence of the universes. With this development, the episodic structure of the police procedural, with every episode-case closed or at least offering a mini-resolution, comes to be pitted against multi-episode arcs and relationship dramas, creating an entangled network of interconnections. From a police procedural the story becomes one of the confrontation or even collision of interdimensional worlds.[25]

What is unique about *Fringe* is that the alternate universe is a concept fundamental to the series, a pivot on which the plot rests, rather than a superficial gimmick or temporary diversion, as in many previous shows.[26] Story arcs in *Fringe* make it possible to build the whole parallel world "with a depth and thoroughness that no other television show has yet achieved."[27] It is the first show that visualizes the physics theory and what it would mean for another universe to exist alongside ours, where it would be located, how it could be experienced, what the relationship between the universes and their inhabitants could be. The two universes in *Fringe*, as theorized by the many-worlds interpretation of quantum mechanics, are variations of each other's events. On the other side of the looking glass the society is more advanced, yet at the same time it is a police state where people have their "Show Me Cards," the World Trade Center is intact after the September 11 attacks, JFK is still alive and Nixon adorns a dollar coin, zeppelins are commonplace over New York, coffee is a rare and luxurious commodity, and so on.[28] The show registers variations

---

24   *Fringe*, "Peter," s. 2 e. 16.
25   Brian McHale differentiates three types of confrontation of worlds: spatially distant, temporally distant and interdimensional. *Postmodernist Fiction* (London and New York: Routledge, 2004), 60–61.
26   Mike Brotherton, "Déjà New. Not Quite the Parallel Universe Story We've Seen Before on TV," in *Fringe Science: Parallel Universes, White Tulips, and Mad Scientists*, ed. Kevin Grazier (Dallas: BenBella Books, 2011), 86.
27   Brotherton, "Déjà New," 86.
28   See also "83 Crazy Differences between *Fringe*'s Alternate Universe and Ours," *Blastr*, December 14, 2012. http://www.blastr.com/2011/01/83_differences_between_fr.php.

in the lives and personalities of the protagonists as well: Olivia's counterpart, Fauxlivia ('false Olivia'), is an easy-going red-head, her mother is still alive while her sister is dead, contrary to Olivia's; and alternate Walter, Walternate, is an emotionless Ministry of Defense who sends cybernetic shape-shifters to infiltrate the prime universe.

The show structurally and thematically underlines speed and multiple temporalities. Time's urgency manifests itself in the way the team is always racing against the clock, forced to make quick decisions in an attempt to outsmart the orchestrator(s) of The Pattern incidents, to gather essential information and to predict a killer's next moves. Contrary to classic television series in which "To watch an episode of 'Dallas' is to be stunned by its glacial pace – by the arduous attempts to establish social relationships, by the excruciating simplicity of the plotline, by how obvious it was,"[29] *Fringe* depicts multiple plot threads at an unrelenting tempo, adhering to the rules of complex narratives by misguiding and deceiving the viewer, introducing unexpected twists and complicating the temporal relations between the characters and plot threads. The show requires active modes of spectatorship, as the rapid cuts and frenetic style of the camera do not allow the viewer to take in all the information. It is only upon a second watching that we can discover the small details that foreshadow the parallel universe conceit: for instance, Olivia experiencing flashes sideways into another universe or Peter remembering the scar on his childhood toy to be on the other side. The densely layered diegesis embeds further details whose importance can only be realized when you pause the film; for example, at the end of "The Same Old Story" (s. 1 e. 2) we can see a man in a hospital bed between two identical men in incubators for two seconds (50:02–50:04), which indicates secret research by Massive Dynamic, a company founded by Walter's coworker, William Bell (Leonard Nimoy); at the end of "Of Human Action" (s. 2 e. 7) a Massive Dynamic employee flips for a couple of seconds through files with identical boys (Tyler 03, Tyler 04, Tyler 05) and different guardians, which alters the meaning of the whole episode. It is advisable then not to even blink while watching, as one might miss crucial information. Every episode also contains details that are not necessary to the comprehension of the plot: Easter eggs which foreshadow the next episode, mysterious glyphs appearing at the interstitial moments between acts and an appearance of the Observer too quick to be registered. In this way *Fringe* is symptomatic of what Tim Blackmore has

---

29   Malcolm Gladwell after Thomas Elsaesser, "The Mind-Game Film," in *Puzzle Films*, ed. Buckland, 33.

called the "speed death of the eye,"[30] generated by the short shot length and visual excess that makes it impossible for the eye to register all images and the brain to process them. However, this feature of the show simultaneously ensures its maximum rewatchability. The subsequent viewings contribute to the pleasure of watching as they allow us not only to digest the plot more slowly, but to appreciate the slight differences between the universes (e.g., different actors playing in famous films advertised on billboards) as well as ontological shocks and "temporal teases."[31]

The "temporal tease," that is, experimentation with serial time that disrupts the linear flow of time, is part of temporal displacement, characteristic of complex narratives. The narrative and imaginative richness of *Fringe* allows the directors to exercise almost all techniques of temporal displacement. First of all, time in the series splits into parallel universes, which becomes the overriding temporal framework of the narrative. Second, the alternate realities proceed along several timelines. Third, the events within those timelines are far from chronological – they are presented by means of such techniques as flashbacks, flashforwards, flashes sideways and time travel. Additionally, the characters navigate a variety of temporal zones, governed by distinct velocities.

4   The Temporal Pyrotechnics in *Fringe*

In its unprecedented temporal complexity the show accommodates various timelines. The First Timeline, never presented on screen, unfolds without any intervention until the Observers are created in 2167. The Original Timeline, with which the show starts, departs from the First Timeline at the moment when a team of Observers, scientists from the future, is sent back in time to witness their beginnings. One of the Observers, September (Michael Cerveris), accidentally prevents Walternate from inventing a cure for Peter and thereby leads to Walter crossing over to the other universe to save the boy. The timeline encompasses two parallel time-streams: the prime universe and the alternate (parallel) universe. The main events begin in 2008 first in prime universe (season 1), then flashes sideways into the parallel universe are introduced before action starts to shuttle back and forth between the two parallel realities

---

30   See Tim Blackmore, "The Speed Death of the Eye: The Ideology of Hollywood Film Special Effects," *Bulletin of Science Technology Society* 27.5 (2007): 367–72.

31   Melisa Ames, introduction to *Time in Television Narrative. Exploring Temporality in Twenty-First Century Programming*, ed. Melisa Ames (Jackson: University Press of Mississippi, 2012), 8.

(seasons 2–4), making it "a cross-time story."[32] Several flashbacks take the viewer back to the 1980s to the time of Peter's illness and Walter's crossing to the other universe, and the aftermath of that decision in both universes. This timeline has several branching futures. One of them is the destruction of the alternate universe in 2011 and Walternate's revenge in 2026. This event is shown by means of a flashforward combined with time travelling when Peter enters the Doomsday Machine and his consciousness rockets forward in time to 2026 to live through several days of this potential future.

In 2026 Walter sends the Doomsday Machine back in time, creating a time loop in the process at least once, while Peter closes the loop in another potential future by making a different choice in the Machine and bridging the two universes. He gets erased from the timeline, a scenario orchestrated by the Observers to rectify the error made by September: "He never existed. He served his purpose."[33] The timeline is rebooted and overwritten with new events and new memories to accommodate the course of events without Peter, giving rise to the Adjusted Timeline. What is in operation here is backward causality – it is the future that influences the past. Contrary to the theory that the future is open to alternatives that the past is closed to,[34] Peter's erasure from time has backward consequences through which the indeterminism, contingency and instability of the future are projected into the past. Paradoxically, both courses of events take place for the characters in this fission-fusion instance of universe branching. The alternate universe timeline gets adjusted as well, and Fauxlivia forgets her son who gets erased. Peter finds himself between universes, "neither here nor there." When he reappears in the timeline ("Subject 9," s. 4 e. 4), his memory is intact – he claims to be Walter's son although for Walter his son had died as a child. One timeline is superimposed upon another then and Peter is simultaneously dead and alive. In this timeline in "Letters of Transit" (s. 4 e. 19) we are taken in a flashforward without any signposts to the year of 2036 for one episode. After that episode time spools back for the rest of the season to suddenly go into fast forward in season 5.

In the Adjusted Timeline the Observers invade in 2015 so Walter devises a plan to get rid of them (season 5). When the plan succeeds, Walter and the Observer child are erased from the timeline as their travel forward to 2167 has created a paradox. Time loops back to 2015 in which the Observer invasion

---

32   Paul J. Nahin, *Time Machines. Time Travel in Physics, Metaphysics, and Science Fiction* (New York: Springer, 1999), 318.

33   *Fringe*, "The Day We Died," s. 3. e. 22.

34   Julian Barbour, *The End of Time. The Next Revolution in Physics* (Oxford, New York: Oxford University Press, 2000), 55. See also Arthur N. Prior, *Papers on Time and Tense* (Oxford: Oxford University Press, 2003), 57.

never takes place, and the timeline is reset again, taking the shape of The Final Timeline.

Within each timeline the narrative intersects a variety of heterogeneous disjunctive speeds, which become further instances of techniques of temporal distortion that constitute narrative complexity. Several episodes feature characters who can predict the future: a girl experiences flashes of future events resonating backwards in time ("Forced Perspective," s. 4 e. 10), after experimental treatment a man becomes super intelligent and runs patterns of future events, and even coordinates a chain of incidents leading to a person's death ("The Plateau," s. 3 e. 3), a former MIT physicist utilizes the theory that "time could be flattened" so that one could see through past, present and future, which allows him to save people from their nightmare futures ("Making Angels," s. 4 e. 11), Agent Astrid Farnsworth (Jasika Nicole) from the alternate universe runs scenarios on the basis of probability theory, Walter's camera registers events before they happen ("Subject 9"), and, finally, the Observers enjoy an out of time perspective, unlimited by the human perception of time. Thanks to the technology implanted in their bodies, the Observers (like Peter later, in season 5), relying on the observation of patterns, are able to trace a system's evolution through time according to the laws of physics.[35]

Yet in this deterministic image of the world they do not seem to take chance into consideration, which is perfectly illustrated in "The Plateau" where an Observer-like protagonist plans Fauxlivia's death, doing complex calculations. However, what he does not predict is that Fauxlivia has been replaced with Olivia, which is the "wild card" that disturbs the predicted effect. Another unpredictable variant the Observers do not consider is the act of observation itself, which in a quantum world always influences what is observed. The most poignant exemplification of this is September distracting Walternate in his lab and the whole chain of events that follows. September discloses: "There are things that I know but there are things that I do not. Various possible futures are happening simultaneously. I can tell you all of them but I can't tell you which one of them will come to pass because every action causes ripples, consequences both obvious and unforeseen" ("Firefly," s. 3 e. 10). Although they can predict the future outcome before it happens, the Observers do not know every detail of what happens, which is confirmed when September tells Olivia that in all possible futures she must die, which is not accurate.

Besides predictions of the future, other disturbances of the linear time continuum involve locally merged space-time of the two universes. In "6B" (s. 3

---

35   See Sean Carroll, chapter 7, *From Eternity to Here: The Quest for the Ultimate Theory of Time* (Oxford: Oneworld Publications, 2011), 119–42.

e. 14) a building becomes a soft spot between the universes after an old lady becomes quantum entangled with her late husband's double in the parallel universe; in "Welcome to Westfield" (s. 4 e. 12) the town of Westfield is merged with its counterpart in the alternate universe and, as a result, people merge with their doppelgängers and receive another set of memories and some duplicate organs; and in "The Consultant" (s. 4 e. 18) several people in the prime universe die after their frequencies have been synchronized with those of their counterparts in the parallel reality.

Many time displacements appear around Peter's erasure from the Original Timeline and his reappearance: desynchronized images of Peter whom Walter intends to capture by bullet time photography seen in *The Matrix*, time slips, déjà vu type incidents (the same event happens twice), frozen moments, time jumps (Peter gets from one moment to another with whole periods swallowed), reversals (to the state from four years before). Some of the incidents appear to be produced by "a time chamber" artificially created by an engineer to return him and his wife to the condition of four years earlier when she was still healthy ("And Those We've Left Behind," s. 4 e. 6).

Further time distortions depicted in the show encompass accelerated time when a man goes from being a fetus to old age and death within several hours ("The Same Old Story") and when Fauxlivia's pregnancy is sped-up ("Bloodline," s. 3 e. 18); time compression, when Peter in the Doomsday Machine lives through several days in 60 seconds; and time dilation in the pocket universe, created by Walter to hide an Observer child ("Through the Looking Glass and What Walter Found There," s. 5 e. 6). A pocket universe is an interdimensional space between the two universes with different laws of space and time. While only five days pass to the man accidentally trapped there, in fact more than twenty years elapse in the prime reality.

Time travelling completes the list of temporal contortions in the series. Apart from Peter's mental time travelling, there are two more instances of that technique of temporal distortion in the show: Walter uses the black hole to send the Doomsday Machine back in time; and the physicist Alistair Peck (Peter Weller) physically travels back in time three times, thus enacting three mutually exclusive versions of events which create time loops and make chronological sequences self-erasing ("White Tulip," s. 2 e. 18).

All the discordant temporalities coexistent in the diegesis of *Fringe* indicate, as Lutz Koepnick points out, "the copresence of various ... beats and rhythms in our temporally expanded moment."[36] In each episode *Fringe*

---

36   Lutz Koepnick, *On Slowness. Toward an Aesthetic of the Contemporary* (New York: Columbia University Press, 2014), 6.

intersects a variety of heterogeneous disjunctive speeds or temporal zones: the enumerated temporal phenomena, the ticking of mechanical clock, the pace of an investigation, chases, time needed for scientific processes, different psychological vectors of the characters' consciousnesses, the zone of an LSD-induced state, and additionally, parallel universes, different timelines, and, finally, the velocity of action and style. Speed relates to these distinct rates of accelerations and velocities then, but also, as Timothy Corrigan insists, to other relationships, such as "the intervals, pauses, and integrals that form the transitions between shifting velocities and accelerations as a 'depth of speed.' Within the movements and experiences of speed as various temporal degrees or 'gears,' in short, there potentially exist spaces of interpellation in which actions, thoughts, decisions, and emotions may adjust, redirect, or even control the flow of speed."[37] In *Fringe* the pauses that constitute a passage between various velocities usually take place in Walter's lab during the team's friendly morning banter or where, apart from working on the current case, Walter engages in scientific experimentation (e.g., growing an ear in an omelet), perfects his recipes (strawberry milkshakes or liquefied meat in an alginate bath), or unwinds taking home-made LSD or listening to music. These vignettes de-dramatize the action and help to build the character and his eccentric personality, contributing to the emotional involvement of the viewer. They also "adjust, redirect, or even control the flow of speed," as Corrigan theorizes, because it is at the moments of rest or cooking that ideas and problems are processed and Walter experiences his epiphanies. These moments thereby become instances of "still speed": "unexpected moments of stillness in the midst of rapid cuts and movements.... Speed is here not opposed to slowness, for it is in stillness that one may be said to truly find speed. And rather than merely going against speed, stillness contains speed and determines its quality."[38] Indeed, Walter must negotiate chunks of "still speed" in the interstices of accelerated time-frames because science, thinking and problem solving cannot be rushed.

In its complex narrative form, *Fringe*, as a postcontinuity narrative, does not depict a coherent world, a fact which is necessitated by its locatedness in quantum universe. Since the show presents two coexistent parallel worlds, there cannot be any spatiotemporal continuity. Nevertheless, even if we consider each of the worlds separately, they are not coherent either, as the passing of time in each of them resists any attempts at linear chronologization. In its

---

37   Timothy Corrigan, "Still Speed: Cinematic Acceleration, Value, and Execution," *Cinema Journal* 55.2 (2016): 120.

38   Trinh T. Minh-ha and Elizabeth Dungan after Corrigan, "Still Speed," 123.

rejection of conventional one-dimensionality of time, the multiverse in *Fringe* resonates with scientific theories and thus brackets together manifold temporal regimes, dissolving temporal continuity by means of breaks of chronology, disjunctive leaps between alternative trajectories, twists and puzzles, along with the dizzying pace of events, all of which contribute to the narrative headspin of the viewer and "moment[s] of temporal disorientation."[39] Additionally, although the universes are alternate, it transpires that the relations between them are distributed laterally and they are actually interdependent linked temporal planes: "our two worlds were inextricably linked. Without one, the other simply cannot exist."[40] In spite of the meticulous world-building (distinct visual styles and colors of credits ascribed to each of the worlds, timelines and time periods, different music for the past and present, different appearance of the characters), a contiguous course of events cannot be construed. The network of interconnected worlds and looping timelines evolves into a multidirectional story whose various branches are threaded by contingency, rather than causality.

5 "You're not you, are you?"[41]

The framework of juxtaposed multiple temporalities becomes the setting for the story of Olivia's manifold identities. Her constant breaching of distinct ontological levels in the interactions with various characters and their worlds contributes to a large extent to the multi-directionality of the narrative. *Fringe* foregrounds character temporality, which Booth cites as one form of temporal displacement in television.[42] He defines it as temporal displacement that features "characters who exist separated from their own temporal context, estranged from their 'normal' time-space arena."[43] One variety of that separation is disembodiment. Olivia is disembodied when she inhabits her FBI partner, John Scott's (Mark Valley), memories with her body in the sensory deprivation tank, and, another time, when Bell's consciousness inhabits her body and Peter and Walter enter her mind in an animated episode inside Olivia's mind in between worlds, in between times ("Lysergic Acid Diethylamide," s. 3 e. 19).

---

39   Booth, "Memories, Temporalities, Fictions," 374.
40   *Fringe*, "The Day We Died."
41   The words Walter addresses to Astrid who appears to be alternate Astrid in "Making Angels," s. 4 e. 11.
42   The other two are extensive flashback and memory temporality. Booth, "Memories, Temporalities, Fictions," 376.
43   Booth, "Memories, Temporalities, Fictions," 381.

Another form of a character's temporal displacement which Booth mentions is their transfer from their own temporal realm to another one. Temporality and spatiality are treated as a space-time amalgam in such cases and "a character's temporal displacement is mirrored by a similar spatial distortion,"[44] manifested when the characters navigate two or more disparate spaces that are dimensionally disconnected from each other. In *Fringe* it is reflected in Olivia's trips sideways in time, an ability she learns in her childhood as a result of being experimented on. This ability is first revealed as flashes from the other side. In "The Road Not Taken" (s. 1 e. 19) Olivia sees two bodies on the site of investigation although there is only one because, as it appears, there are two bodies in the same investigation in the parallel universe. After that she experiences further momentary intrusions into another realm, in which spatio-temporal continuity is violated: in her boss's, Broyles (Lance Reddick), office she involuntarily breaches dimensions and talks to the alternate Broyles who vanishes when the prime universe Broyles enters the room (which mirrors her childhood experience of switching to the other universe and meeting Walternate in "Subject 13," s. 3 e. 15); she answers the phone and reads a file in the alternate universe and after the switch she's doing the same thing in her own universe; and she has a momentary glimpse of the other universe's bleak cityscape on fire. Walter calls her experience "protracted déjà vu": "a momentary glimpse to the other side," "the road not taken."[45] Perhaps the most dramatic example of Olivia's loss of individual temporal continuity is her meeting with William Bell in his office in the World Trade Center ("A New Day in the Old Town," s. 2 e. 1). She drives a car and then has a meeting with Bell during which she experiences slippages of time because she is "out of sync" with that universe. In the meantime, to the people who look at her car, she has been in a car crash and then disappeared from her car with the door still locked and the seatbelt fastened. While the Fringe team are looking at the empty car, Olivia suddenly crashes through the windscreen and tumbles onto the ground, as if continuing her accident after visiting Bell in the alternate universe. Although her body in fact does not participate in the accident, she suffers from its consequences; as Bell explains: "I pulled you out of a moving car. Momentum can be deferred, but it must always be paid back, in full … As I once said to Walter, physics is a bitch" ("Momentum Deferred," s. 2 e. 4). Olivia is thus extracted from her own temporal realm to be temporarily interpolated into another temporal frame.

Olivia's worldline (the path which she follows in space through time) is neither continuous nor unified; it oscillates between various dimensions and

---

44   Booth, "Memories, Temporalities, Fictions," 381.

45   *Fringe*, "The Road Not Taken," s. 1 e. 19.

timelines, accelerates and slows down, branching, taking detours and coiling into overlaid timescales. Her existence at the intersection of plural dissonant temporal frames imposes the unstructured quality of personal time and problematizes the sense of her own temporality. Her slipping in and out of time disturbs her sense of self, causing confusion about who she is, about her 'true' identity.

The complex character temporality realized in the character of Olivia Dunham subverts traditional notions of subjectivity, and is conditioned by the rejection of linear temporality. If the problem of identity over time involves a continued sense of self, of memory, we cannot speak of a continued self here because traditionally it is connected with the notion of linear time, while Olivia traverses various temporal planes and various sets of memories. In consequence, the psychological continuity of her identity undergoes interferences, fluctuations and interruptions. The psychological discontinuities are first caused by John Scott's memories being intercepted by her consciousness during the shared dream state. Several of his memories blend with hers to such an extent that she perceives them as her own. Other discontinuities involve the aforementioned instance of Bell's inhabiting Olivia's body with her own consciousness suppressed; the shared psychic connection with other Cortexiphan subjects thanks to which they can combine energy to cross to the parallel universe ("Over There. Part 1," s. 2 e. 22); a close bond with Nick Lane (David Call) who helps her to tap into alternate Nick's mind ("Worlds Apart," s. 4 e. 20); and, finally, the character's twenty year encasement in amber (a substance used to contain fringe events).

Olivia's identity undergoes further disintegration and metamorphosis in the Adjusted Timeline. Like the others, she forgets Peter and her memories become reassigned to the new timeline, but when Peter returns, she gradually regains the old memories. The two timelines compete for ontological supremacy and with time, when September neglects to permanently erase Peter, Olivia fully regains her memories while the adjusted life story becomes hazy, like a dream: "it's like looking at a photo album of somebody else's life."[46] Her functioning in the Adjusted Timeline becomes a bit problematical as it is only Olivia, Peter, and later Walter who preserve the memory of the Original Timeline while other people continue psychologically in the Adjusted Timeline.

The most significant breaks in the psychological continuity of the protagonist are connected with displacements and misrecognitions with her double from the alternate universe, Fauxlivia. The two characters are established as being both similar and very distinct, both separate and connected, like the two

---

46   *Fringe*, "The End of All Things," s. 4 e. 14.

universes themselves. The similarities are first pointed out by Olivia: pleading with Fauxlivia to help her, she says: "You've got to trust me. I'm you" ("Over There: Part 2," s. 2 e. 23); while reading Fauxlivia's diaries she hopes to find clues to her secret agenda: "We're the same – the other Olivia and me. I should be able to think the same way so maybe I'm gonna be able to find a pattern.... We speak the same language, we use the same phrases. I'm reading this and I'm thinking this is how I make sense of things. Make sense of feelings. Of Peter ..." ("Reciprocity," s. 3 e. 11). Although Olivia has claimed with such conviction that she and Fauxlivia are "the same," simultaneously, however, she cannot forgive Peter that he does not have some sixth sense ability to differentiate between Fauxlivia and her: "She wasn't me. How could you not see that?"[47] Olivia thus expects Peter to recognize in her some elusive property through which her true self is disclosed just by looking into her eyes.

On the other hand, while similar, the two Olivias have different personalities, styles, and ways of conduct. This is corroborated by external observers as well: when asked by Fauxlivia what Olivia is like, Peter first replies, "She's a lot like you," but after reflection, he finally states, "maybe she's nothing like you at all" ("Over There: Part 2"). In her analysis of the complex relationship between alternate selves in *Fringe*, Eleanor Sandry concludes that Fauxlivia is established as a separate character, rather than just the other side of Olivia.[48] Manuel Garin supports this claim, arguing that "[t]he essential mechanism of *Fringe* lies in this exorcism of the double ..., which transforms the traditional doppelgänger – the shadow, the other – in a gradual coexistence of two real, serialized heroines. The series literally splits the character in two," featuring Olivia from the prime universe and Fauxlivia from the parallel one.[49] Exploiting this assumption, the creators of the show implement a series of displacements between the two heroines which blur the spatio-temporal and mental boundaries, and indicate the instability of identity.

First, Walternate engineers an exchange of Olivias to have Fauxlivia infiltrate the prime universe while Olivia is supposed to perform her counterpart's duties in the alternate reality. Contrary to Olivia, Fauxlivia preserves the continuity of identity as she only pretends to be her double without any pharmacological support. She adopts Olivia's hair color and way of walking but does not have the knowledge extensive enough to be a perfect impersonator. This

---

47   *Fringe*, "Marionette," s. 3 e. 9.
48   See Eleanor Sandry, "Same ... Yet Other. Interpersonal Communication Across Alternate Worlds," in *The Multiple Worlds of Fringe: Essays on the J. J. Abrams Science Fiction Series*, ed. Tanya R. Cochran et al. (Jefferson, North Carolina: McFarland, 2014), 77–92.
49   Manuel Garin, "Truth Takes Time: the Interplay Between Heroines, Genres and Narratives in three J. J. Abrams' Television Series," *Communicacion y Sociedad*, 26.2 (2013): 52.

is summed up by unknowing Peter's words: "With all the little differences ever since you got back from the other side … it's like you are a completely different person" ("Do Shapeshifters Dream of Electric Sheep?," s. 3 e. 4).

Olivia, on the other hand, is brainwashed into believing that the gaps in her memory have been caused by a head trauma and that she is in fact Fauxlivia. The psychological brainwashing is supported by the pharmacological procedure of "transferring memories" whose purpose is to overwrite her own memories with Fauxlivia's. Walternate's chief scientist, Brandon (Ryan McDonald) explains: "She needs to be completely immersed in Agent Dunham's life. Over time she will reach a plateau and her new identity will become fixed" ("The Plateau"). When the memory transfer kicks in, Olivia indeed treats Fauxlivia's memories as her own, and she even adopts her way of conduct, becoming thereby psychologically continuous with Fauxlivia, while her own identity is receding. This internalization of Fauxlivia's memories, knowledge and traits is enough for Brandon to consider her Fauxlivia: "for all intents and purposes she is our Olivia now,"[50] thus equating identity with memory. Although the effect of the procedure is able to fool the coworkers, it fools Olivia herself only temporarily, because the hallucinations of Peter surfacing from her unconscious help to reject the false memories and hold on to her true identity. Her knowledge also exhibits small gaps, which can become a matter of life and death ("The Plateau"). The gaps and hallucinations prove that the new identity never fully coagulates but it merely disrupts the 'true' identity, staging a break in the psychological continuity.

The blending of the two Olivias and their fluctuating identities in impossible times and spaces establishes a zone of plural truths and problematic identifications. Switching between the universes, between the timelines, and between herself and her double, the character of Olivia stages the confusion between and overlapping of ontological levels. In consequence, her identity comes to be defined by the unstable temporalities and memory shifts as she strives to negotiate a self in the temporal movements whose acceleration leads to the blurring of spatio-temporal horizons.

Olivia's psychic fragmentation and disjointed movement through time, emulated by the formal fragmentation and the accelerated rhythm of the narrative, participates in "a mapping of the epistemological uncertainties of modernism onto the ontological indeterminism of quantum science and science fiction. The crisis of subjectivity is repositioned as an ontological crisis, rather

---

50   *Fringe*, "Olivia," s. 3 e. 1.

than simply as an epistemological one."[51] With her centrifugal self, Olivia epitomizes this crisis of subjectivity; the inability to describe her identity over time because it is not ascribed to any unified time or to any fixed temporal location. If self-identification is traditionally connected with a reference in time, the culture of acceleration has disintegrated temporal horizons, and, as a result, has brought about a situation where identity lacks a point of reference.[52] Time's urgency thereby leads to the deconstruction of identity as it destabilizes its very determinants.

References

"83 Crazy Differences between *Fringe*'s Alternate Universe and Ours." *Blastr*, December 14, 2012. http://www.blastr.com/2011/01/83_differences_between_fr.php.

Ames, Melisa. Introduction to *Time in Television Narrative. Exploring Temporality in Twenty-First Century Programming*, edited by Melisa Ames, 3–24. Jackson: University Press of Mississippi, 2012.

Aronofsky, Darren, dir. *Requiem for a Dream*. USA: Artisan Entertainment and Thousand Words, 2000.

Barbour, Julian. *The End of Time. The Next Revolution in Physics*. Oxford, New York: Oxford University Press, 2000.

Blackmore, Tim. "The Speed Death of the Eye: The Ideology of Hollywood Film Special Effects." *Bulletin of Science Technology Society* 27.5 (2007): 367–372.

Booth, Paul. "Memories, Temporalities, Fictions: Temporal Displacement in Contemporary Television." *Television&New Media* 12.4 (2011): 370–388.

Bordwell, David. "Intensified Continuity: Visual Style in Contemporary American Film." *Film Quarterly* 55.3 (2002): 16–28.

Bordwell, David. "Unsteadicam Chronicles." *Observations on Film Art Blog*. http://www.davidbordwell.net/blog/2007/08/17/unsteadicam-chronicles/.

Bordwell, David. *The Way Hollywood Tells It: Story and Style in Modern Movies*. Berkeley, Los Angeles, London: University of California Press, 2006.

Bress, Eric and J. Mackye Gruber, dir. *The Butterfly Effect*. Canada, USA: BenderSpink, FilmEngine, Katalyst Films and Province of British Columbia Production Services Tax Credit, 2004.

---

51 Scott Bukatman, *Terminal Identity. The Virtual Subject in Postmodern Science Fiction* (Durham and London: Duke University Press, 1993), 174–75.

52 Sadeq Rahimi, "Identities Without a Reference: Towards a Theory of Posthuman Identity," *M/C Journal* 3.3 (2000). http://journal.media-culture.org.au/0006/identity.php.

Brotherton, Mike. "Déjà New. Not Quite the Parallel Universe Story We've Seen Before on TV." In *Fringe Science: Parallel Universes, White Tulips, and Mad Scientists*, edited by Kevin Grazier, 81–96. Dallas: BenBella Books, 2011.

Buckland, Warren, ed. *Hollywood Puzzle Films*. New York, London: Routledge, 2014.

Buckland, Warren, ed. *Puzzle Films: Complex Storytelling in Contemporary Cinema*. Malden and Oxford: Blackwell Publishing, 2009.

Bukatman, Scott. *Terminal Identity. The Virtual Subject in Postmodern Science Fiction*. Durham and London: Duke University Press, 1993.

Carroll, Sean. Chapter 7. In *From Eternity to Here: The Quest for the Ultimate Theory of Time*, 119–142. Oxford: Oneworld Publications, 2011.

Corrigan, Timothy. "Still Speed: Cinematic Acceleration, Value, and Execution." *Cinema Journal* 55.2 (2016): 119–125.

Denson, Shane. "Crazy Cameras, Discorrelated Images, and the Post-Perceptual Mediation of Post-Cinematic Affect." In *Post-Cinema: Theorizing 21st-Century Film*, edited by Shane Denson and Julia Leyda. Falmer: REFRAME Books, 2016. http://reframe.sussex.ac.uk/post-cinema/2-5-denson/.

Elsaesser, Thomas. "The *Mind-Game Film*." In *Puzzle Films: Complex Storytelling in Contemporary Cinema*, edited by Warren Buckland, 13–41. Malden and Oxford: Blackwell Publishing, 2009.

Fincher, David, dir. *Fight Club*. USA: Fox 2000 Pictures and Regency Enterprises, 1999.

Fringe, 100 episodes. Written by J. J. Abrams, Alex Kurtzman and Roberto Orci. USA: Fox, 2008–2013.

Garin, Manuel. "Truth Takes Time: the Interplay Between Heroines, Genres and Narratives in three J. J. Abrams' Television Series." *Communicacion y Sociedad* 26.2 (2013): 47–64.

Gladwell, Malcolm. "Brain Candy. Is Pop Culture Dumbing Us Down or Smartening Us Up?." *The New Yorker*, May 16, 2005. http://www.newyorker.com/magazine/2005/05/16/brain-candy.

Gleick, James. *Faster. An Acceleration of Just about Everything*. New York: Vintage Books, 2000.

Koepnick, Lutz. *On Slowness. Toward an Aesthetic of the Contemporary*. New York: Columbia University Press, 2014.

Kracauer, Siegfried. *Theory of Film: The Redemption of Physical Reality*. Princeton, NJ: Princeton University Press, 1960.

McHale, Brian. *Postmodernist Fiction*. London and New York: Routledge, 2004.

Mittell, Jason. "Narrative Complexity in Contemporary American Television." *The Velvet Light Trap* 58 (2006): 29–40.

Nahin, Paul J. *Time Machines. Time Travel in Physics, Metaphysics, and Science Fiction*. New York: Springer, 1999.

Niccol, Andrew, dir. *In Time*. USA: Regency Enterprises, 2011.

Nolan, Christopher, dir. *Memento*. USA: Newmarket Capital Group, Team Todd, I Remember Productions and Summit Entertainment, 2000.

Piepiorka, Christine. "You're Supposed to Be Confused. (Dis)Orienting Narrative Mazes in Televisual Complex Narrations." In *(Dis)Orienting Media and Narrative Mazes*, edited by Julia Eckel, Bernd Leiendecker, Daniela Olek and Christine Piepiorka, 183–202. Bielefeld: Transcript, 2013.

Prior, Arthur N. *Papers on Time and Tense*. Oxford: Oxford University Press, 2003.

Rahimi, Sadeq. "Identities Without a Reference: Towards a Theory of Posthuman Identity." *M/C Journal* 3.3 (2000). http://journal.media-culture.org.au/0006/identity.php.

Sandry, Eleanor. "Same ... Yet Other. Interpersonal Communication Across Alternate Worlds." In *The Multiple Worlds of Fringe: Essays on the J. J. Abrams Science Fiction Series*, edited by Tanya R. Cochran, Sherry Ginn and Paul Zinder, 77–92. Jefferson, North Carolina: McFarland, 2014.

Shaviro, Steven. "Post-Continuity: full text of my talk." http://www.shaviro.com/Blog/?p=1034.

Stork, Mattias. *Chaos Cinema Part 1*. https://vimeo.com/28016047.

Stork, Mattias. *Chaos Cinema Part 2*. https://vimeo.com/28016704.

Thomas, David Dylan. "Paranormal is the New Normal." In *Fringe Science*, edited by Grazier, 1–16.

CHAPTER 15

Foresight and Urgency: The Discrepancy between Long-Term Thinking and Short-Term Decision-Making

Kerstin Cuhls

Abstract

Foresight is the long-term view into the future or different futures, defined as the more action-oriented 'structured debate about complex futures.' The academic pendant is Futures Research dealing with possible, probable and desirable future developments. Even if *the* future cannot be predicted, major developments emerge already today in their basics. The guardrails of the possible, probable and desirable can be determined in this sense by scientific methods and in social discourses. Foresight is thus a concept to prepare for futures and *avoid* urgent reactions, quick and un-reflected, reactive answers to problematic situations or sudden occurrences. Methods are available to work with the different time horizons (e.g., Delphi surveys), to work with different long-term scenarios in preparation or decision-making, or even to travel in time as thought experiments. But although time scales up to 30, 40 or even more years have to be considered when e.g., investing in new infrastructures, technologies or to change the behavior of people, decision-making is often still ad hoc and does not take the time to think about the consequences. It remains in reaction to urgency. In Foresight and Futures Research, a very linear time concept is still in the forefront, although the thought experiments make it possible to go back and forth in time thinking, prepare for different futures, or even shape 'the' preferable future with visioning processes (mainly in innovation research but also in transformative studies). This contribution demonstrates examples from empirical research mainly in 'government Foresight' but also 'Strategic Foresight' of companies, associations or others with the aim to avoid urgency situations. It tries to explain why both long- and short-term time considerations are so important and what long- and short-term means for the different stakeholders (relativity of time considerations).

1 Introduction

In the 1990s, Foresight gained attention on a national scale in many countries. Foresight not only looks into the future by using all instruments of futures research, but includes utilising implementations for the present. What does a result of a Foresight study or process mean for the present? Foresight is not planning, but Foresight results provide 'information' about the future and are therefore one step in the planning and preparation of decisions (Coates 1985).

As there are many misunderstandings and different definitions around (e.g., Martin 1995 and 2010; Cuhls 2003, Sardar 2010, Marien 2010), the first part of the paper explains our understanding of Foresight in applied research at the Fraunhofer Institute for Systems and Innovation Research, Competence Centre Foresight, which was developed in 2009 to clarify that Foresight has a process and a time component. At the beginning of the paper, I explain this definition, what it means for our understanding and perception of 'time.' In the second part, I explain some methods to deal with the dilemma of 'time' in futures thinking and how Foresight is used to prepare for Futures to avoid stress or urgency reactions when certain avoidable or expected developments occur.

The third part of the paper describes some Foresight processes, which are performed within this dilemma – between long-term view and short-term action-orientation (urgency). Some ways out of this dilemma are shown in the last section. A way to address this issue is to introduce long-term thinking at every level of decision-making even if we cannot predict 'the future' and understand Foresight as a process to bridge the gap between present and future.

2   What Is Foresight?

Foresight takes the long- and medium-term view. It is not planning per se, but a step on the way to planning (see Coates 1985 or Coates et al. 2010 when discussing the term 'Strategic Foresight') and it is clear that the prediction of future events in time often goes wrong. Therefore, if prognostics or time predictions are made, direct evaluations of right or wrong prognosis are nonsense. It rather should be asked what the view ahead evoked and changed for the future. It is for example more important to discuss the different science and technology or societal topics, make them available for discussion and with this discussion start their realization or stop undesired developments. It is possible to work with assumptions and different options, the possible, probable and preferred (desirable) future or even a vision. In German language, Grunwald (2012) uses the word 'Technikzukünfte' (technical / technology futures), and the fact that the plural use of 'futures' is increasingly accepted demonstrates that there are always different options (see e.g., the journal *Futures*) from which to choose. The choice has an influence on the future that unfolds.

There is no 'theory' about Foresight (or futures studies or futures research), even if some attempts are made (e.g., Bell 2009) and quality criteria are discussed (e.g., in Gerhold et al. 2015 or Kuusi et al. 2015). In any Foresight, we learn more about today and we have the possibility to shape the future – at least partly. Therefore, the idea of defining Foresight as 'using the future to discuss and decide in the present to shape the future' is very appealing.[1] Foresight thus takes the future into the present to find solutions, make things happen or avoid undesired developments. It thus gives us the 'time' to prepare for certain futures.

Foresight is thus a mix of need-orientation and the development of drivers. In technology Foresight, it combines technology push and pull, which can start from both ends. Assessments of the issues identified are made with regard to certain criteria such as importance for the economy or sustainable development. Foresight is needed first to select the areas that should be examined in more detail via forecasting methods later. In these cases, forecasting can be a component of Foresight (for further explanation see Cuhls 2003). Foresight assumes that there is always an alternative, the future is not fixed.

The further examination of one of the anticipated futures can ask questions like: What does this option mean for today? Does something have to be changed? If yes, what? Who has to do it? Why? What does that evoke? Where is the change necessary? And how will the future option change when these measures are taken? If changes are unnecessary, will the option really develop like assumed before? Who is affected? What impacts does it have? In reality, Foresight often tells us more about the present than about the future. But it also tells us about the time we still have to decide and to act until the issue becomes an urgent one, evoking stress, being an unsolvable problem or cannot be developed anymore, e.g., because there is already a different solution or product on the market.

In reality, decisions are often made, ad hoc, without looking into the long-term and all the consequences. If the long term is considered, to make a real choice is sometimes a time-consuming task, especially as in organizations there is often not a single decision-maker but many individuals and groups are involved. In many cases, the 'new' targets are not newly identified for the decisions but they are recognized as an adopted priority. In these cases, the facts were already implicitly known but not yet recognized as priorities explicitly. The target can be modified, but once it is set, it cannot totally be abolished

---

1   Thanks to Robin Bourgeois, who brought me back to this thinking during the discussions of a European Commission/SFRI Workshop on "Democracy 2.0 – Foresight for better R&I policy," June 8, 2016.

without revising an explicit decision. At this point, either new forecasts concerning the target are needed, or if the forecasts are already made and available, the part Foresight can play ends, and planning for the future ('forward planning' or 'strategic planning') or the definite implementation of the decision starts. With Foresight tools and methods the priorities can clearly be identified. The involvement of different stakeholders gives legitimation or helps to make the priorities become real, not to predict a single future.

The approaches in (technology) Foresight can have many objectives (Cuhls 1998). Most of them help to overcome urgency in the sense that they open up the view and help us to be less surprised if a development or issue occurs. Foresight cannot meet all of the objectives, specific targets have to be set, selected from the full set of possibilities. In the context of policy-making and decision-making, the most important are:

- to enlarge the choice of opportunities, to set priorities and to assess impacts and chances,
- to anticipate the impacts of current (research and technology) policy,
- to ascertain or test new needs, new demands and new possibilities as well as new ideas,
- to focus selectively on economic, technological, social and ecological areas as well as to start monitoring and detailed research in these fields,
- to define the definition of desirable / undesirable or preferred futures and
- to start and stimulate continuous discussion processes.

In reality, most areas of the future are largely shaped by underlying reciprocal influences. The development of one field influences another field often more and more indirectly than expected, e.g., developments in electronics influence biotechnology, developments in biotechnology influence society and the healthcare sector. At first sight, these mutual influences are often unseen, but with anticipating different paths, the overlaps become obvious. At even if at present they cannot be assessed fully, the view ahead gives us time to consider the impacts and prepare for them. Only the visible or acknowledged parts, structures or framework conditions can be understood or partially influenced. If the knowledge in systems analysis theory is also taken into account, the mutual influences of systems and rules, in which the actions of persons are embedded, must also be reckoned with. Uncertainty was perceived in futures research when new experiences of the chaos theory emerged (Steinmüller 1995). The new thoughts about Foresight that emerged in the 1990s (starting with Irvine and Martin already in 1984) did not say that the future cannot be influenced directly, but made it clear that the influence on future developments is strictly limited and that the impacts can only partially be estimated. Nevertheless, the future can be 'prospectively monitored' so that urgent situations are limited. The accelerating changes that a person has to adapt to socially and psychologically make it necessary to anticipate these changes before they become reality (Helmer 1967). Therefore, meanwhile a lot of 'horizon scanning' activities are performed to monitor what is lying ahead (Cuhls 2015) and detect weak and strong signals. Most of them are part of Foresight processes. Some are stand-alone activities and there are also fully automated futures searches making more and more use of the possibilities the internet and Big Data approaches provide.

There is a link from Foresight to (strategic) planning. Some parts even overlap. Even if Foresight is only a step in planning, it can be used for strategic purposes and is thus often called 'Strategic Foresight' (Coates, Durance and Godet 2010; Godet 2000), for example in an expert commission the author was involved in (European Commission Expert Group 'Strategic Foresight for R&I Policy in Horizon 2020,' SFRI). In our Competence Center Foresight at Fraunhofer ISI, we prefer the order of 'Foresight-first, strategy based on it' and call the respective business area 'Foresight for Strategy Development' working out visions and strategies mainly for companies, but also public organizations. Scott (2001) argues similarly for strategic planning: 'Strategic planning involves creating a vision of the business the company is in or wants to be in, setting the company's goals, and determining resource allocation and other actions to pursue those goals.' This is to avoid urgency situations.

All the above and other definitions have in common that the goal of planning is a plan, be it the combination of information, milestones and measures related to fixed dates or the arrangements and allocation of resources in advance. This gives time to prepare for deadlines – and avoids stress and urgency situations, in which ad hoc action is needed. Plans describe the arrangements to reach targets (Montgomery and Porter 1996). 'Scenario planning' is intentionally both: Foresight and planning (Ringland 1998, 2002; Fink et al. 2001). Here, the borders are fluid.

Foresight is acting in the long-range view on the one hand, but has to deliver the preparation for decisions in the short-term on the other hand. This sounds very logical, but reality shows that most people do not think long-term for different reasons: they cannot do it, they do not want to do it, they are lazy, they do not have time, they want to flee ... When performing workshops, we notice different reasons. An explanation from neurosciences is: in a case of urgency, our brain still reacts like in the Stone Age: run to survive – means, we are still in the habit of fleeing or hectic activity instead of waiting, thinking and then acting (Burnett 2016). Instead of acting, most people rather react in a case of urgency. Only if they have a strategy in mind or prepared, they can go on acting in the prepared way which does not make the same stress in cases of urgency.

The most famous example are the Shell scenarios preparing for the oil crises (Schwartz 1996). As Shell had different scenarios and options prepared, they were much better able to cope to the new situation even though the situation of the oil crisis was much more severe than expected in any of the scenarios.

The problem is that most people are in a steady situation of urgency: reading e-mails, using social media, being called on the phone, just asked by someone, etc. There is no time to think long-term or even consider the long-term view. Therefore, Foresight provides the opportunity to give a place and time (better time-out) for considering futures, also thinking in alternatives – and thinking without any urgency. In the usual day, one has to be reminded that there is something further away in the future and that there is always an alternative (we do not have to like the alternative but it exists). Some methods support this.

3   Methods and Time

There are a lot of Foresight methods available (see overview in Cuhls 2008) but not all Foresight methods have to do with time or are applied to avoid situations of urgency. There are methods to unfold different futures and work with scenarios, and there are methods, like Delphi surveys, to assess future statements. These are specific surveys in two or more rounds with feedback estimating in most cases the time of realization of specific issues, but nowadays enhanced with different questions, argumentations, giving realtime feedback, etc. (see Cuhls 1998, Aengenheyster et al. 2017, Gordon and Pease 2006). Their major purpose is to estimate the time horizon and the feasibility of occurrence and gain thus 'information about the future.' There are methods to find out what people want, for example with visioning approaches, and most of them are combined with workshop concepts. They are action-oriented and try to make use of the knowledge in the room, to communicate this knowledge about the future and to make the participants 'ambassadors of the future' or multipliers who spread the ideas and thus avoid things to become urgent.

In some cases, the length of time it takes is problematic because when the projects or processes are finished, the results are sometimes out-dated, overrun by sudden events, which again create a new type of urgency. The problem remains that we cannot foresee in the sense of 100% prediction. We can only estimate roughly what is lying ahead.

Linear thinking still underlies most of the methods – especially in extrapolations and quantitative futures research, so-called 'forecasting,' one only analyses one path into the future and extrapolates from history and the present. This helps in Foresight to go to extremes, but one does not know where the curve breaks or when sudden events (or Wild Cards in scenarios) completely change the direction of the path into the future. In scenarios, different futures are looked at but thinking linear remains the same. One can escape this dilemma a bit with open explorative methods going much further than brainstorming, e.g., with Mental time travel to open up minds and thinking (Cuhls 2016).

Thinking about the time horizon in Foresight, the so-called Three Horizon Model going back to Curry and Hodgson (2008) of differentiating time between the next 5 years (analysis), the following 10 years as research, and the following years (imagination) is used (Hobcraft 2016). This is a rather practical model but does not explain time as such. It structures different futures and clarifies that the timelines in the sense of the years ahead are often not linked to the time horizons. But human beings' imagination can also have leaps and is not necessarily bound to linear extrapolation as some of the Delphi surveys have shown (Grupp 1995). Human beings in their daily thinking tend to follow the linear thoughts; however, when asked, they can also imagine something completely different, not linearly linked to the present (Cuhls 2016).

The Three Horizon Model is quite helpful to demonstrate our dilemma and it is reflected when applying Foresight methods – but three horizons do not seem to be enough. To understand the dilemma between the long-term view and urgent current decision-making some examples are given.

4   Examples from Foresight Activities to Overcome 'Urgency'

The following examples explain real Foresight projects. They were performed for shaping the future instead of predicting it. Their general objective behind the formal agenda is always to avoid future situations of urgency. Three projects try the view ahead in order to be earlier than others with developing something new in science and technology. One project tries to look into the future to deal with obvious trends in demography and to avoid their negative impacts.

The first two are purely public projects on a national level, the other two are private projects, one by a research organization in applied sciences (the Fraunhofer Society) and one by a private organization of a Federal State in Germany (Rhineland Palatinate). The author of this paper was involved in all of them, e.g., in the first one as the general project manager.

4.1   The BMBF Foresight Process (Which Ran from 2007 to 2009, and Was Later Referred to as 'Cycle 1')

The BMBF is the German Federal Ministry of Education and Research and it funded a fully-fledged Foresight activity for two years. The process was very broad, complex and made use of different methods, partly in parallel, partly successively. The starting point for this Foresight process was the 17 thematic fields of the German government's High-Tech Strategy and ongoing Foresight activities in the departments of the BMBF, i.e., the BMBF's portfolio. By mid-2009, a set of advanced methods of futures research had been developed to identify new research and technology focuses in 14 selected *established future fields*. This resulted in the so called *future topics*.

The future topics were analyzed in several steps. A topic coordinator and a sparring partner were responsible for each future topic. They conducted searches supported by bibliometrics, interviewed experts, discussed with national and international experts in a more open way and selected single topics that were labeled as 'new to BMBF,' were still undergoing research in a time-frame of more than 15 years, and somehow relevant. From among this set of topics, the most relevant were extracted and described in papers with a given standardized format. These future topics were evaluated in an online survey (single round). 19,365 science and technology experts were invited by e-mail to take part in the survey on the online platform. 2,659 individuals participated. The science and technology topics were given in a short form and participants were asked, for example, to judge items on their importance for economy, quality of life, scientific and technological importance, whether or not the issue will still be on the research agenda, the time of realization in steps and some other question like problems. They could give comments. In the analysis, it became obvious that most of the topics were regarded as sufficiently important (rate higher than 50%).

The same criteria were applied to the so called New Future Topics, which were interdisciplinary topics combining two or more of the original fields and, in most cases, also touching upon societal issues. During the course of this evaluation, questions were asked about the extent to which the research prospects and structure of the future topics were stable or still in flux. The topics were then selected, after being measured against the questions formulated at the outset, and seven additional very interdisciplinary future fields were identified. The major challenge here was the formulation of the topics (e.g., Human-Technology Interaction or Time Research). The results are documented in Cuhls et al. (2009a, b, c).

As a result of the whole process, a report with 14 future fields and 7 new (interdisciplinary) futures fields were defined and described (Cuhls et al. 2009a, b). One of the new future fields was 'Time Research,' described as *investigating time-related technologies and time-critical processes in new depth* (Cuhls et al. 2009a, b, c). The factor time is not yet adequately understood and is therefore a bottleneck in many developments. For the project managers and sponsors of this Foresight, this was a surprising result but could be explained. Research into time is the central aspect of a future field that extends into many applications, including issues such as the chronological sequence of complex processes in making applications faster, more efficient, cost-effective and intelligent, or in parallelising and synchronising processes (e.g., internet servers, production processes). The issue of dynamic and chronological development on various time scales, especially of non-linear processes, can also only be dealt with in the long term.

One very dynamic future topic in time research is chronobiology (research into natural rhythms and biological clocks), which produces findings on precisely-timed medication delivery, for example. Chronobiology could boost developments in automated pharmaceutics ('missile drugs,' depots 'targeted drugs'), possibly helping to alleviate the side-effects of nocturnal or irregular working hours. It could also be used to identify optimal times for learning, for example. For the financing agency, the German BMBF, 2009, the end of the project would be a good time to start a programme making the transfer from basic to clinical chronobiological research – this was one of the results of the project (Cuhls et al. 2009b, 97–113) because the issue met all the criteria asked for. It is a long-term issue, there are new projects and actors in Germany, there is a dynamic in the field and it will have a huge impact on the economy, on science and technology and the quality of human life (Cuhls et al. 2009b, 106–10) New implications will emerge from these new findings for society and the time management of individuals (e.g., in shift work or security monitoring by people) in the long term. Urgency in development would be avoided when starting soon. And as there were not many national activities in other countries (Cuhls et al. 2009b, 106–11), BMBF would set itself in the front of research and be ahead of others, thus rather creating urgency and being the actor than suffering from urgency and reacting.

Central research aspects of time research include understanding and specifically being able to 'control time' to be more efficient with the help of time efficiency research, precise time measurement (e.g., for GPS applications in precision agriculture, remote maintenance of machines) and time resolution (e.g., 4D precision).[2] This could optimize existing technologies as well as lead to completely new time technologies. (Cuhls et al. 2009b, 111–13).

---

2   For example in 3D printing, there are materials, printed, an approach of additive manufacturing. The fourth dimension is time. Then it is called 4D printing. The product can reshape its form after a specific time.

4.1.1   From Long-Term Searches to the Integration into the
             BMBF Portfolio

Time was thus an issue selected as a research result, but time was also a factor
when the implementation of the project results was tried. Here, we saw differences for the different future fields. Without assuming/creating a sense of
'urgency,' it was not possible to implement results. The field Man-Technology
Cooperation was very close to the present development (Cuhls et al. 2009b,
15–17). It was clear from the beginning of the BMBF Foresight Process that a
topic like this (at first called 'borders between humans and technology are
more and more blurred') would be one of the selected future topic fields. But
in this case, the formulation was difficult, because it was feared that journal-
ists would mock about it. After the presentations of all new future fields in
the respective, potentially responsible BMBF departments, this was the most
successful field because a division in BMBF was founded, called 'Referat 524:
Demographic Change and Man-Technology Cooperation' (later: changed to
Interaction). The reason was that there were already signs for urgency in dealing with this future field as such because other countries started to develop
science and technology for it. The race was already started – that means it was
easy to argue that German research and development needs programmes on it
to remain in the race and on the same level as other countries in science and
technology.

     Contrary to that, another result identified as important, aging research (un-
derstanding the molecular processes of aging) and time research did not get
the same attention. These fields were regarded as more long-term, far in the future, and *no urgency* was felt to become active in pushing science and technology solutions for it. In the end, there were no specific programmes dedicated
to these fields.

4.2   *BMBF Foresight Cycle II*

The BMBF started its second cycle in 2012 with a new two years search
and analysis phase. While the first cycle put emphasis on possible future
technological developments ('technology push'), the focus of this second cycle
was on a demand perspective ('demand pull'). Again, issues were searched
for to 'gain time' in science, technology and innovation to meet new demand
before the demand gets urgent. The project was not meant to find solutions
for urgent problems, rather for wicked problems.[3] The project was carried out

---

3   A wicked problem is a social or cultural problem that is difficult or impossible to solve for
    as many as four reasons: incomplete or contradictory knowledge, the number of people
    and opinions involved, the large economic burden, and the interconnected nature of these

by two organizations, the VDI Technology Center GMBH and the Fraunhofer Institute for Systems and Innovation Research (ISI).

This project searched for societal trends (both hidden and open trends) on the one hand, and an update of most of the science and technology fields on the other hand in order to see if the priorities are still in line or if certain developments have become more urgent. When described in papers and a mind map, the results of both approaches were presented in a workshop and 'innovation seeds' were worked out. There were structured descriptions (in templates) for all trends and technologies. From them, so-called 'Stories from the Future' were told (Zweck et al. 2015a, b, c). These narratives are told as if they are already realized (in present tense, like scenarios) and similar to hindsight, it is often explained how this situation developed and which hurdles had to be overcome. This process was published in the reports but also 'living documents' in the intranet of the BMBF were used for comments and changes. This means, the translation work of naming future issues, developments and technologies goes on. This is an important part of creating urgency.

The problem of this process was again that there was no pressure, or better: no urgency, to make use of the results. As in many Foresights (starting from the first Delphi surveys in Germany or Japan (Cuhls 2001, 2003, 2005, 2008, 2017), the results were provided in publications for anyone who would like to use them – but the problem in these cases is: There is no *urgency*, no driving forces, no promoter. There might be the demand for the new topics or solutions to a problem, but there is nobody who is guiding the solutions, so there is no direct use of Foresight results. This leaves the results underexploited – if they were urgently needed, exploitation would have been taken place. The reason is again the gap between the long-term view that is necessary for decisions, and the decisions themselves – but the decisions are often postponed to a later time. These 'later times' might be too late for the development of a technology (because someone else developed it or because the problem is solved in a different way) or for the problem solution as such. Often, postponing is the reason why problems remain unsolved, even if they are known (e.g., we know that chronobiology in medication can play a large role, but we still take our pills regularly and three times a day instead of selecting the right timing for best effectiveness).

---

problems with other problems. Poverty is linked with education, nutrition with poverty, the economy with nutrition, and so on, see e.g., https://www.wickedproblems.com/1_wicked_problems.php. The expression goes back to Rittel and Webber 1973.

4.3   *Fraunhofer Future Markets*

The Lund Declaration (European Commission 2015) formulated a recommendation that European research should focus on the Grand Challenges of our time, moving beyond current rigid thematic approaches. This enhanced the existing discussion about challenges and need-oriented approaches versus science and technology push and stressed the point that challenges should be touched upon before they become very urgent. So-called 'Grand Challenges' are discussed at many levels, for instance, the EU, the regions, nations, cities, organizations. The EU Framework Program 'Horizon 2020' stresses programs that are based on 'social challenges' (European Commission 2011, 5ff).

The Fraunhofer Society with its mission of contributing to science, technology and societal question is Europe's largest non-profit contract research organization. Fraunhofer wanted to take the demand for solving some of the Grand Challenges problems seriously by adapting its corporate process for defining and developing research themes across institutes. Fraunhofer had at that time more than 60 institutes, and 6 groups of institutes with similar technological scope. The strategic planning activities of Fraunhofer's headquarter were complemented by a process to identify and strategically develop research themes across institutes (Fraunhofer Future Topics). Fraunhofer performs this process iteratively every three years (Klingner and Behlau 2008).

In order to differ from the rather technology-driven processes of the past, a new approach was sought. This new strategy process was supposed to orient itself more towards demand-driven questions following the principles of corporate social responsibility and developing new ways for Fraunhofer research markets of the future. The idea fits well into the mission of Fraunhofer, which is to conduct innovation-oriented research for the benefit of private and public enterprises, as well as society in general. But the first question to ask was: What are the global challenges and questions of the future? Only if they are understood, some projects to meet demand and thus avoid urgency in meeting this demand can be launched.

The rationale started with the assumption that there are obviously science- and technology-driven approaches that make use of long-term thinking, but that the opposite, needs-driven approaches, are rare. Some of these needs can be defined by the Global/Grand Challenges. This does not mean that the whole research landscape should be focussed only on Global/Grand Challenges, but that a more active part is needed here. The Lund Declaration (European Commission 2010) made clear that 'European research must focus on the Grand Challenges of our time moving beyond current rigid thematic approaches. This calls for a new deal among European institutions and Member States, in which European and national instruments are well aligned and cooperation builds on transparency and trust.' Therefore, every research organisation and every actor in the innovation system has to position itself or himself in this new arena and consider what the Grand Challenges means for itself or himself. It is obvious that the Fraunhofer headquarter has to position the whole Fraunhofer Society in this context.

For Fraunhofer itself, this question was already posed earlier because inside the Fraunhofer Society with its 60 institute profiles, there is a broad portfolio with a huge variety of scientific disciplines, applications and knowledge in general available. Therefore, with intelligent cooperation, global challenges can be addressed, and for Fraunhofer, an add-on can be identified by directing these cooperations towards something that is supposed to deliver early results.

The idea was to define pragmatic areas in order to foster this cooperation among Fraunhofer Institutes and solve some of the problems technically before the demand becomes urgent. Global or Grand Challenges were therefore regarded as a 'means' to direct Fraunhofer's collaborative research into a direction with societal impact. The Global Challenges that could be addressed by Fraunhofer institutes in general were identified, and projects to actively promote solution-finding were called for. In order to support these projects, budgets were provided only for projects spanning the knowledge domains. This approach left the scientists enough freedom to find their own solutions. On the other hand, the – often technically minded – researchers are forced to think out of their normal box. The aim of the 2010 process was that, ultimately, each future topic would be promoted and developed by at least one dedicated (and centrally funded) R&D project of significant size. That means 'real' prototypes or results developed in a collaborative manner were expected from the projects performed across Fraunhofer knowledge domains represented by the institutes and groups of institutes. A needs-oriented approach and cross-institute problem-solving should open up new contract research markets in a 3–7 year perspective, i.e., an actual market perspective of 5–10 years.

The process had a first top-down part, in which global societal challenges were analyzed and adapted to Fraunhofer-specific challenges. The specific challenges served as a framework for the second, bottom-up part of the process. Within a competitive call, institutes teamed up to develop technological solutions to the challenges in the form of collaborative project proposals. The most convincing projects were internally funded.

A generalist team extracted the technological solution approaches for each sub-challenge. Each technological solution approach was rated in terms of its fit with the Fraunhofer R&D portfolio and clustered into solution fields. In a series of workshops, the generalist team, together with additional Fraunhofer experts, drafted a long list of Fraunhofer challenges by combining the sub-challenges of step 1 and solution field with sufficiently matching Fraunhofer objectives and mission. Each head of the above mentioned groups of institutes was asked to prioritise the entries in the long list. With this input, the generalist team formulated the final list of Fraunhofer challenges. To summarise the process, it can be stated that the guiding principles when designing and performing the process were to:

- Find the right challenges for Fraunhofer: look at global challenges and ask which of them are crucial for Fraunhofer or should be tackled with Fraunhofer projects and thus avoid urgency and preparing projects in a hurry;
- Involve the intellectual resources of many Fraunhofer scientists: include different people from Fraunhofer in the process so that in the end there is support for the future projects of Fraunhofer. Thus one is trained to work in teams over institution borders and know better who knows what to avoid urgency in other cases;
- Bring together different knowledge domains: the projects that should be funded are supposed not to be technology-driven, but bring together different disciplines and backgrounds in order to solve problems – some urgent and some more long-term problems;
- Perform dedicated technological R&D projects: the projects should be supported by technologies from Fraunhofer and need to produce results that really offer a solution for a part of the problem;
- Ensure broad acceptance through a transparent process: the process is performed in an open and transparent way so that every institute has a chance to apply and participate. This is necessary for acceptance at an early stage when the problems are not yet urgent.

The Fraunhofer challenges served as a framework for an internal competitive call. Institute consortia consisting of at least 4 institutes could propose collaborative projects to tackle the challenges. They had to explain their understanding of the challenge and which aspect of the challenge their project would provide a solution for. The anticipated impact of the project on the challenge had to be quantified, as well as the market potential for Fraunhofer that would be opened up through the project. A jury consisting of senior Fraunhofer experts and external experts evaluated the proposals. The most convincing proposals received substantial funding for 3 years. A second round of calls had also been performed (for details see Cuhls et al. 2012).

In this case, the future topics were derived from Grand Challenges with the assumption that there is already a specific need for problem solutions. But this would not make a solution 'urgent,' it just makes it necessary. Only the internal funding (relatively high) made the projects really interesting for the institutes to apply for the money. Real urgency only came up when it was clear that in the three years the projects were running, a problem has to be solved (by a real application, a prototype). This made the 'long-term challenge' an urgent one for the respective institutes.

4.4   *Future Radar 2030*

In 2002, the work of the Commission of Enquiry on 'Demographic Change' (Enquete Commission 2002) sparked the awareness that population developments will change all areas of society and the economy in Germany. There was first awareness that the changes might have an impact that may become urgent at a certain point in time – even though it was still early enough to change some of the impacts or prepare for them. This led to a Foresight process discussing future impacts of demographic change in a German Federal State, which was based on statistics on the one hand, and workshops, including a survey, on the other hand. The State Statistical Office of Rhineland-Palatinate published a regional model calculation about population change through 2050 for the first time in 2002. Regional population data were used to estimate the development of different regions ('Landkreise': rural districts) and towns ('kreisfreie Städte': urban districts) in Rhineland-Palatinate. Different model calculations were derived from varying the basic assumptions (State Statistical Office Rheinland-Pfalz 2004).

Whereas the 'new' federal states of former East Germany had already experienced drastic changes in their population's composition since re-unification in the form of a high (westward) migration and an increasingly ageing population, this development will only reach some of the 'old' federal states in former West Germany from about 2015. Based on the assumption that immigration and birth rates continue unchanged, some calculations were made in 2002 for Rhineland-Palatinate through 2050. These prognoses were based on an extrapolation of the existing demographic data. No variations or alternative scenarios were calculated, because the intention was to show what kind of developments people in the state could expect if nothing were changed. The following 'extrapolations' were a kind of 'shock' to the participants:
- the population will decline at a moderate rate in the medium term and strongly (up to 21%) in the longer term;
- the demographic ageing process will continue undiminished, the average age of 42 at the time of the study will increase to 51;
- the share of over 60-year-olds rises from 20% to 32%;
- at the same time, the share of under 20-year olds will drop from 20% to 15%;
- the number of those gainfully employed will shrink by 26%;

- the change varies greatly according to district: districts near towns and with good transport infrastructure will be less affected than more rural ones;
- the largest regional population decline will be in the town of Pirmasens (ca. 34%);
- the smallest in the rural district (*Landkreis*) of Mainz-Bingen (ca. 4%) (State Statistical Office Rheinland-Pfalz 2007).

These figures developed a certain pressure to act ('urgency'), to do something about the long-range development, which was at the time already well known – but not really of interest to anyone. A citation of one of the participating mayors (one with a city really decreasing in numbers, and looking really shocked) illustrates the normal attitude: 'In 2050, I am not mayor, anymore.' But then he noticed, that there is a kind of urgency, that the problems will not only start during his time as mayor but they also have to be addressed soon in order not to become really big problems (left houses as ruins in the city centre, unpayable infrastructures, e.g., water infrastructure, no internet for the few people, no kindergarden, etc.).

Demographic research expects that the developments described can no longer be changed, but only mitigated. The trends described above will have far-reaching consequences for the future development of the German federal state of Rhineland-Palatinate and prompted the board of the 'Future Initiative Rhineland-Palatinate' (Zukunftsinitiative Rheinland-Pfalz, ZIRP) to address this topic in the lead project *'Zukunftsradar 2030'* (*Future Radar 2030*). The ZIRP is a public-private partnership (PPP) supported by about 70 individuals, enterprises and organizations from industry, politics, science and the cultural scene. As demographic change is relevant to every area of public and private life, as well as the economy, the problem was regarded as very complex and needed

- multi-disciplinarity,
- a broad spectrum of expertise,
- a broad spectrum of topics,
- the inclusion of all relevant stakeholders and institutions in the region.

The general public was included in the ZIRP process in five different ways:

1. Experts from different institutions and different regions of the federal state were selected to be 'ambassadors' for the topic of demographic change. The experts were drawn from different age groups, had different backgrounds (ranging from mayors to students, from teachers to company staff) and were selected because they acted as multipliers.
2. An internet platform dealing with the topic made the results available to all interested persons in the Rhineland-Palatinate.
3. The press and media were included to inform decision-makers, especially in companies and institutions – and motivate them to discuss the topic.

4. Regional workshops for different target groups were intended to not only inform the participants, but also motivate them to actively debate the topic.

5. The population was informed via press releases and media reporting, but also by topic-oriented travelling exhibitions and local citizens' events to achieve a broad sensitization for the topic of demographic change.

The interdisciplinary collaboration of experts with different experiences can be regarded as one of the greatest advantages of Foresight processes. The conjunction of a broad spectrum of experiences and mentalities on one topic makes it necessary for the organizers to first reach a common basis of fundamental information and an understanding of fundamental terminology. It is difficult to persuade these very different individuals to work on future issues – and work together without hierarchical or disciplinary thinking.

It was decided to address four thematic complexes:

1. Demographic change as a challenge for local authorities;
2. Demographic change as a challenge for the world of work;
3. Generations cooperating together in demographic change and;
4. New market opportunities resulting from demographic change.

The 'expert talks' held during joint workshops were the first methodological challenges in the project. The participants were very heterogeneous, from different backgrounds and different places all over the federal state of Rhineland-Palatinate. There were students, teachers, mayors of cities or villages, industry representatives, entrepreneurs, architects, managers, physicists, scientists and policy-makers, etc. Each workshop group had a different setting, most groups had about 15–20 participants, each workshop was performed with 3 to 5 groups at different locations in the state.

To gain these individuals' support, mental time trips (as described in Cuhls 2016) were used as mind-openers at the beginning of the first workshops (Future Team sessions). The aim of these workshops was an introduction to the topic, 'shocking' the participants with the relevant data and extrapolative prognosis, the search for focus topics and the development of 'pictures of future developments' reflecting the subjective imaginations of the participants. The latter was the stage, at which mental time travelling was applied. None of the external facilitators had had any experience with mental time travel before the workshops started. Therefore, the experiment started with different ways of performing mental time travel into the year 2030. A description of the whole project can be found (Cuhls et al. 2012).

To sum up, Rhineland-Palatinate is one of the 'old' federal states of Germany, expected to be influenced early on by demographic change and a shrinking society. Looking at the pure extrapolative numbers created a certain 'urgency' on the policy side to act instead of wait and react. The state was much better prepared than others due to its Future Radar 2030 project. One reason is that the public were prepared and informed at a relatively early stage in the development, and participants in Future Radar 2030 acted as 'multipliers,' spreading information and even launching their own activities in the different regions and on a smaller scale. The preparation of measures does not go so far to be called 'deliberate democracy' (as in Fishkin 2018 or Stirling et al. 2018) because citizens are participating in the preparation of gaining information, but the decisions are still made elsewhere. And even if – as is so often the case – measures were postponed because there 'was still time to do something about it,' the mind-open multipliers continued to spread information and were enthusiastic because they had 'seen' or 'felt' some of the developments (like empty houses in villages, long journeys to the nearest school). This federal state started the relevant discussions and measures much earlier than the federal government, which addressed the topic 10 years later in the 'year of demographic change' (in 2013).

5   Do We Need a Different Perception of Time, Timing and Urgency?: A Summary

There is always something more urgent than the long-term. But to plan for today, we need the long-term view, different perspectives on it and to plan for the time in-between. That is the ideal and roadmaps, often regarded as a Foresight method, can be based on it. In reality, we have rather static reactions concerning time and timing. In urgency cases, we often forget the long range implications.

We see permanent problems or challenges, we see larger challenges that accumulate, get worse slowly, but suddenly turn into severe urgent problems. Sometimes, this is called 'Frogboiler' as frogs are boiled in cold water that gets warmer – and suddenly the frogs are dead. Challenges like this might be climate change or demographic changes (see example Future Radar 2030 above). We quite well anticipate these changes and that they may bring problems; others, e.g., from societal development, are less well-known and occur unnoticed. These problems always force a societal (and political) reaction. Nevertheless, at a certain point in time, urgency to deal with the problems comes up. What we ignore is that there is a time-lag between (potential) perception and occurrence of first events, a time we can use for preparation (pre-action/ pro-active measures).

Another difficulty is dealing with sudden events. The normal reaction is: There are permanent problems with greater or lesser urgency and quick extreme and not fully thought through reactions; there are grand challenges and there is classic policy thinking. Suddenly, when a wild card occurs, an un-expected event – that was completely ignored or regarded as impossible or maybe regarded as possible but the exact timing was completely unknown. Then, hectic reactions occur – *urgency* to do something, to act is felt. This can be avoided with Foresight by working with wild cards, with expected and un-expected but imaginable occurrences. Having at least some scenarios in mind (for creating Plan A, Plan B, Plan C ...) in case something similar might happen, even if we cannot predict and do not know what and when it might occur.

The third difficulty we face in Foresight is concerned with science, techno-logy and innovation. Technical solutions, scientific experiments or prototype developments as well as certain policies have unclear impacts and reactions towards them. Even in these cases, politics react, sometimes in panic – and feel urgency. Sometimes, the quick reactions even make things worse. With better impact anticipation, which can be a part of Foresight, at least some of the cases can be identified, see e.g., BMBF Foresight I above where proactive action was taken in a new division for human-technology interaction. If it is possible to re-guide and change the development, give it up or even stop it (by law, etc.) is the question of implementation and decision. Sometimes it is also the question of participation in the processes: The citizens are involved, the more information is spread, the more informed decision-making can be. There are more possibilities than we normally imagine. Thus, the poten-tial needs to be used much more than in classic technology assessment. The same argumentation can be found in 'deliberate democracy' attempts, which was discussed during a workshop of the European Commission expert group 'Strategic Foresight' (SFRI) in June 2016. Similar arguments can be found in Fishkin 2018; Smith and Stirling 2018 and Stirling et al. 2018.

Again, there is the problem of attention: Many developments are well-known, but in times of information overflow, they do not get any attention. We human beings ignore rather the 'negative developments' because we do not like them, and from neurosciences we know that it is easier for our brain to ignore negative emotions (Burnett 2016). *What is not urgent, is not seen or ignored.* Normally, people wait until something is urgent (or on the agenda, deadline, something serious is happening) – only then, they (re-)act. The other attitude is to wait without acting – because 'in the long run, we are all dead.' That means, we adults are not touched upon, anymore, only our children and the coming generations. Often, we leave it to them to solve the problems.

Thus Foresight – although seen as necessary – is difficult for human be-ings. It is not against human 'nature' but challenges human usual behavior. To act and be really implemented, Foresight also seems to *need a certain sense of urgency*. Therefore, sometimes, we Foresighters shock our participants in the workshops with questions about outcomes: What happens, if there is no more oil? What happens if your town only has 50% inhabitants, anymore? What comes after a real BREXIT?... Ideally, we would think ahead, use our knowledge for the present decision and create a solution or something new or just different, that we really implement. As systems are complex, this means navigating in the unknown and dealing a lot with complexity (Nowotny 2016).

This leads to conclusions that link Foresight much more to planning, develop new planning and timing procedures, which are more open and more flexible. Planning procedures are different in different cultures – especially in Germany they are rather fix. Therefore, I am a bit critical about games and scenarios in e.g., wargaming as Foresight experiences or for planning procedures. Experience with German Armed Forces, where they have exact planning, exists and sometimes the soldiers are lost if 'the plan' does not work. How are you going to behave in war where the enemy does not act as foreseen in your plan and as you are prepared for? What is Plan B, Plan C, Plan D? Therefore, we need Foresight for the long run and more tests with methods that are flexibly preparing us for futures thinking and for thinking in alternatives. Some call it 'thinking for storage,' to store the ideas for a long time until they are needed.

Equally important is timing, to be there at the right place and point in time with the ideas and with this kind of thinking in place. It is often said that we have problems to think about something as long as it is uncoded, unnoticed, and we have no word for it. In German language, we even do not have a word for 'timing,' but when it is needed, we use English (or better: Denglish). Even without a correct wording, we need to try it. It takes time to convince people to accept such an inexact science and it needs a long time and patience to see the fruits of the work done now.

Strategic planning thus needs to start from the vision, the goal, and work back and forth, taking the time factor and the human time factor into account. This is especially important for innovation research, where you cannot plan for an 'idea': you can create the surrounding for being creative, but the more you enforce it, the more you are under time pressure and the less creative you are (ideas get more and more limited). Urgency is really counter-productive, here.

Currently, we have a debate about 'resilience,' defined as (e.g., an individual's) the ability to successfully adapt to life tasks in the face of social disadvantage or highly adverse conditions. Resilience is one's (or the environment's) ability to bounce back from a negative experience with 'competent functioning.' But if a change of a system is needed, then resilience means falling back into the previous stage and that is not enough. We need resilience in our lives to cope with time, not to be overloaded with work or to organize ourselves as human beings. But if real innovations (also social innovations) are to be integrated or added, or to change something, resilience hampers the disruptiveness that is necessary in the sense Schumpeter (creative destruction, 1912/1934) described it. Resilience is therefore very good for adaptation and makes the system sustainable for a longer period of time. But if real change is necessary, a kind of disruption is needed. In former times, this was often combined with war, but war is no solution. Examples, where these changes are needed are: retirement systems in the age of demographic aging, health systems, political systems or organizations of companies.

The clear answer to my question if Foresight can help to avoid situations of urgency is therefore: yes, we need Foresight to counter urgency and we need a partly circular (or better: spiral) thinking of *time* and a more parallel thinking of *timing*. In a world of multitasking, multi-events, etc., things are not only going on in parallel, they have to be combined at certain points in time (e.g., innovation have to be on the market at the 'right' time, if their ideas are too early, they cannot be taken over). We still need to experiment with the approaches and method combinations – we have examples like the ones mentioned above, but there is no 'evidence' in the classic scientific sense. One of the first things (or "thinks"?) in Foresight to do is thus to *create some urgency* in challenging cases. This is necessary when there is no obvious activity asked for because the problem still lies far in the future. Yet, we need to keep the long-term view to understand the impacts of current activities – otherwise our activities can be counter-productive and induce more problems instead of finding answers to challenges.

References

Aengenheyster, Stefan, Kerstin Cuhls, Lars Gerhold, Maria Heiskanen-Schüttler, Jana Huck, and Monika Muszynskae. 2016. "Real-Time Delphi in practice – A comparative analysis of existing software-based tools." *Technological Forecasting and Social Change.* Available online at http://dx.doi.org/10.1016/j.techfore.2017.01.023 DOI: 10.1016/j.techfore.2017.01.023.

Bell, Wendell. *Foundations of Futures Studies: History, Purposes, and Knowledge. Human Science for a New Era.* 2 vols. Volume I. New Brunswick and London: Transaction Publishers, 2009.

Burnett, Dean. 2016. *The Idiot Brain: A Neuroscientist Explains What Your Head is Really Up To.* London: Guardian Books / Faber and Faber.

Coates, Joseph, Philippe Durance, and Michel Godet. 2010. 'Strategic Foresight Issue: Introduction.' *Technological Forecasting and Social Change* 77, no. 9: 1423–25. Accessed June 21, 2016. doi:10.1016/j.techfore.2010.08.001.

Coates, Joseph F. 1985. 'Foresight in Federal Government Policymaking.' *Futures Research Quarterly*, no. 1: 29–53.

Cuhls, Kerstin. 2016a. Mental Time Travel in Foresight Processes – Cases and Applications. *Futures*, Special Issue "Experiencing Futures"; DOI: http://dx.doi .org/10.1016/j.futures.2016.05.008. http://www.sciencedirect.com/science/article/pii/ S001632871630129X.

Cuhls, Kerstin. 2016b. Shaping the Future: Science and Technology Foresight Activities in Japan, with Special Consideration of the 10th Foresight, *Asien*, Nr. 140: 103–30.

Cuhls, Kerstin. 1998. *Technikvorausschau in Japan: Ein Rückblick auf 30 Jahre Delphi-Expertenbefragungen.* Technik, Wirtschaft und Politik. Schriftenreihe des Fraunhofer ISI 29. Heidelberg: Physica-Verlag.

Cuhls, Kerstin. 2001. 'Foresight with Delphi Surveys in Japan.' *Technology Analysis & Strategic Management* 13, no. 4: 555–69.

Cuhls, Kerstin. 2003. 'From Forecasting to Foresight Processes – New Participative Foresight Activities in Germany.' In 'Special Issue on Technology Foresight.' Kerstin Cuhls and Ahti Salo. Special issue, *Journal of Forecasting* 22, 2–3: 93–111.

Cuhls, Kerstin. 2005. 'The Delphi Method.' In *Delphi Surveys: Teaching material for UNIDO Foresight seminars*, edited by Kerstin Cuhls, 93–112. Wien: UNIDO. http:// www.unido.org/fileadmin/import/16959_DelphiMethod.pdf.

Cuhls, Kerstin. 2008. 'Foresight in Germany.' In *The Handbook of Technology Foresight: Concepts and Practice*, edited by Luke Georghiou et al., 131–53. PRIME Series on Research and Innovation Policy. Cheltenham, UK; Northampton, MA, USA: Edward Elgar Publishing Ltd.

Cuhls, Kerstin. 2008. 'Japanese S+T Foresight 2035: The Eighth National Japanese Science and Technology Foresight.' In *The European Foresight monitoring network: Collection of EFMN briefs*, edited by European Commission, 139–42. Community research Case studies 23095 EN. Luxembourg: Office for Official Publ. of the Europ. Communities.

Cuhls, Kerstin. 2008. *Methoden der Technikvorausschau – eine internationale Übersicht.* Stuttgart: Fraunhofer IRB Verlag, 2008.

Cuhls, Kerstin, Walter Ganz, and Philine Warnke, ed. 2009. *Foresight Process -New Future Fields.* Karlsruhe/Stuttgart: IRB Publishers. http://www.isi.fraunhofer.de/ bmbf-Foresight.php.

Cuhls, Kerstin, Walter Ganz, and Philine Warnke, ed. 2009. *Foresight-Prozess im Auftrag des BMBF. Etablierte Zukunftsfelder und ihre Zukunftsthemen.* Karlsruhe/Stuttgart: IRB Publishers. http://www.isi.fraunhofer.de/bmbf-Foresight.php.

Cuhls, Kerstin, Alexander Bunkowski, and Lothar Behlau. 2012. 'Fraunhofer future markets: From global challenges to dedicated, technological, collaborative research projects.' *Science and Public Policy*, no. 39: 232–44. doi:10.1093/scipol/scs018.

Cuhls, Kerstin, Amina Beyer-Kutzner, Walter Ganz, and Philine Warnke. 2009a. 'The Methodology Combination of a National Foresight Process in Germany.' *Technological Forecasting and Social Change* 76, no. 9: 1187–97. doi:10.1016/j.techfore.2009.07.010.

Cuhls, Kerstin, Walter Ganz, and Philine Warnke. 2009b. *Foresight Process on behalf of the German Federal Ministry for Education and Research: New Future Fields.* Karlsruhe, Stuttgart: IRB Verlag. www.bmbf-Foresight.de.

Cuhls, Kerstin, Walter Ganz, and Philine Warnke. 2009c. *Foresight-Prozess im Auftrag des BMBF: Etablierte Zukunftsfelder und ihre Zukunftsthemen.* Karlsruhe, Stuttgart: Fraunhofer IRB Verlag, 2009.

Cuhls, Kerstin, Walter Ganz, and Philine Warnke. 2009d. *Foresight-Prozess im Auftrag des BMBF: Zukunftsfelder neuen Zuschnitts.* Karlsruhe, Stuttgart: Fraunhofer IRB Verlag, 2009.

Cuhls, Kerstin, Heinz Kolz, and Christoph M. Hadnagy. 2012. 'A regional Foresight process to cope with demographic change: Future radar 2030 (Zukunftsradar 2030).' In 'Special Issue on Foresight and New Trajectories.' Edited by Kerstin Cuhls. Special issue, *International Journal of Foresight and Innovation Policy* 8, no. 4: 311–34.

Cuhls, Kerstin, Annelieke van der Giessen, and Hannes Toivanen. 2015. *Models of Horizon Scanning. How to integrate Horizon Scanning into European Research and Innovation Policies.* Report to the European Commission, Brussels. https://www.isi.fraunhofer.de/content/dam/isi/dokumente/ccv/2015/Models-of-Horizon-Scanning.pdf.

Curry, Andrew, and Anthony Hodgson. 2008. 'Seeing in Multiple Horizons: Connecting Futures to Strategy.' *Journal of Futures Studies* 13, no. 1: 1–20.

Enquete Commission 'Demographic Change' of the German Parliament. 2002. *Publication no. 14/8800.* Berlin.

European Commission. 2010. 'Lund Declaration.' *European Planning Studies.*

European Commission. Communication from the Commission to the European Parliament, the Council, the European Economic and Social Committee and the Committee to the Regions. Horizon 2020: The Framework Programme for Research and Innovation: COM (2011) 808 final. Unpublished manuscript.

Fink, A., O. Schlake and A. Siebe. *Erfolg durch Szenario-Management. Prinzip und Werkzeuge der strategischen Vorausschau [Success through Scenario Management. Principles and Instruments of Strategic Foresight].* Frankfurt/ New York: Campus, 2001.

Fishkin, James. 2018. *Democracy When the People are Thinking: Revitalizing Our Politics through Public Deliberation.* Oxford, UK: Oxford University Press.

Gerhold, Lars, Dirk Holtmannspötter, Christian Neuhaus, Elmar Schüll, Beate Schulz-Montag, Karlheinz Steinmüller, and Axel Zweck, ed. 2015. *Standards und Gütekriterien der Zukunftsforschung: Ein Handbuch für Wissenschaft und Praxis.* Zukunft und Forschung 4. Wiesbaden: Springer Verlag.

Godet, Michel. 2000. 'The Art of Scenarios and Strategic Planning:Tools and Pitfalls.' Technological Forecasting and Social Change, no. 65: 3–22.

Godet, Michel, and Philippe Durance. 2011. *Strategic Foresight for Corporate and Regional Development.* Paris: UNESCO. http://www.laprospective.fr/dyn/traductions/2dunod-unesco-strategic-Foresight-ext-veng.pdf.

Grunwald, Armin. 2012. 'Technikzukünfte: Vorausdenken – Erstellen – Bewerten.' In *acatech IMPULS*, edited by acatech. Heidelberg: Springer; Vieweg.

Grunwald, Armin. 2012. *Technikzukünfte als Medium von Zukunftsdebatten und Technikgestaltung.* Karlsruhe: KIT Scientific Publishing.

Helmer, Olaf. 1967. *Analysis of the Future: The Delphi method.* Santa Monica: Rand Corporation.

Hobcraft, Paul. 'The Innovation Bunker Series on Cognitive Traps'[blog post]. Accessed June 23, 2016. https://paul4innovating.files.wordpress.com/2015/06/the-innovation-bunker-series-on-cognitive-traps.pdf.

Irvine, John, and Ben R. Martin. 1984. *Foresight in science: picking the winners.* London/Dover: Pinter Publishers.

Klingner, R., and L. Behlau. 2008. 'Fraunhofer Future Topics' Paper presented at Third International Seville Conference on FTA, as part of the strategic planning of a distributed contract research organisation, Seville, Spain, 16–17, 2008.

Kuusi, Osmo, Kerstin Cuhls, and Karlheinz Steinmüller. 2015. 'Quality criteria for scientific futures research.' *Futura*, no. 1: 60–77.

Kuusi, Osmo, Kerstin Cuhls, and Karlheinz Steinmüller. 2015. 'The futures Map and its quality criteria.' *European Journal of Futures Research* 3, no. 1. Accessed August 3, 2016. doi:10.1007/s40309-015-0074-9.

Marien, Michael. 2010. 'Futures-thinking and identity: Why "Futures Studies" is not a field, discipline, or discourse: a response to Ziauddin Sardar's 'the namesake."' *Futures* 42, no. 3: 190–94. Accessed August 3, 2016. doi:10.1016/j.futures.2009.11.003.

Martin, Ben R. 2010. 'The origins of the concept of 'Foresight' in science and technology: An insider's perspective.' *Technological Forecasting and Social Change* 77, no. 9: 1438–47.

Montgomery, C. A., and M. E. Porter, ed. 1996. *Strategy: Seeking and Securing Competitive Advantage.* Wien: Ueberreuter (Manager Magazin Edition).

Nowotny, Helga. 2016. *The Cunning of Uncertainty.* Cambridge: Polity Press.

Ringland, Gill. 1998. *Scenario planning: Managing for the future.* Chichester: John Wiley & Sons.

Ringland, Gill. 2002. *Scenarios in public policy.* Chichester: John Wiley & Sons. http://www.gbv.de/dms/sub-hamburg/339906855.pdf.

Sardar, Ziauddin. 2010. 'The Namesake: Futures; futures studies; futurology; futuristic; Foresight – What's in a name?' *Futures* 42, no. 3: 177–84. Accessed August 3, 2016. doi:10.1016/j.futures.2009.11.001.   http://ac.els-cdn.com/S001632870900175X/1-s2.0-S001632870900175X-main.pdf?_tid=83b49584-5953-11e6-b624-00000aabof6b&acdn at=1470212770_8a52997d1e610e023dcc6949add6258.

Rittel, Horst W. J., and Melvin M. Webber. 1973. 'Dilemmas in a General Theory of Planning.' *Policy Sciences.* 4: 155–69. doi:10.1007/bf01405730.

Schumpeter, Joseph A. 1912/1934. *The Theory of Economic Development: An Inquiry into Profits, Capital, Credit, Interest, and the Business Cycle.* Transaction Publishers.

Schwartz, Peter. 1996. *The Art of the Long View: Planning for the Future in an Uncertain World.* New York: Bantam Doubleday Dell Publishing.

Scott, George. 2001. 'Strategic Planning for High-Tech Product Development.' *Technology Analysis & Strategic Management* 13, no. 3: 343–64. doi:10.1080/09537320120088174.

State Statistical Office Rheinland-Pfalz. 2004. *Rheinland-Pfalz 2050: Vol 2: Impacts of theDemographic Development.* Bad Ems.

State Statistical Office Rheinland-Pfalz. 2007. *Rheinland-Pfalz 2050 – second regionalized population estimation/zweite regionalisierte Bevölkerungsvorausberechnung (Basisjahr 2006).* Bad Ems.

Steinmüller, Karlheinz. 'Drei Beiträge zu Grundfragen der Zukunftsforschung.' Unpublished manuscript.

Smith, A., and A. Stirling. 2018. 'Innovation, Sustainability and Democracy: An Analysis of Grassroots contributions.' *Journal of Self-Governance and Management Economics* 6(1), 2018, pp. 64–97. doi:10.22381/JSME612018.

Stirling, Andy, Cian O'Donovan, and Becky Ayre. 2018. 'Which Way? Who Says? Why? Questions on the Multiple Directions of Social Progress' on Technology's Stories [blog post dated May 24, 2018]. http://www.technologystories.org/which-way-who-says-why/. Accessed August 23, 2018.

Wright, G., and P. Goodwin. 1998. *Forecasting with Judgment.* Chichester: John Wiley & Sons.

Zweck, Axel, Dirk Holtmannspötter, Matthias Braun, Kerstin Cuhls, Michael Hirt, and Simone Kimpeler. *Forschungs- und Technologieperspektiven 2030: Ergebnisband 2 zur Suchphase von BMBF-Foresight Zyklus II.* 3 vols. Bd. 2. Düsseldorf: VDI Technologiezentrum GmbH, 2015.

Zweck, Axel, Dirk Holtmannspötter, Matthias Braun, Lorenz Erdmann, Michael Hirt, and Simone Kimpeler. *Geschichten aus der Zukunft 2030: Ergebnisband 3 zur Suchphase von BMBF-Foresight Zyklus II.* 3 vols. Bd. 3. Düsseldorf: VDI Technologiezentrum GmbH, 2015.

Zweck, Axel, Dirk Holtmannspötter, Matthias Braun, Michael Hirt, Simone Kimpeler, and Philine Warnke. *Gesellschaftliche Veränderungen 2030: Ergebnisband 1 zur Suchphase von BMBF-Foresight Zyklus II.* 3 vols. Bd. 1. Düsseldorf: VDI Technologiezentrum GmbH, 2015.

CHAPTER 16

"More Than Watchmen": Dante on Urgency in Ritual

Dennis Costa

There is no 'urgency' without time, though it is also the case that some kinds of urgency distend or interrupt time. In Book xi of his *Confessions*, Augustine famously chose to de-emphasize any sense of time as substantial and cosmic in favor of a psychosomatic temporal model according to which the human person may be accurately characterized by its urgency, here and now, to open itself up to something other than itself, by its repeated resistance to mere chronology, by its interruption of time's conventional measurements. Augustine understood a person's being-in-the-world in terms of his or her *distentio animi* – the intentional act (which may become an habitual act) of 'stretching out' one's 'mind' or human powers. As Robert Jordan has very carefully described Augustinian temporality, "Time is a relation, with a foundation in successive states of finite or limited being, whose measurement is a cognitive act, terminating in the 'distentio' of the mind."[1] I have long thought it important to re-focus Augustine's point of view in Book xi somewhat, by emphasizing that individual human psyche finds itself immersed in the time that is always relative to the being or substance of finite matter, not only by its embodied gesture towards another human being – the chronotype, as it were,[2] of the ethical field – but also by its no less urgent gesture towards the rest of space-time – the chronotype of cosmology and the natural sciences, of art and, just as importantly, the chronotype of contemplation and of cult.[3] Human beings do not just wait

---

1   See his "Time and Contingency in Augustine," in *Augustine: A Collection of Critical Essays*, ed. R. A. Markus (New York: Doubleday & Company, 1972), p. 272.

2   The term was coined for medical research and normally refers to the phenomenon and effects of a person's (or animal's) own circadian rhythms. It seems to me an elegant term for referring to a particular person's aptitude for making the gesture Augustine calls *distentio animi:* a scientist's scientist, we might say, or a truly virtuous person, one whose aptitude for enacting the good is honed by repetition, as Aristotle says in the *Nicomachean Ethics*. The expression also has a natural affinity with biblical typology, since the 'types' of Christ in the Hebrew Bible (Solomon, David, Isaiah's 'suffering servant of Yahweh') are persons who were, among others of their attributes, most apt to be urgent for others.

3   I would add that the normal translation of *distentio animi* as a particular 'stretching out of the *mind*' should also be amended to be inclusive of Augustine's clear sense of the human

© KONINKLIJKE BRILL NV, LEIDEN, 2019 | DOI:10.1163/9789004408241_018

upon a succession of things expected. They actively hope for (or hope to observe or to represent, in the sciences and in art) things unexpected. In cultic acts, they may also experience a kind of discernment that is, as Augustine dared to say, a synchronous being 'present' with a future in the precise terms of a past, the enactment of which discernment controverts the supposedly sequential 'flow' of time.[4] Cultic acts, if conceived as rhetorical structures, very often enact the figure of *hyperbaton*, according to which the 'normal' grammatical flow of subject-predicate-object gets purposefully distorted. The discerning activity of hoping may be considered a different order or *taxis* for the measurement of time; its exactitude is simply different from that of the atomic clock.[5] For Augustine and later for Dante, urgency that is hopeful, as well as being idiosyncratic, always verges upon being communal in nature.[6]

---

as the admixture of physical, mental and spiritual powers. David van Dusen, who describes Augustinian temporality as "sensuous-affective," would omit the word 'spiritual.' Van Dusen corrects the mistake – which is also the mistake of Cartesian dualism – of seeing Augustine's sense of time as being only 'mental.' See David van Dusen, *The Space of Time: A Sensualist Interpretation of Time in Augustine, Confessions X to XII*. Leiden: Brill, 2014.

4 This is the case in Augustine's report of his conversation with his mother, Monica, on the seashore at Ostia in *Confessions* x. van Dusen, who studies Augustinian *distentio* writ large – as Time itself in all its dimensions – states that "... it is only as *distentio* that a presence of past things endures. And it is only as *distentio* that a presence of future things is traversed, transposed, ..." Given van Dusen's desire to find other, material contexts for Augustine's 'mystical' moments, he also adds to the above sentence "... diminished and exhausted in a *provenience* of sense-imaginal or sense-affective *impresence*." I would disagree. See van Dusen, *op.cit.*, p. 311. See also Augustine's *Sermons* 169.15, 18, a special case of what I am calling temporal discernment that concerns conversion: "If you stretch yourself out towards what is in the future, if you consider future realities, forget past things, do not turn back to look again at them so as not to stop yourself there where you have glanced." This new temporal perspective is a central part of Dante's definition of the process of purgation in the afterlife; it characterizes the cult or "piety" – what Dante terms "la religione de la montagna" (*Purg.* XXI. 41–42) – on his Mt. Purgatory. Augustine's Easter sermons specifically conjoin ritual urgency with hope on behalf of the about-to-be-baptized. See Philip T. Weller, ed., *Selected Easter Sermons of St. Augustine*. London: B. Herder, 1959.

5 A curious sign posted (at least in the year 2016) on the door of the parish church in the town of Pisacq, Perú, in the Sacred Valley of the Incas, reads: "La hora cristiana es la hora exacta." This startling announcement would seem to imply much more than simply the pious believer's habit of showing up in church 'on time.'

6 For the embrace of the other as part and parcel of Augustinian *distentio*, see the entire study by Luigi Alici, *L'Altro nell'io: dialogo con Agostino*. Roma (Città Nuova), 1999. Hope, in these senses, is a God-given energy; it is not the same thing as the entirely human and temporal experience of *exspectatio* which can certainly – to refer again to van Dusan's analysis (see fn 4, above) – be "diminished" and "exhausted," and is always subject to the vagaries of affect and imagination.

It seems important for all these reasons to pay attention to urgency as a factor in ritual and also to think about the related question of the ritualization of urgency.

In Book x of the Confucian *Analects*, when Master K'ung (or "the gentleman") crossed the palace's sacred threshold he "bent over, as though the gate were not big enough to admit him." When he passed by the sacred space of proclamation, "his face took on a look of concentration and his pace was solemn."[7] "When carrying the [sacred] tablet of jade, he seems to double up, as though borne down by its weight. He holds it at the highest as though he were making a bow, at the lowest, as though he were proffering a gift. His expression ... changes to one of dread and his feet seem to recoil ..."[8] If Confucian*ism* means knowing what actions are appropriate in any situation, it is precisely the urgent proximity to cult that makes Master K'ung take appropriate action in the profoundest senses and in his own person.

The Mayan shaman in Chiapas, Mexico makes great haste, immediately following the occurrence of the vernal equinox, to chant prayers "at the threshold, at the altars," and right next to the newly emerging cornstalks. Once arrived 'there,' however, he recites prayers of impressive length that depend for their effect on the force of anaphora or repetition: "The holy hour has struck, has struck ... Your lowly children arrived with anxious heads/ arrived with anxious hearts ... Will the holy words be now in unison? / will the holy lips move now in unison?... So may it take place / Your great festival ... For this I come kneeling / For this I come bending low ... For this come our humble candles / For this comes our chunk of incense / For this comes our cloud of smoke."[9] I dutifully report the translator's rendering, "anxious," but would include a caution with respect to it and a preference for 'uncertain' rather than 'anxious.' It is characteristic of what one scholar has termed the "apostrophic voice" in liturgy[10] that urgent petition always recognize the reality of uncertainty, of the dreadful (including the dread of performing the rite incorrectly) and even of danger, but

7   Burton Watson, tr., *Analects of Confucius*, (New York: Columbia UP, 2010), p. 66.

8   Arthur Waley, tr., *The Analects of Confucius*, (New York: Vintage Books, 1989), p. 147. A particularly excellent, and more recent study of the larger relationships between Chinese ritual, both Confucian and Daoist, and nature is the chapter by James Miller in R. L. Nadeau, ed., *The Wiley-Blackwell Companion to Chinese Religions*, (Chichester: Wiley-Blackwell, 2012), chapter 16.

9   The prayer is translated from the Tzotzil Maya by Robert M. Loughlin. See D. E. Breedlove and R. M. Loughlin, *The Flowering of Man: a Tzotzil Botany of Zinacantán* (abridged edition), (Washington and London: Smithsonian Institution Press), pp. v and 49ff.

10   See Catherine Pinkstock's brilliant study of the rhetoric of the Tridentine Mass, *After Writing: on the Liturgical Consummation of Philosophy*, (Oxford: Blackwell Publishers, 1998), part II, chapter 4, "The Apostrophic Voice."

that it also avoid any nervous anxiety in a clinical sense. The shaman's questions are real, not merely rhetorical; they require that his subsequent, urgent statement be an affirmation, without anxiety, but one that is also qualified by the subjunctive mood. "So may it take place / Your great festival!"

In *Psalms* 130:5–6 – and in the title of this essay – an individual human soul (as metonymic, surely, for all the faithful of Israel) both hopes for and waits for God "more than watchmen wait for the morning, wait for the morning" – which is to say that the soul hopes and waits with a kind of moral urgency. But according to one ancient explication of this psalm, that urgency *ipso facto* extends itself to a community: communally designated dawn-watchers, perhaps even Levites, wait upon the appearance of the sun during the course of what we can call diachronic time, in order to give others the designated signal to perform communal cult, "to offer up the morning sacrifice," to initiate something like what we have to call synchronic time.[11] Urgency is itself ritualized in every element of the Passover Seder, including in the special bread that is unleavened because the community is typologically already on the road, in great haste, towards its experience of divine rescue; and yet it still must 'take the time' to sit down, at some length, at table. During the Orthodox Christian liturgy of St. John Chrysostom, there are usually at least four litanies or iterations of the voice of petition – the "apostrophic voice." But one of these is curiously named "the insistent litany" because the community's response to each petition, *Kyrie eleison / Lord, have mercy,* is sung not once, as in the other litanies, but in rapid-fire repetition, *Kyrie eleison Kyrie eleison Kyrie eleison Kyrie eleison ...* as it were, a hundred times, or urgently. In both Judaism and Christianity, messianic expectation has it that the divine providence may never be reduced to being only the object of individual or even tribal urgency – no matter how powerfully such urgency may get ritualized. It is, rather, the entire created order that is believed to be urgent within its own terms, that is to say, within the terms of finite space-time. In Paul's epistle to the Romans: "For creation's urgency [*apokaradoxia / expectatio*] is waiting for [*apekdechetai / expectat*] the unveiling of the children of God." That is a sentence in which a noun and a verb, expressing in Greek an 'earnest' or 'urgent' kind of waiting, have as their common referent what came to be used as both a synchronous and a transtemporal expression: '*apokalypsis / revelation.*' Paul's 'apocalypse' points to a present 'unveiling' of the entire created order – in the terms of its past (even its cosmological or geological histories) – that is final. The epistle

---

11   This is *Psalms* 129:5–6 in the Vulgate Bible. The Targumic interpretation is reproduced in A. F. Kirkpatrick, ed., *The Book of Psalms* (Cambridge: Cambridge UP, 1906), vol 18, part 3, p. 760, fn 6.

goes on to figure cosmic urgency as a process of parturition, as birth-pangs: "For we know that the entirety of creation has been groaning and in birth-pains up until now ... and we too are groaning within, urgent for the deliverance [*ten apolutrosin*] of our body" (*Rom.* 8:19, 22–23).[12] The biblical *Apocalypse*, similarly, posits an ongoing cosmic or heavenly liturgy, according to the terms of which any earthbound liturgy must be modeled. "And they had no rest day and night, saying 'Holy, Holy, Holy'" (*Apoc.* 4:8). Richard of St.-Victor's reading of this passage highlights the urgent relationship in Christianity between 'now' and 'then' and also nuances the figure of sacred journey – the figure also at the center of Dante's fiction – from the one to the other, as well as what Richard posits as the real communion/communication between the one and the other. "For when it says, 'They have no rest,' it hardly means physical labor. Never ceasing from giving fervent praise to the Creator is without doubt the highest and perfect repose ... On the way home, they have a foretaste ... of the joyfulness of heavenly harmony, with which they will be totally saturated once they arrive home."[13]

Traditional Christian liturgies have always aimed at making these kinds of temporal urgency palpable. The German poet Hölderlin wanted to invoke something similar with the opening line: "As when on festival-day ..." / "Wie wenn am Feiertage ..." For Hans-Georg Gadamer, "It is of the nature of the festival that it should proffer time, arresting it and allowing it to tarry."[14] The 'proffer' that is ritual is offered openly to others, and with an urgency that is simultaneously tarrying. According to Ivica Žižič, "... the feast is given as an

---

12   I choose the English 'deliverance' punningly, and to denote that Paul composes a mixed metaphor. The Greek noun in question has nothing to do with childbirth; it means deliverance from a debt or, literally, redemption. The economic figure for Christ's saving acts is, of course, dominant in Paul. The noun for 'urgency' or 'eager expectation' in verse 19, above, is 'apokaradoxia,' literally 'sticking one's head and neck out.' It is not inconceivable that Dante knew the word when he characterizes those "few" who are fit to read the *Paradiso* as those who "over time have stretched out their necks" towards the new and unexpected, whether in Scripture or in the cult of the Eucharist. See *Par.*11.10–11. Another neotestamentarian expression for 'urgency' is 'prothumia' (*Acts* 17:11, *Vulgate* Latin 'aviditas'), the noun that is used to characterize an eager, messianic expectancy among the members of the synagogue in Berea that was visited by Paul and Silas.

13   Richard of St.-Victor, *In Apocalypsim libri septem*, in J.-P Migne, ed., *Patrologia Latina*, vol. 196, col.752: "Quod autem dicit, non habebant requiem, minime designat laborem. Hoc est nimirum summe et perfecte quiescere [NB: Augustine's verb in *Confessions* i.1] a Creatoris pia laude numquam cessare ... Et jam in via ... praegustant coelestis harmoniae jucunditatem qua plenarie saturabuntur in patria."

14   *The Relevance of the Beautiful and Other Essays*, ed. Robert Bernasconi (Cambridge UP, 1986), p. 22.

encounter where subjectivity is surpassed and where intersubjectivity is enacted."[15]

The setting of Dante-pilgrim's first evening and night in purgatory superficially resembles Virgil's evocation of the Elysian Fields in *Aeneid* VI, where his Anchises is "reviewing with urgent attention" (*studio recolens*, vi.681) the faces of the famous who have died, and where a kind of secular praise-singing is heard (*choro paeana canentis*, vi.657). But in *Purgatorio* VII–VIII both the urgency and the singing are explicitly liturgical. The fictive Virgil's explication of his own situation of being 'hung up' in the Limbo, as in a place of intellectual "sighs" (*sospiri*, VII.30), is designed by Dante-poet to be weighed over and against the purgatorial souls' "sighing" hymn ("ad te suspiramus")[16] to the Virgin Mary, who will help speed their salvation. That famous hymn, the "Salve, Regina," is the first one the pilgrim hears on this evening, even though it appears last in the order of service for the liturgical office of Compline, the elements of which office Dante will continue to quote or paraphrase throughout these canti. In one sense, the souls are waiting here in terms of diachronic time, that is, until the light returns the following morning. This is clearly the sense given by the pilgrims' escort, the Lombard poet Sordello: "… and we'll wait over there for the new day" ("… e là il novo giorno attenderemo," VII.69). But it becomes clear in the opening of canto VIII that the souls' waiting (*aspettando*, VIII.24) is a synchronic experience of time, a ritualistic keeping vigil. The hymn sung and heard next in Dante's text is the *Te lucis ante terminum*, the hymn sung first at Compline. The rhetorical figure, again, is *hyperbaton*; the ordinary order of the service is skewed in Dante's fiction in order to reflect the extraordinary simultaneity of temporalities knowable in ritual.[17] Dante has it that, although the twilight hour may evoke a kind of poignancy – when "the day on which they said to their sweet friends *addio*" turns the

---

15   Žižič reads both Gadamer and Heidegger's reception of Hölderlin's poem in "Interrupting Time: Feast as Play and Art," in Davor Džalto, ed., *Religion and Realism* (Newcastle upon Tyne, Cambridge Scholars Publishing, 2016), p. 166.

16   Among the souls pointed out to the pilgrim here, two are paired by gerunds: one is sighing (*suspirando*, VII.108) and the next, simultaneously, is singing (*cantando*, VII.113). The gerund that follows these two epitomizes an attitude of prayer: looking up (*guardando in suso*, VII.134).

17   Paul Rorem's study of liturgical references in Dionysios the Areopagite makes the same point: "In the liturgical realm, the timeless is expressed by the sequential plurality of extensions in temporal actions." See his *Biblical and Liturgical Symbols within the Pseudo-Dionysian Synthesis* (Toronto: Pontifical Institute of Mediaeval Studies, 1984), p. 119. Because liturgical acts-in-time are many and also have manifold spatial extension, the order of their 'sequence' is not at all necessarily *a, b, c, d*, but may often be *d, b, c, a*. The good liturgist, in other words, intends to enact the simultaneous.

heart's desire of travellers *back* – the same time of day in purgatory prompts in one soul, the soul who intones the hymn *Te lucis*, an experience of what this essay has described as sacred urgency. The purgatorial singer looks *forward*, toward the east, "as though he were saying *a Dio*: 'I'm calmed by nothing else.'" Dante has it punningly (a poignant *addio* over and against the urgent *a Dio*) that religious ritual, here performed both individually and communally, has nothing whatsoever to do with nostalgia.[18] When a threatening snake arrives on the scene, Dante-pilgrim is the only one who is in any way 'anxious' – he is terrified – because he is not as yet fully a part of the purgatorial community; in the fiction, he has not as yet fully embraced the seeming contradiction between urgency and fulfillment, as have the rest of the souls – "that gentle army" (*quell'esercito gentile*, VIII.22).[19] The snake's "assault" (VIII.110) apparently occurs every night. Its real threat is quelled, every night, liturgically.

In the twenty-third canto of the *Paradiso*, Dante compares his Beatrice to a mother bird whose urgency for her brood is 'prevenient'; her waiting anticipates a future that is already realized in part. The bird wakes in the middle of the night and "on an exposed branch, anticipates the hour / and with ardent desire waits for the sun" ("... previene il tempo in su aperta frasca, / e con ardente affetto il sole aspetta," *Par.*XXIII.7–8). Fully, "aware" (*attenta*, l.11), fully "erect" (*eretta*, l.10), "suspended, eager" (*sospesa e vaga*, l.13), just like the bird's expecting to feed her young, Beatrice introduces the pilgrim into a ritual moment, in the terms of which he becomes "like someone who, full of desire, wanting something – is satisfied by the hoping for it" (*Par.*XXIII.14–15).[20]

In the fiction of a visionary experience in the heaven of the Sun as being para-liturgical – cantos X through XIV of the *Paradiso* – Dante and Beatrice are described as having already moved beyond our terrestrial, sub-lunar, world, and as moving subsequently within and in terms of the velocities of the heavens of Mercury and Venus, which bodies-in-motion are among the nine, overlapping mensurations of fourteenth-century space-time. As the pilgrim crosses over the threshold of the wider, faster solar sphere, his body speeds up.

---

18   This is not the place for a complete reading of Dante's presentation of the office of Compline, including specific references to the three proper psalms (the first referenced, in VIII.10, occurs last in the order of service), the hymns and the formal oration, which urges the community at prayer to "keep vigil" (*vigilate*).

19   The oxymoron, entirely irenic in character, goes all the way back to St. Paul, in *Ephesians* 6:14–17. Its application to the Christian community is corroborated – and not necessarily only from the Pauline source – as early as Clement of Alexandria, who writes of "pacific soldiers" and of an "army that does not shed blood" in *Protreptikos* 116. See Alieto Pieri, tr., and ed., *Protreptico ai Greci* (Roma: Edizioni Paoline, 1966), p. 281.

20   "fecimi qual è quei che disiando / altro vorria, e sperando s'appaga."

His temporal reckoning, as the poet writes it, is somewhat confused by the fact that he is also and simultaneously moving at the speed of discursive thought.[21] But the fiction has it that the unique opportunity for him at this moment is that Beatrice is moving even faster and far more urgently – at the speed of insight. "Beatrice is she who super-conducts [my energy] / from the good to the better, so very fast / that her act has no temporal extension" ("È Beatrice quella che sì scorge / di bene in meglio, sì subitamente / che l'atto suo per tempo non si sporge," *Par.*x.37–39). Her urgency for him is both intellectual and moral; she wants him to give thanks that the solar sphere and its precise path through the universe and its precise velocity and luminescence are all part of a created order, and therefore of a taxonomy in which every existent is really (not just theoretically) connected and connectable to every other existent. The effect of her urgency is that the pilgrim's "heart" (metonymic, here, for something akin to the Augustinian being-in-the-world), a heart fully aware of its mortality in the near term, makes so genuine a religious act of love that "it eclipsed Beatrice in forgetfulness" ("che Beatrice eclissò ne l'oblio," *Par.*x.60). He forgets about her! And then – I take this to be a truly radical moment – "She didn't mind ... but she smiled at it, such that the shining of her smiling eyes split up my unified mind into further differences" ("Non le dispiacque, ma sì se ne rise, / che lo splendor degli occhi suoi ridenti / mia mente unita in più cose divise," *Par.*x.61–63). Beatrice becomes the proximate occasion of the fact that the genuinely religious, even mystical experience must, over finite time, necessarily give way to the discursive mind's asking all kinds of practical, scientific and philosophical questions. Again, urgent, personal gesture verges upon a community. A dozen philosophers and theologians now, unexpectedly, surround Dante-pilgrim. All of them are saints, which is to say that the questions they have are always in the process of being fully answered. Dante's fiction of the appearance of saints in this place, within the limitations of space-time, and on behalf of a living person's "divided" mind, allows him to dramatize a distinction – central to late medieval philosophy – among three types of time: 1) ordinary time, along with its conventional chronographic markers, 2) extraordinary, transtemporal, 'sempeternal' or aeveternal time – what I have been glossing as the urgent temporality of cultic action (in late-antique and medieval texts, something akin to the temporality of angels) – and 3) eternity or timelessness. Ritual only makes sense, is only a "meet, right, just and healthy thing" to do, when it is mediated by the cosmos of time-bound substances (by the stars, by human bodies, by

---

21   "... but I didn't perceive the ascent, any more than one perceives a thought's coming before it begins"; "... ma del salire/ non m'accors'io, se non com'uom s'accorge, / anzi 'l primo pensier, del suo venire" (*Par.*x.36–39).

that Mayan "chunk of incense") and when ritual occurs within the course of the myriad aeons of the world's finite existence. "Glory" – a proper theological word for timeless worship – needs no mediate terms. Thus the daring logic of the Christian doxology, in which believers (including Dante's saints within the physical sphere of the sun) are urgent to sing "Gloria!" "now and always, unto aeons of aeons!" That is, believers are urgent to sing "Gloria!" right now, even in terms of their mixed experience of time.[22] Those twelve encircling theologians (including, it turns out, Thomas Aquinas, Boethius, Dionysius the Areopagite and Albertus Magnus) are now moved to start singing and to dance a circle-dance around the pilgrim. The poet's simile likens their being so moved to the verge-escapement movement of a modern (fourteenth century), double-geared clock that marks time as it "draws and drives," "tira e urge," and triggers an alarm, so that "God's spouse gets up to sing a pre-dawn song to her husband so that she may love him" (Par.x.139–148).[23] Dante's figure for sacred urgency could not be more explicitly erotic: rapid movement back-and-forth creates a tension, imaged as the 'tumescence' of all those who are spiritually urgent: the verb "turge" ('swells') rhymes with "urge" ("drives"). (English "urgency" is rooted in the Latin urgere, to push, to drive.) The tension built up by pulling and pushing, drawing and driving, inevitably triggers, leads inevitably to ... morning prayer, the celebration of the liturgical office of Matins. The entire passage (Par.x.139–148) is worth quoting:

> Then, like a clock that rings / the hour at which God's spouse rises / to sing a dawn-song to her husband, so that she may love him, / as one part draws and drives the other, / ringing tin tin in tones so sweet / that a

---

22   ... εἰς τοὺς αἰῶνας τῶν αἰώνων – the final phrase of the doxology.

23   "... draws and drives" is Allen Mandelbaum's admirable translation. The verge-escape-ment mechanism 'drives' (pushes, "urge") a weighted bar, the foliot, which oscillates back and forth; as it rotates from the foliot, the escapement 'draws' (pulls, "tira") one part of the verge and pulls the other, creating a consistent movement. See Carlo Cipolla, Clocks and Culture: 1300–1700, New York: W.W. Norton & Co., 2004. In his finely detailed essay, "Miraculous Syllogisms: Clocks, Faith and Reason in Paradiso 10 and 24" (Dante Studies, 117 (1999), 59–84), Cristian Moevs reads what I have called an erotic figure quite otherwise. For Moevs, the clock mechanism that pulls and draws is a figure for the successful (that is to say, according to Moevs, ultimately "miraculous") operation of a syllogism in the operation of the human intellect. Moevs also cites Par.xxiv.10–12, where what I have called herein the circling saints' ritual urgency – love's triggering ritual movement in time – is inscribed participially: fiammando (their "flaming forth") and volte (their having "turned 'round").

person well-disposed[24] will swell with love; that's how I saw the glorious circle move itself and answer voice to voice with a melody and a sweetness unknowable / except in that place where enjoying is eternalized.[25]

("Indi, come orologio che ne chiami
ne l'ora che la sposa di Dio surge
a mattinar lo sposo perché l'ami
che l'una parte e l'altra tira e urge,
tin tin sonando con sì dolce nota
che 'l ben disposto spirto d'amor turge;
così vid'io la gloriosa rota
muoversi e render voce a voce in tempra
e in dolcezza ch'esser non po' nota
se non colà dove gioir s'insempra.")

The contemporary physicist Étienne Klein calls such a moment an intersection between the cosmic course of time and a particular "arrow" of time, in which Beatrice's and the saints' urgent acts – time's 'contents' or arrow – have their own temporal orientation.[26] Beatrice will go on to epitomize the erotic trope in *Paradiso* XXX by using the very same rhyme-words, at the moment in which she invites the pilgrim to inspect the revelation of Itself that God is about to make on his behalf:

---

24   As for *'l ben disposto spirto*, the modern reader's temptation is to hear the phrase as indicating a general moral disposition or rule of moral constraint. But here "spirto" ('mind') is best rendered as 'person,' making the phrase more holistic and indicating an urgency that is moral, intellectual, emotional, physical and temporal.

25   The similitude of a sung Matins is made more 'verisimilar,' as it were, by its comparison to twelve circling saints. As Thomas Aquinas introduces them to the pilgrim, one by one, they are each imaged as lights or flames. One of them, Dionysios the Areopagite, is even a "candle" (*cero*, X.115). They are collectively a "chorus" (*coro*, X.106). In the solemn celebration of Matins, twelve lit candles in a candelabrum are successively extinguished at the end of each 'nocturn' or section of the office. In the Byzantine tradition, which Dante could have known (making his word-choice, *coro*, into a pun), the liturgical chandelier or candelabrum, called a "choros," is round; when it is pushed or swung, its circling of lights imitates in the physical church the annual movement of the stars (of the twelve zodiacal houses) from a given terrestrial point of view. Dante's clearest reference to the "choros"-figure comes in *Par.*XI.13–15, in which each of the saints' circling comes full round and then stops, "like a candle in a candelabrum"/ *come a candellier candelo.*

26   See his "What Does the Arrow of Time Mean?," *KronoScope / Journal for the Study of Time*, 16.2 (2016), 187–98.

That great desire which right now sets you on fire and draws you [the verb is 'urge'], / towards having some idea of what you're looking at, / gives me more pleasure the more it swells up ['turge']. (*Par.*XXX.70–72)

"L'alto disio che mo' t'infiamma e urge,
d'aver notizia di ciò che tu vei,
tanto mi piace più quanto più turge."

Many other moments of the poem could be adduced here. One way to synthesize their import would be to say that, in Dante's poem, hell is finally mere repetition, with no urgency whatsoever. (A possible exception would be Dante's fiction of the damned souls' *desire* to stand in relation to an other as being occasioned by the singular presence of a living person – Dante-pilgrim – in hell.) The conceivable gesture towards community always fails. Purgatory, for Dante, is the extending of temporal urgency as *we* know it, until such time as individual persons come to desire to be better and come to learn, as well, that the opposite of self-centered lethargy, or 'wasting time,' is, precisely, liturgy. The many acts of ritual worship in the *Purgatorio* are urgent – to quote Catherine Pinkstock – to "receive time from eternity as a gift and ... offer time back to eternity as a sacrifice."[27] Paradise is without time. Dante devotes many words to saying that discursive language is "defective" with respect to its reality. At the poem's close, he compares himself and his art to the intellectual urgency of a mathematician who attempts, and still fails, to find an adequate formula for squaring the circle.

---

27   C. Pinkstock, *op.cit.*, p. 118.

CHAPTER 17

Time's Redeeming Urgency

Walter Schweidler

At first glance, time's urgency seems to intensify the existential challenge with which we as human beings are confronted by the knowledge of our finiteness. Given the limited amount of time we have, it seems to us that the speeding up of its course means that we lose it faster than we otherwise do when *tempus fugit*. In the philosophical perspective, however, this defeatist view turns out to be only the consequence of a wrong understanding of finiteness based on what Hegel called the "bad" or "spurious infinite", i.e., the illusion that the flight of time could be overcome by the extension of its course. The opposite insight, i.e., that we can cope with time's flow by never extending or stopping but only by turning it against itself, is incorporated in a way hitherto unsurpassed into Proust's *Recherche du temps perdu*. It is first and foremost this insight which makes Proust's work a principal legacy for the philosophy of time.

Many times in this unique kind of novel the author emphasizes its philosophical claim and dimension. The *Recherche* which is incorporated into this book is not a search for memory but for truth; it is called a search for time only, as Gilles Deleuze has formulated very tersely, in as far as time is in essential relation with truth.[1] That relation, however, is a unique one: If we want to understand it we must see that truth is not simply the topic or the theme of this novel but that it is incorporated into it as it is incorporated virtually into the book which is given into our hands. And we must recognize that truth is a matter not of theory but of redemption. For Proust, the beginning and the completion of the work that is incorporated into this book witnesses the release from the deepest existential threat of which we can think. They are the evidence of a race run with death and of its happy end. The decisive aspect of time which was at stake in this race and which is the real issue of this book is the turning of its direction against its natural course or, as we shall learn, our time's transition into a life whose end becomes the cause of its beginning.

"Maintenant je peux mourir", *now I can die*: These were Proust's words, spoken with a smile and with glowing eyes, in the morning after he had set the word "Fin" under the *Recherche du temps perdu*.[2] Spoken in the context of that

---

1 Gilles Deleuze. *Proust und die Zeichen* (Frankfurt am Main, Berlin and Vienna: Ullstein, 1978), 16.
2 Jean-Yves Tadié. *Marcel Proust* (Frankfurt am Main: Suhrkamp, 1987), 430.

moment, these words meant something categorically different from a worker's release after having finished his job. They expressed not relief but redemption, they witnessed the final and irreversible escape from the deepest existential threat. It was a threat that has nothing to do with the problem of bringing a literary work to its end and preparing it for publication. Céleste Albaret, Proust's housekeeper to whom he spoke those words has reported that with the same breath he conceded that there was still much to do;[3] from her we know also that months later he complained that many passages of the novel were yet to be completed. His last notes before his death were written into the typescript of the *Prisoner*. And it is well known that his publishers constantly were in despair about his never-ending efforts of reworking his manuscripts. So, the point of Proust's final experience of redemption is of a very different and somewhat unique kind.

An author's task and striving to finish his job is something that is made necessary by the moment everybody has to cope with in his or her life: The moment when one has to die. But the moment which was at stake on that spring morning was not the moment when Proust would have to die; it was, exactly as he spoke, the moment when he *could* die, i.e., the moment in which not his work but he himself had reached his end – and lived to see it. Therefore, at least at our present *point de depart*, from where we try to understand the kind of threat the escape of which is incorporated in the *Recherche*, we must distinguish clearly the sense in which Proust said that he "could" die now from any kind of final reckoning of one's life. Such an expression of this kind of reckoning may take place, for instance, when a father sees his children attain a good standard of living and expresses his content by the statement that his work is done and he can go. This is something very honorable but it is philosophically irrelevant. The point of difference is that in such a life-reckoning statement a person's insight that he can "go" now is largely tantamount to the following: that he can "disappear" or, to put it brutally, that soon he can *be* ignored. That means, it is an essentially passive sense which is expressed here by means of the word "can", whereas Proust uses the word in an absolutely different meaning, which actually is a highly active one. It is the sense in which one says, for example, that one "can" read German or that one can play chess now. What by this aspect of potentiality comes to the fore is the horizon of *disposition*. Therefore, the philosophically decisive question is exactly that which in Proust's dictum is meant by and contained in the dispositional aspect of the expression "I can die". To answer this question, we will have to go through at least three dimensions of death which are constitutive for what is the meaning

---

3   Ibid.

of "can" in the context of the expression "I can die". Let me briefly point out or at least indicate these three dimensions by reference to some passages mainly in the first and the last volume of the *Recherche*, i.e., *Swann's Way* and *Finding Time Again*.

Firstly, death's *revelation*: It belongs to the deepest constitution of personal existence that only the end of one's life can reveal what it has been in its whole and final course. That insight contains much more than the prudent but somewhat superficial warning that we should not praise the day before the evening. It also contains more than the ethical wisdom to which Aristotle referred when he said that *eudaimonia*, felicity, should not be ascribed to anyone's life before it has ended. Again, in order to understand the existentially revealing dimension of death, we have to keep away from any aspect of a reckoning of life. The decisive point is, at first glance, much simpler; death actually is the principle of truth not in respect to life's reckoning but only to life's end. But *simplex sigillum veri*, "simplicity is the sign of truth": It is the point at which we face a fundamental paradox. I mean the following: On the one side, death, and that means the very concrete moment of death, belongs to one's life. Is it not perhaps the most comforting insight one can reach that if one did not have to die one would not exist? Since a person has his identity essentially through his relation to all other persons, if I were not the person I am then I would not be there. If I did not have to die, if I had before me an infinite future expanse in which I would eventually realize all the possibilities of any other person, then I would not be the being which I am, and that is to say: I would not exist. But on the other side, we all know what Wittgenstein laconically stated in the *Tractatus*: "Den Tod erlebt man nicht."[4] It is death, the core of one's personal existence, of which one has absolutely no experience in one's life. Indeed, it is even ontologically impossible that one could have such an experience: Since, as Aristotle said, *vivere viventibus est esse*,[5] for a being whose existence is constituted by its life it is *per definitionem* impossible that it, this very being, could ever experience its death after having died. So, what are we to make of this paradox, the paradox that death belongs to my life more than anything and at the same time death is necessarily beyond everything my life will ever encompass?

In this limited context I can only present it as my thesis that with this question we have formulated the existential and, which for him meant the same, the philosophical problem to which Proust dedicated his *Recherche du temps*

---

4 Ludwig Wittgenstein. *Tractatus Logico-Philosophicus* (London: Routledge & Kegan Paul, 1960), 6.4311.

5 Cf. Aristotle. *De Anima*, in *Opera omnia: Græce et latine*, vol. 3. (Paris: Ambroise Firmin-Didot, 1854), II, 4, lines 2–3.

perdu and, which for him meant the same, his life. The answer to this question is what, according to Proust, every human being's life demands to be given. It is the really and essentially *universal* problem of personal existence. And it is the answer which the *Recherche du temps perdu* demands his author to give – and to give in a unique, incomparable way. These two aspects, the universal and the individual, stand in a relation to one another which, as I would like to argue, is the decisive principle of the philosophical project in Proust's book of his life. I would characterize this relation as a *metonymic* one. What I mean by that can be explicated only by turning our attention to the place in the novel where its author presents the definite clarity he has reached about these two aspects and about their connection – and, exactly thereby, about his task. That place is nothing but the very end of the novel where the author notes:

> ... the fact that we occupy an ever larger place in Time is something that everybody feels, and this universality could only delight me, since this was the truth, the truth suspected by everybody, that it was my task to try to elucidate. (6, 355)[6]

In the last lines of the novel, immediately before the "Fin", he repeats that task of describing the persons in it as beings with an "almost infinitely extended" place in Time (6, 358), adding however one condition, namely

> if enough time was left to me to complete my work. (Ibid.)

Seemingly, there is a contradictory aspect between the "Time" in which we can attain our almost infinitely extended place and the "time" the end of which threatens to run out before the author can finish his work. Thinking about the time left to him to fulfill his task the author faces that aspect of uncertainty which for us marks the first indication of time's urgency. In one of the great existential metaphors in the book that aspect is characterized by essentially artistic scenery:

> My life had been like a painter who climbs up a road overhanging a lake that is hidden from view by a screen of rocks and trees. Through a gap he glimpses it, he has it all there in front of him, he takes up his brushes. But the night is already falling when there is no more painting, and after which no day will break. (6, 345)

---

6   See Marcel Proust. *In Search of Lost Time. Volume 6: Finding Time Again*, trans. Ian Patterson (London: Penguin Books 2003).

Here we come to notice that the revealing and the threatening dimension of death are as inseparable as the two sides of a page of a book. However, and what we are given to understand here is decisive for us at this point, exactly like two inseparable sides of a page they are also infusible. Here we come to the decisive aspect which radically distinguishes the task Proust is going to teach and to save us from any kind of existentialistic pathos in reclaiming or uncovering the "absurdity" or the "foundering" constitution of personal life. The task we have to solve is revealed by death, but it is this only because *death itself, in its true meaning, has been revealed by the fulfillment of this task*. In other words: The task which is ontologically constituted by the fact and by the concrete moment of one's death can be at the same time *revealed and fulfilled*. This is what happens in the very instant in which we discover what Proust means by the term "lost time" and the truth to which the search for it has led. What is at stake here is the time in which the moment of death's revelation can be turned into the instant of life's fulfillment. That is what brings us to our next step.

Secondly, death's *contingency*: The problem which we face by the paradox of death's revelation is, as we saw, that the only moment which can reveal the whole course of one's life cannot at the same time belong to that life as a part of it which one can experience. The task, though, which we now have indicated as a possible way to deal with that problem, is the following: to turn our attention to a certain kind of moment which we know, which we have experienced and will experience further, but in which we nevertheless find fulfilled what the moment of death restricts us from. The poetic presentation, the literary reformulation of that kind of moment is what Proust's *Recherche du temps perdu* is famous for more than for any other of its treasures and through which it even has gained a certain amount of popularity. The essence of it, developed most paradigmatically in the immortal *madeleine* episode in *Swann's Way*, is quite obvious: *involuntary memory*, *déjà vu* in the most philosophical role it has ever played in the theatre of human ideas. The revelation of the whole past of "Combray", emanating from the scent of a small cake, half-dissolved in a cup of tea, is certainly the key scene in *Swann's Way*. In the course of the whole *Recherche*, however, the late – and explicit[7] – counterpart to it in *Finding Time Again*, the scene when the kaleidoscope of the crucial experiences and memories of his whole life emerges from a footstep on two unevenly laid paving-stones, is even more important for what the author has to tell us

---

7   Cf. ibid., 175: "Just as at the moment when I tasted the madeleine, all uneasiness about the future and all intellectual doubt were gone."

about the crucial relation between the moment of death and the instant of the fulfillment of life. The decisive point is what we could call the objective side of involuntary memory, i.e., the *contingency* of finding again what otherwise would be lost forever:

> But sometimes it is just when everything seems to be lost that we experience a presentiment that may save us; one has knocked on all the doors which lead nowhere, and then, unwittingly, one pushes against the only one through which one may enter and for which one would have searched in vain for a hundred years, and it opens. (6, 174)

Here the metaphor of the painter getting his glimpse of the face of death is obviously anticipated, but more than this: Its meaning is here already explicated, namely by the immediate touch between the described instant of involuntary memory and "a second intimation" which, as the author says,

> occurred to reinforce the one which the two uneven paving-stones just had given me and to extort me in the perseverance of my task. (6, 176)

This is exactly the point where the deepest passivity is turned into its dialectical counterpoint, the highest activity that constitutes the artist's task. Exactly here we find the solution to the paradox of death's revelation and life's fulfillment, and the philosophical principle of this solution is explicated as the representative, substitutive relation in which the artist who turns involuntary memory into his task stands for time itself. "Time, the artist", he says, had "rendered" all the models of his work

> in such a way that they were still recognizable, but they were not likenesses, not because he had flattered them, but because he had aged them. He is also an artist who works extremely slowly. (6, 243)

With this we have reached the point where the deepest core of what Proust means by "lost time" can be pointed out. The scheme which explicates the contiguity between death's revelation and life's fulfillment is the turning of involuntary memory into the truth of the most voluntary work we can conceive, the work of art. So, the relation between lost time and time found again, the explication of time turned back out of its end into the beginning of its whole course, has to be grounded on the difference between involuntary and voluntary memory. The author characterizes this difference in *Swann's Way* as the difference between the life and death of past itself:

The fact is, I could have answered anyone who asked me that Combray also included other things and existed at other hours. But since what I recalled would have been supplied to me only by my voluntary memory, the memory of the intelligence, and since the information it gives about the past preserves nothing of it, I would never have had any desire to think about the rest of Combray. It was all really quite dead for me. (1, 46)[8]

At this point, marking the essence of the difference between dead and living past, he invents the key category in the relation between death and truth: The essentially contingent *chance* they open for us:

Dead for ever? Possibly. There is a great deal of chance in all this, and a second sort of chance, that of our death, often does not let us wait very long for the favours of the first. (1, 46)

Now the contradictory aspect, the tension between death and truth, has been transformed into the temporal difference between two moments of chance, thereby opening a gap which can allow us to fill it with what constitutes the fulfillment of our life. Immediately following, augured only by the wondrous reference to an old "Celtic belief", we find the anticipation of what 3000 pages later is described as the revealing experience of the uneven paving-stones which finally will allow him to persevere in his task.

It is the same with our past. It is a waste of effort for us to try to summon it, all the exertions of our intelligence are useless. The past is hidden outside the realm of our intelligence and beyond its reach, in some material object (in the sensation that this material object would give us) which we do not suspect. It depends on chance whether we encounter this object before we die, or do not encounter it. (1, 47)

Here now we have reached the threshold of our third and crucial step.

Thirdly, death's *urgency*: Following the experience of the kaleidoscope of all the moments standing for his whole life, the author describes the decisive, redeeming insight into the chance he has found now, the chance which is given to him by the small amount of time between it, this insight, and his death. What he can do in order to fulfill his life, i.e., to include in it everything that belongs to it, is to transform it into a book, the book of his life:

---

8   See Marcel Proust. *In Search of Lost Time. Volume 1: The Way by Swann's*, trans. Lydia Davis (London: Penguin, 2003).

Finally, this idea of Time was valuable to me for one other reason, it was a spur, it told me that it was time to start, if I wanted to achieve what I had sometimes sensed during the course of my life, in brief flashes ... which had made me feel that life was worth living. How much better life seemed to me now that it seemed susceptible of being illuminated, taken out of the shadows, restored from our ceaseless falsification of it to the truth of what it was, in short, realized in a book! (6, 342)

That is Proust's philosophical solution of the paradox of death's revelation and life's fulfillment: To transform life into the book of life allows its writer to turn its end into its beginning – given that he has enough time to complete the plot, i.e., to reach the point where the described life has led him to the insight that it can be transformed into the book in which this point is reached now. If he manages to reach this point, then his life lies in front of him as its own book, containing its end. And that he had reached this point was what Proust expressed by the words "I can die now". We will not grasp the real meaning of these words as long as we hear them as an expression of relief from a poetic task which has been finished by the artist instead of understanding them as the opposite, i.e., as a message of redemption which is told to us by time itself which has gotten rid of anything seemingly outside of or beyond it. It is the message which reaches us from the incomparable, unique place where one lives to look at one's end. In this horizon, the words "I can die now" are related to the old philosophical tradition of the *ars moriendi*, the idea of turning life from its natural temptation to flee from death into the opposite, i.e., into a self-fulfilling way of getting acquainted with death and thereby integrating it, illuminating it as its part instead of excluding it as its despised shadow.

To underline and concretize this we have to dive once more into the depths of the meaning of the word "I can". In one of the most important passages of his *Metaphysics*, in which Aristotle analyzes the relation between *dynamis* and *energeia*, he refers to two different kinds of "goings-on" in the world, movement and actuality. For the latter it is characteristic that the end to which it leads is already present in everything belonging to it; only actuality is what we call in its full sense "action". What is only movement, however, does not bear its end in itself but is only

relative to the end, e.g. the removing of fat, or fat-removal, and the bodily parts themselves when one is making them thin are in movement in this way (i.e. without being already that at which the movement aims)...; but that movement in which the end is present is an action. E.g. at the same time we are seeing and have seen, are understanding and have understood, are thinking and have thought (while it is not true that at the same time we are learning and have learnt, or are being cured and have been cured). At the same time we are living well and have lived well, and are happy and have been happy.[9]

This is of the highest importance for the philosophical dimension in which Proust's contribution to the tradition of the *ars moriendi* must be seen: It is not simply a psychological or perhaps an ethical dimension, but the ontological dimension in its deepest sense. Here we have to go one step beyond what we said in the beginning: We must hear the words "I can die now" not only in the horizon of expressions like "I can speak German now" or "I can play chess now" but also in the sense of what we say when we recognize "I can see him now" or "I can understand you now". These expressions show how we in our ordinary life and speech are fully aware of that deep paradox at the core of personal existence: As a rational being I am identical with myself in the most eminent sense when I am in action, and to act means to make the experience of a reality which has its end in itself, i.e., of a past which is integrated in what is going on presently. To see somebody means to become aware of having seen him, to understand means to become aware of having understood, and as far as we can turn life into action to live means to become aware of having already lived. There is no moment of one's life, including its beginning and its end, which we could experience and which would not reveal to us that it has been already lived. That means: There is the sound of *reincarnation* in the voice which tells us that we can turn life into action, and I think that the fact that this sound is obviously present in Proust's *Recherche* supports my thesis that turning life into action, into the literary work of a novel describing the process that in the end leads to its beginning, is exactly Proust's program in that book of life. In the *Prisoner*, immediately following the description of the narrator Bergotte's death, the author initiates a beautiful piece of reflection on the possibility of reincarnation with the same words we have read in *Swann's Way* when he asked the question if the past dies with the present:

> He was dead. Dead for ever? Who can say? Certainly spiritualist experiments provide no more proof than religious dogma of the soul's survival. What we can say is that everything in our life happens as if we entered in bearing a burden of obligations contracted in an earlier life [...] All these

---

9   Aristotle. *Metaphysics*, trans. W. D. Ross, in *The Complete Works of Aristotle: The Revised Oxford Translation*, ed. Jonathan Barnes, vol. 2 (Princeton, NJ: Princeton University Press, 1995), I, 6, 1048b.

obligations which do not derive their force from the here-and-now seem to belong to a different world founded on goodness, conscientiousness, sacrifice, a world quite different from this one which we leave to be born onto this earth, and to which we perhaps shall return ... (5, 170)[10]

I do not want to say that any kind of belief in reincarnation is what Proust is going to support in the *Recherche*, although the reflection in the quoted passage leads to the conclusion

> that the idea that Bergotte was not dead for ever is not all implausible. (Ibid.)

But to go much further in that direction would be misleading because Proust's solution to the paradox does not consist in any belief but in action. Turning life into action means to live it once again in the writing its book, the plot of which leads to and ends with the decision to write it. This kind of action, not any belief underlying it, constitutes what Proust calls and reclaims as the truth of his work and of his life. In as much as the author of his life's book manages to turn his life's time into the narration of the way to the end where he decides to become what he is, the author and the subject he writes about will coincide.

I cannot go further into this fascinating philosophical project, but let me just conclude with the remark that with it Proust has also given a deep contribution to the theme of "time's urgency". It is based on Proust's decisive insight that the *ars moriendi*, that our only chance to "learn to die" does not consist in any strategy to extend the time we have before our death but, on the contrary, to lead life to an end which anticipates and thereby integrates death into its course. In the light of this insight, time's urgency becomes the closest ally in our enterprise to forestall life's end by the symbolic repetition of its beginning in itself, i.e., in time. The decision to integrate this renewal of its beginning into its course is tantamount to the exclusion of any possible extension beyond the moment of that decision. It is this moment which, in order to anticipate death, we must reach again, and as swiftly as possible. The urgency of our return to it is made clear in the length of the detour it necessitates, which we once more must make in the form of its narration. Thus, the urgency of life's time and the chance to reach its aim essentially converge. Given that life coincides with the process of composition, we constantly experience a life that is already lived. After having begun to turn life into the plot of its book, any slowdown of our

---

10   See Marcel Proust. *In Search of Lost Time. Vol. 5: The Prisoner and the Fugitive*, trans. Carol Clark (London: Penguin, 2003).

action would be tantamount to the most dangerous delay we can face in our race run against unforestalled death. It is the most urgent aim for us to reach the point where we can say: "I can die now." So, time's urgency becomes the scheme of release from the threat of death. As a human being, it belongs to my nature that I cannot want to die. But as an *animal symbolicum*, as an apprentice of the artist Time, I want to become somebody who can die.

Index of Authors

Index of Subjects